Sustainability Science

Harald Heinrichs • Pim Martens
Gerd Michelsen • Arnim Wiek
Editors

Sustainability Science

An Introduction

Springer

Editors
Harald Heinrichs
Institute of Sustainability Governance (INSUGO)
Leuphana University
Lüneburg, Germany

Gerd Michelsen
Faculty of Sustainability
Leuphana University
Lüneburg, Germany

Pim Martens
Maastricht University
Maastricht, The Netherlands

Arnim Wiek
School of Sustainability
Arizona State University
Tempe, AZ, USA

ISBN 978-94-017-7241-9 ISBN 978-94-017-7242-6 (eBook)
DOI 10.1007/978-94-017-7242-6

Library of Congress Control Number: 2015957804

Springer Dordrecht Heidelberg New York London

Printed on acid-free paper

Springer Science+Business Media B.V. Dordrecht is part of Springer Science+Business Media (www.springer.com)

Contents

Contributors

David J. Abson Faculty of Sustainability, Leuphana University, Lüneburg, Lüneburg, Germany

FuturES, Leuphana University, Lüneburg, Germany

Maik Adomßent Faculty of Sustainability, Leuphana University, Lüneburg, Germany

Rimjhim M. Aggarwal School of Sustainability, Arizona State University, Tempe, AZ, USA

Karoline Augenstein Wuppertal Institute for Climate, Environment and Energy, Wuppertal, Germany

Günther Bachmann German Council for Sustainable Development, Berlin, Germany

Matthias Barth Institute for Integrative Studies, Leuphana University of Lüneburg, Lüneburg, Germany

Sjouke Beemsterboer International Centre for Integrated assessment and Sustainable development (ICIS), Maastricht University, Maastricht, Netherlands

Michael J. Bernstein School of Sustainability, Arizona State University, Tempe, AZ, USA

Frank Biermann Institute for Environmental Studies, VU University Amsterdam, Amsterdam, Netherlands

Chiung Ting Chang Institute of Public Affairs Management, National Sun Yat-sen University, Taiwan

James Clark York Green Chemistry Centre of Excellence, University of York, York, UK

Ron Cörvers International Centre for Integrated assessment and Sustainable development, Maastricht University, Maastricht, The Netherlands

Joop de Kraker International Centre for Integrated assessment and Sustainable development, Maastricht University, Maastricht, The Netherlands

Management, Science and Technology, Open Universiteit, Heerlen, The Netherlands

Daniel Fischer UNESCO Chair Higher Education for Sustainable Development, Leuphana University Lüneburg, Lüneburg, Germany

Bastien Girod Department of Management, Technology, and Economics, Swiss Federal Institute of Technology (ETH) Zurich, Zürich, Switzerland

Jasmin Godemann Agricultural, Nutritional and Environmental Science, Justus-Liebig University Gießen, Gießen, Germany

Aaron Golub Toulan School of Urban Studies and Planning, Portland State University, Portland, OR, USA

Erik G. Hansen Centre for Sustainability Management, Leuphana University of Lüneburg, Lüneburg, Germany

Werner Härdtle Institute of Ecology, Faculty of Sustainability, Leuphana University, Lüneburg, Germany

John Harlow School of Sustainability, Arizona State University, Tempe, AZ, USA

Harald Heinrichs Institute of Sustainability Governance (INSUGO), Leuphana University, Lüneburg, Germany

Maud M.T.E. Huynen International Centre for Integrated assessment & Sustainable development (ICIS), Maastricht University, Maastricht, The Netherlands

Beatrice John Institute of Ethics and Transdisciplinary Sustainability Research, Leuphana University Lüneburg, Lüneburg, Germany

René Kemp International Centre for Integrated assessment and Sustainable development (ICIS), Maastricht University, Maastricht, Netherlands

Sonja Klinsky School of Sustainability, Arizona State University, Tempe, AZ, USA

Pius Krütli ETH Zurich, USYS Transdisciplinarity Laboratory (TdLab), Sonneggstrasse, Zurich, Switzerland

Klaus Kümmerer Sustainable Chemistry and Resources, Institute of Sustainable and Environmental Chemistry, Faculty of Sustainability, Leuphana University of Lüneburg, Lüneburg, Germany

Daniel J. Lang Faculty of Sustainability, Institute for Ethics and Transdisciplinary Research, Leuphana University of Lüneburg, Lüneburg, Germany

Stefan Lechtenböhmer Wuppertal Institut für Klima Umwelt Energie, Wuppertal, Germany

Heather Sealy Lineberry Arizona State University Art Museum, Tempe, AZ, USA

Gesa Lüdecke Institute for Sustainability Communication, Leuphana University Lüneburg, Lüneburg, Germany

David Manuel-Navarrete School of Sustainability, Arizona State University, Tempe, AZ, USA

Pim Martens Maastricht University, Maastricht, The Netherlands

Leuphana University, Lüneburg, Germany

Darryn McEvoy Climate Change Adaptation Program, Global Cities Research Institute, RMIT University, Melbourne, Australia

Gerd Michelsen Faculty of Sustainability, Leuphana University, Lüneburg, Germany

Andreas Möller Institute for Environmental and Sustainability Communication, Leuphana University Lüneburg, Lüneburg, Germany

Jens Newig Institute for Sustainability Communication, Leuphana University Lüneburg, Lüneburg, Germany

Lars J. Nilsson Adjunct Professor, Department for Technology and Society/ Environment and Energy Systems, Lund University, Lund, Sweden

Nils Ole Oermann Faculty of Sustainability, Leuphana University Lüneburg, Lüneburg, Germany

Marjan Peeters Leuphana University of Lüneburg, Lüneburg, Germany

Marco Rieckmann Department I / Education and Pedagogical Sciences, University of Vechta, Vechta, Germany

Niko Roorda Roorda Sustainability, Tilburg, Netherlands

Stefan Schaltegger Centre for Sustainability Management, Leuphana University of Lüneburg, Lüneburg, Germany

Uwe Schneidewind Wuppertal Institute for Climate, Environment and Energy, Wuppertal, Germany

Thomas Schomerus Leuphana University of Lüneburg, Lüneburg, Germany

Daniel Schulz Institute for Sustainability Communication, Leuphana University Lüneburg, Lüneburg, Germany

Mandy Singer-Brodowksi Wuppertal Institute for Climate, Environment and Energy, Wuppertal, Germany

Heiko Spitzeck Sustainability Research Center, Fundação Dom Cabral, São Paulo, Brazil

Michael Stauffacher ETH Zurich, USYS Transdisciplinarity Laboratory (TdLab), Sonneggstrasse, Zurich, Switzerland

Jan H. Stel International Centre for Integrated assessment and Sustainable Development (ICIS), Maastricht University, Maastricht, The Netherlands

Han van Son School of International Studies, Avans University, Breda, Netherlands

Michael von Hauff TU Kaiserslautern, FB Wirtschaftswissenschaften, Kaiserslautern, Germany

Leuphana University, Lüneburg, Germany

Goddert von Oheimb Institute of General Ecology and Environmental Protection, Technische Universität Dresden, Tharandt, Germany

Henrik von Wehrden Centre for Methods, Leuphana University, Lüneburg, Germany

Institute of Ecology, Faculty of Sustainability, Leuphana University, Lüneburg, Germany

FuturES, Leuphana University, Lüneburg, Germany

Olaf Weber School for Environment, Enterprise and Development (SEED), University of Waterloo, Waterlooo, Canada

Annika Weinert Faculty of Sustainability, Leuphana University Lüneburg, Lüneburg, Germany

Annika Weiser Institute of Ethics and Transdisciplinary Sustainability Research, Leuphana University Lüneburg, Lüneburg, Germany

Arnim Wiek School of Sustainability, Arizona State University, Tempe, AZ, USA

Chapter 1
Introduction

Harald Heinrichs, Arnim Wiek, Pim Martens, and Gerd Michelsen

Abstract Three hundred years after defining sustainable development in forestry and 25 years after conceptualizing sustainability as a societal guiding vision and regulative idea, the necessity for further operationalizing and realizing sustainability is greater than ever. The textbook at hand provides a state-of-the-art overview of key areas of sustainable development. Like a mosaic, the chapters compose building blocks, which assemble an encompassing perspective on sustainability science. We hope to contribute with this textbook to the further establishment of sustainability science and to enable the next generation of sustainability experts to get a "grip" on the challenging and exciting "centenary topic" of sustainable development.

Keywords Sustainable development • Sustainability science • Transformation • Inter- and transdisciplinarity

The necessity for sustainable development was first documented in 1713, in the book *Sylvicultura Oeconomica* by German chief miner Hans Carl von Carlowitz (2013). He asserted that "sustainable forestry" is key for long-term success in mining and related livelihoods. He argued that the demand for trees for heating, building, brewing, mining, and smelting activities could only be met if a balance between harvesting and growing/restoring trees would be reached. Nowadays, 300

H. Heinrichs (✉)
Institute of Sustainability Governance (INSUGO), Leuphana University, Lüneburg, Germany
e-mail: harald.heinrichs@uni.leuphana.de

A. Wiek
School of Sustainability, Arizona State University,
PO Box 875502, Tempe, AZ 85287-5502, USA
e-mail: arnim.wiek@asu.edu

P. Martens
Maastricht University, Maastricht, The Netherlands
e-mail: p.martens@icis.unimaas.nl

G. Michelsen
Faculty of Sustainability, Leuphana University, Scharnhorststraße 1,
21335 Lüneburg, Germany
e-mail: michelsen@uni.leuphana.de

H. Heinrichs et al. (eds.), *Sustainability Science*, DOI 10.1007/978-94-017-7242-6_1

years later, concepts of sustainable development and sustainability have reached far beyond the realm of forestry or natural resource management and have entered discourses and practices in urban development, chemical industry, tourism, policy making, and education, to name a few. Sustainability has become an important reference point for safeguarding the future across societies worldwide. The broad dissemination of sustainability as a societal guiding principle can be ascribed to the Brundtland Commission Report (WCED 1987) and the succeeding United Nations Conference on Environment and Development in Rio de Janeiro in 1992 (United Nations 1993).

In 2012, 25 years after the Brundtland report and 20 years after the Rio conference, representatives from governments, business leaders, NGO activists, and engaged researchers met again in Rio de Janeiro to take stock and discuss the future of sustainable development. The outcome of this latest sustainability summit was mixed – at best. Despite some progress on sustainability issues in government, business, and civil society, the world continues on unsustainable pathways (e.g., United Nations 2011; UNEP 2012; WWF 2012). The global community is still far away from realizing inter- and intra-generationally just development that balances ecological, social, and economic needs. For example, CO_2, the key driver of anthropogenic climate change, continues to increase despite international climate policy instituted in 1990; biodiversity loss is accelerating; global poverty reduction is lagging behind its goals; and social inequality has intensified over the past 30 years through economic globalization – in some cases passing critical tipping points (Rockström et al. 2009). Thus, it was of utmost importance that the Rio+20 conference agreed on next steps – such as developing global sustainability goals by 2015 or establishing a global sustainability council – to accelerate progress toward sustainable development.

Despite the ongoing debate about the form and shape of the "Great Transformation" toward sustainability (Raskin et al. 2002; WBGU 2011), it is obvious that sharply altered and improved decision-making and action are necessary to secure a better future for humankind and the planet. Next to decision-makers in politics, business, media, and civil society, as well as citizens and consumers, academia has to play an important role in this endeavor. Through research and teaching, higher education institutions are prime places for exploring and shaping the future. However, the traditional academic disciplines, which operate as if "society has its problems – universities have their disciplines," are not adequately equipped for the enormous challenges ahead (Van der Leeuw et al. 2012). The disciplines that aim at contributing effectively to sustainable development need to switch their modi operandi toward transformational and solution-oriented research and education (Wiek et al. 2012; Miller et al. 2014; Wiek and Kay 2015). Beyond interdisciplinary collaboration (working across disciplinary boundaries), transdisciplinary research projects are needed in which researchers and practitioners collaborate in problem-solving efforts (Lang et al. 2012). Over the past two decades, many inspiring approaches and projects have advanced sustainability science (Kates et al. 2001; Clark and Dickson 2003; Komiyama and Takeuchi 2006; Jernecke et al. 2011; Wiek et al. 2012, 2015; Clark et al. in press). There are now numerous academic journals, conferences, study programs, professorial positions, and so forth devoted to sustainability science. Despite these initiatives, there is still a lack of textbooks providing a broad overview of sustainability science efforts to students specifically. The textbook at hand aims at helping to fill this gap.

The 28 chapters compiled in this textbook address a wide spectrum of topics relevant to sustainable development and sustainability, ranging from justice, science policy, art, and business to mobility, oceans, international development, health, energy systems, and education. We have deliberately abstained from imposing a meta-structure to the book, in order to overcome the limits of disciplinary perspectives on sustainability. Like a mosaic, the individual chapters represent building blocks, which assemble an encompassing perspective on sustainability science. All chapters are written for students and early-career professionals entering the field of sustainability with different interests and backgrounds. All chapters provide introductory level information and indicate further readings. To support a solution-oriented approach to sustainability science, we have asked the authors to cover the specific sustainability challenge (why is this relevant to sustainability science), currently offered solution options to this challenge (what has been achieved), and open issues (what is still needed) of the respective topic.

We hope to contribute with this textbook to the further establishment of sustainability science and to enable the next generation of sustainability experts to get a grip on the challenging and exciting "centenary topic" of sustainable development.

References

Clark WC, Dickson NM (2003) Sustainability science: the emerging research program. Proc Natl Acad Sci USA 100:8059–8061

Clark WC, Tomich TP, van Noordwijk M, Guston D, Catacutan D et al (in press) Boundary work for sustainable development: natural resource management at the Consultative Group on International Agricultural Research (CGIAR). Proc Natl Acad Sci USA

Jernecke A, Olsson L, Ness B, Anderberg S, Baier M, Clark E, Hickler T, Hornborg A, Kronsell A, Loevbrand E, Persson J (2011) Structuring sustainability science. Sustain Sci 6:69–82

Kates RW, Clark WC, Corell R, Hall JM et al (2001) Sustainability science. Science 292:641–642

Komiyama H, Takeuchi K (2006) Sustainability science: building a new discipline. Sustain Sci 1:1–6

Lang DJ, Wiek A, Bergmann M, Stauffacher M, Martens P, Moll P, Swilling M, Thomas C (2012) Transdisciplinary research in sustainability science – practice, principles and challenges. Sustain Sci 7(Supplement 1):25–43

Miller TR, Wiek A, Sarewitz D, Robinson J, Olsson L, Kriebel D, Loorbach D (2014) The future of sustainability science: a solutions-oriented research agenda. Sustain Sci 9(2):239–246

Raskin P, Banuri T, Gallopin G, Gutman P, Hammond A, Kates R, Swart R (2002) Great transition: the promise and lure of the times ahead. Stockholm Environment Institute, Boston

Rockström J, Steffen W, Noone K, Persson A, Chapin FS et al (2009) A safe operating space for humanity. Nature 461:472–475

United Nations (1993) Agenda 21 – programme of action for sustainable development. Agreements negotiated by Governments at the United Nations Conference on Environment and Development (UNCED), 3–14 June 1992, Rio de Janeiro, Brazil. New York

United Nations (2011) The millennium development goals report 2011. United Nations, New York
United Nations Environment Programme (2012) Global Environment Outlook 5 (GEO5) – environment for the future we want. United Nations Environment Program, Nairobi
Van der Leeuw S, Wiek A, Harlow J, Buizer J (2012) How much time do we have? Urgency and rhetoric in sustainability science. Sustain Sci 7(Supplement 1):115–120
von Carlowitz HC (2013) Sylvicultura Oecnonomica – Haußwirthliche Nachricht und Naturmäßige Anweisung zur Wilden Baum-Zucht [Economic news and instructions for the natural growing of wild trees]. Oekom Verlag, München
WBGU – Wissenschaftlicher Beirat Globale Umweltveränderungen [German Advisory Council on Global Change] (2011) Welt im Wandel. Gesellschaftsvertrag für eine große Transformation [World in transition: a social contract for sustainability]. WBGU, Berlin
Wiek A, Kay B (2015). Learning while transforming – solution-oriented learning for urban sustainability in Phoenix, Arizona. Curr Opin Environ Sustain 16:29–36
Wiek A, Ness B, Brand FS, Schweizer-Ries P, Farioli F (2012) From complex systems analysis to transformational change: a comparative appraisal of sustainability science projects. Sustain Sci 7(Supplement 1):5–24
Wiek A, Harlow J, Melnick R, van der Leeuw S, Fukushi K, Takeuchi K, Farioli F, Yamba F, Blake A, Geiger C, Kutter R (2015) Sustainability science in action – a review of the state of the field through case studies on disaster recovery, bioenergy, and precautionary purchasing. Sustain Sci 10:17–31
World Commission on Environment and Development (WCED) (1987) Our common future. Oxford University Press, Oxford
WWF (ed) (2012) Living planet report 2012: biodiversity, biocapacity and better choices. WWF International, Gland, Switzerland

Chapter 2
Sustainable Development – Background and Context

Gerd Michelsen, Maik Adomßent, Pim Martens, and Michael von Hauff

Abstract The debate about sustainability can be traced back into the eighteenth century. It was revived following the publication of the Brundtland Report "Our Common Future" (1987) and the United Nations Conference on Environment and Development in Rio de Janeiro (1992). Since then, interest has been focused on developing new concepts between the seemingly opposing paradigms of strong and weak sustainable development as well as on their application in practice. Moreover, sustainable development entails normative implications that affect inter- and intra-generational justice.

Keywords Sustainable development • Brundtland Report • Strong and weak sustainability • Inter- and intragenerational justice

1 The Idea and Historical Overview

Sustainable development is a concept that, since the 1992 Earth Summit in Rio de Janeiro and the *Agenda 21* adopted in its wake right up to the present, has been used, misused, and sometimes abused. In the concept of sustainable development, a number of social visions based around the ideas of justice, the frugal life, freedom and autonomy, the welfare of humankind, and responsibility for the future can be found with varying degrees of importance. Governments, companies, and non-governmental organizations, as well as national and international conferences, have all emphasized the importance of sustainability. However, one result of sustainability being considered in such widely different contexts is that the concept has become plagued by inaccuracies, ambiguities, and contradictions.

G. Michelsen (✉) • M. Adomßent
Faculty of Sustainability, Leuphana University, Scharnhorststraße 1, 21335 Lüneburg, Germany
e-mail: gerd.michelsen@uni.leuphana.de; adomssent@uni.leuphana.de

P. Martens
Maastricht University, Maastricht, The Netherlands

M. von Hauff
TU Kaiserslautern, FB Wirtschaftswissenschaften, Gottlieb-Daimler-Str., Geb. 42, D-67663 Kaiserslautern, Germany

H. Heinrichs et al. (eds.), *Sustainability Science*, DOI 10.1007/978-94-017-7242-6_2

In his book *Sustainability: A Cultural History*, Ulrich Grober carefully investigates our understanding of sustainability, beginning with the following question (Grober 2012: 15 f.):

> But what is 'sustainable?' The Dictionary of the German Language published in 1809 by Joachim Heinrich Campe, Alexander von Humboldt's teacher, defines Nachhalt (the root of nachhaltig, the German word for 'sustainable') as 'that which one holds on to when nothing else holds any longer'. That sounds comforting. Like a message in a bottle from a distant past, for our precarious times. Another message in a bottle, this one from the famous 1972 report The Limits to Growth (Club of Rome) says: 'We are searching for a model that represents a world system that is: 1. sustainable without sudden and uncontrollable collapse; and 2. capable of satisfying the basic material requirements of all people.'
>
> In both cases, sustainability is an antonym to "collapse". It denotes that which stands fast, which bears up, which is long-term, and resilient. It is immune to ecological, economic or social breakdown. What is striking is that the two terms, from such different epochs, are remarkably congruent. They both locate "sustainability" in the basic human need for security.

The book offers a rewarding insight into how the discussion of sustainability has developed over the centuries and which aspects played a role.

1.1 Beginnings of the Discussion About Sustainability

The origins of the concept of "sustainability" go back 300 years, when in 1713 the German mining director Carl von Carlowitz wrote a treatise on forestry, *Sylvicultura Oeconomica* (cf. Peters 1984; Schanz 1996; Di Giulio 2004). Carlowitz called for "continuous, steady and sustained use" of the forest. Sustainable forest management was to be based on the principle that only as many trees as would allow a continuous replenishment of an equivalent number of mature trees should be cut down in a single year, allowing the forest to be maintained and managed over the long term.

This principle of sustainability unites an economic criterion (e.g., maximum timber production securing the continuing existence of an individual business enterprise or livelihoods) and an ecological one (e.g., preserving a particular ecosystem). From an economic perspective, we can also derive the principle of living from the "interest" of capital (the annual growth in logged timber) and not from the capital itself (the forest). This principle was legally codified in German forestry at the end of the eighteenth century. Since then, sustainable forestry has, however, been reinterpreted a number of times.

At the beginning of the twentieth century, the concept of sustainability in the form of "maximum sustainable yield" was introduced to the fishing industry, and for similar reasons. Conditions were to be created that would allow maximum yields in relation to the size of the fish populations. For over 200 years then, the principle of sustainability, to the extent it was made use of at all, was limited to the timber and fishing industries. It had very little influence on other sectors of the economy. The business principle of "allowance for depreciation" comes closest to the goal of conservation of living from the yield and not from capital.

By the mid-eighteenth century, the first economic analyses focused on nature as a factor of production (in the sense of resources or land) had already appeared. Some 50 years later, the works of important economists such as David Ricardo and Thomas Malthus, as well as of the philosopher John Stuart Mill in the mid-nineteenth century, were premised on the idea of the limited carrying capacity of nature. Malthus, living in a time of extreme population growth in England, diagnosed an imbalance between the resources in a habitat and the size of its population. He predicted starvation, epidemics, and wars would follow. Today, these works are often considered the first systematic studies of the ecological limits on growth in a finite world and are credited with being an early source of critical sustainability. This work was given little attention at the time, however, as environmental problems on a national scale, much less a global, were not part of the political or social discourse at the time.

From the emergence of industrialization at the end of the eighteenth century until the mid-twentieth century, for most people, development was largely about economic and social issues. Questions of survival and improving work conditions were more urgent than what we would today call environmental problems. In addition, new methods in agriculture and food industries improved food supplies and, in spite of greater opportunities for consumption, the population grew more slowly or even remained stable. Malthus' pessimistic thesis was given less attention or even considered out of date. As a result, for more than 150 years, neoclassical economic theory and practice largely ignored nature as a factor in the analysis of production processes. It was not until the 1960s that economists such as Boulding (1966), Ayres and Kneese (1969), Georgescu-Roegen (1971), Ayres (1978), Daly (1973, 1977), and others put nature and the environment, and so, at least indirectly, sustainability, back on the economic agenda. In the wake of a series of environmental catastrophes that could no longer be disregarded, environmental protection became an issue of growing public concern. Winter smog in London and New York, devastating mercury poisoning in Japan, a tanker oil spill are only a few examples. The book *Silent Spring* by Rachel Carson, published in 1962 in the United States, had a very strong impact on the discussion of the risks of chemical pesticides on the environment. In 1972, the Club of Rome commissioned the report *Limits to Growth* (Meadows et al. 1972) and thrust the question of resources into the heart of environmental debates in more-developed countries. The report was based on work done by scientists at the Massachusetts Institute of Technology (MIT), who used computer programs to simulate different scenarios of the Earth's future. The most alarming forecast, and so the most widely reported in the media, was that the Earth would not be able to sustain a continuation of resource-intensive growth policies. Most scenarios show an eventual and significant decline in population and in the standard of living (Meadows et al. 2005). In Fig. 2.1, expected advances in extraction technologies for nonrenewable resources that might be capable of postponing the onset of increasing extraction costs are shown. Also evident is the alarming rise of pollution levels (even

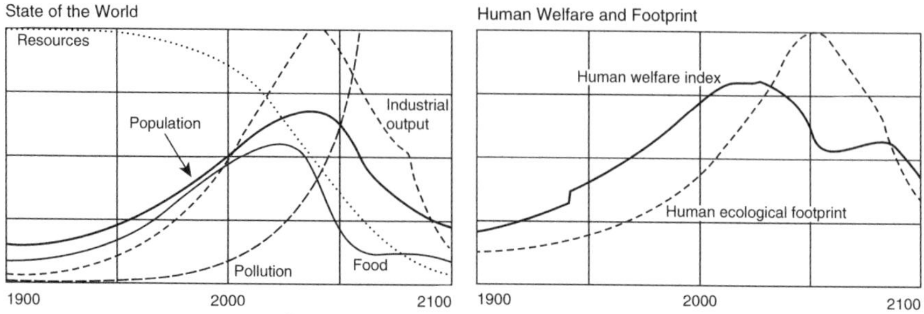

Fig. 2.1 Scenario 2: more abundant nonrenewable resources – one of 10 different scenarios for the future, through the year 2100 that were generated by of the World3 computer model (cf. Meadows et al. 2005). For detailed explanation see text

exceeding the borders of the graph!), to be followed by depressing land yields and requiring huge investments in agricultural revitalization. As a fatal consequence, the population will decline as a result of food shortages and negative health effects from pollution. The report started a largely scientific and political discussion of the relationships between the social means of production and lifestyles, economic growth, and the availability or finiteness of resources. Following the publication of the *Limits to Growth*, Scandinavian countries and the United States started an initiative to have environmental protection taken up by the United Nations.

- ***Task:*** *Find two current examples of both overexploitation and sustainable management of natural resources and describe one positive and one negative example in detail.*
- ***Question:*** *List the most important milestones in sustainability discourse and outline their meaning.*

1.2 Initiatives of the United Nations and Other Organizations

1.2.1 The Stockholm Conference and Its Consequences

In 1972, the first United Nations Conference on the Human Environment took place in Stockholm. The main political interest of the countries of the northern hemisphere was to head off an imminent environmental catastrophe by reaching an agreement on measures to limit industrial pollution and protect the environment. Contrastingly, on the priority list of the developing and undeveloped countries were items such as the eradication of poverty, the establishment of education and vocational training, access to clean water, and medical care – in short, social, and economic development. These were the first interest conflicts between the two goals of "environment" and "development" (Di Giulio 2004). The countries of the southern hemisphere – meaning the less-developed and undeveloped countries in the world – wanted to overcome their "backwardness" through rapid industrialization. Environmental problems were, to the extent that they were recognized at all, accepted as inevitable and were to be dealt with at a later point in time.

Nevertheless, there was a first rapprochement at the Stockholm Conference. The more-developed countries were able to persuade the developing and undeveloped countries that drought, flooding, and inadequate hygienic conditions were also environmental problems and that there was no contradiction between environmental protection and development. It was in this discussion that the formula "poverty is the biggest polluter" emerged. This made it possible for developing and undeveloped countries to become engaged in environmental protection without having to make compromises regarding their development goals. Furthermore, it became clear that the environmental problems recognized in the 1972 Conference (e.g., the destruction of the rainforest or pollution of the oceans) could not be solved without taking social and economic perspectives into account.

The *Action Plan for the Human Environment* adopted by the UN General Assembly in 1972 included:

- Measures for the collection of environmental data, for environmental research, and for monitoring and exchanging information
- Agreements on environmental protection and the efficient use of resources
- Establishment of environmental administration and management agencies
- Programs for the education, training, and information of the public

To implement this action program, the United Nations Environmental Program (UNEP) was established with headquarters in Nairobi, Kenya.

1.2.2 Other Environmental and Development Initiatives

Following the Stockholm Conference, the UNEP created concepts for alternative environmentally and socially acceptable paths of development. Under the heading "eco-development," the economic and consumption patterns of more-developed countries were criticized as models for other nations.

The 1974 Cocoyoc Declaration, the final statement of one of the joint conferences organized by UNCTAD (the United Nations Conference on Trade and Development) and the UNEP and held in the Mexican city of Cocoyoc, together with the 1975 *What Now* report by the Dag Hammarskjöld Foundation, introduced the problematic state of "overdevelopment" alongside the problem of underdevelopment. The demand that basic human needs must be met as an answer to poverty-related overpopulation and environmental destruction was contrasted with the call for a reduction of the exploitation of environmental resources by wealthy countries. A stable ecological and social balance can only be achieved by taking both aspects into account. In this context, issues of power and the distribution of wealth on both the international and the national level were identified as problems.

The concept of "eco-development" was first intended as a developmental approach for the largely rural regions of developing and undeveloped countries. Its theoretical framework, however, allowed it to be expanded to redefine growth and prosperity. Essential elements of this approach were:

- Meeting basic needs using a country's own resources and without imitating the consumption patterns of industrial countries
- Developing a so-called satisfactory social ecosystem, which includes employment, social security, and respect for other cultures
- Anticipatory solidarity with future generations
- Measures for the efficient use of resources and environmental conservation
- Participation of all parties
- Accompanying and supportive educational programs

(Haborth 1991)

The Bariloche Report *Limits to Poverty* (Herrera et al. 1977), published by the Argentinian foundation of the same name, took a more radical position and clearly rejected the thesis of limits to growth. Briefly, it was not economic growth but consumption by more-developed countries that was approaching its limits. These countries should restrict their consumption and make the resulting resources available to developing and undeveloped countries. Economic growth does not necessarily lead to increased environmental pollution, as there are technological solutions to this problem. What is decisive is that there is a comprehensive transfer of technology from north to south so that both development and environmental problems could be solved. Due to the intensifying global environmental situation, the ecological dimension was given greater priority in the subsequent international debate.

The International Union for Conservation of Nature (IUCN) published, together with the UNEP and the UNESCO, the *World Conservation Strategy*. This was the first time the term "sustainable development" was used in a contemporary context. Its core thesis was that without preserving ecological functionality (above all, agricultural, forest, coastal, and freshwater ecosystems), there would be no economic development. Sustainable development was understood as a concept in which the protection and conservation of nature would ensure the preservation of natural resources. Ecological issues (efficient use of resources, protection of species diversity, preservation of ecosystem functions) were given priority. There was less said about the political and socioeconomic conditions that were some of the main causes of the dangers facing the ecosystems.

In the 1980s, the view of ecological problems shifted somewhat from a focus on resources to the sink problem, that is, the threatened capacity of the ecosystem to absorb and process wastes. In addition, it became more widely understood that the production methods and lifestyle of the more-developed countries could not be transferred to the rest of the world – i.e., roughly 80 % of the world population. Linked to this insight, the more-developed countries were, due to their role in a majority of environmental and socioeconomic problems, given the main responsibility for finding a solution to these problems. The so-called Brandt Report (1980) and the subsequent *Palme Repo*rt (1983) – both the result of work done by the North–South Commission of the United Nations – were among the first international documents that dealt with this topic extensively. On the 10th anniversary of the Stockholm Conference in 1982, the United Nations Conference on the Human Environment met in Nairobi, Kenya, to develop a new and long-term strategy for the environment and development.

The concept "sustained livelihood" was introduced to the discourse of the environment and development by women's movements in developing and undeveloped countries (Wichterich 2002: 75). This approach focuses on "the local conditions of life, livelihood security and everyday experiences of women" (Wichterich 2002: 75). Livelihood is defined as the basis of existence, that is, "the capabilities, assets (including both material and social resources) and activities required for a means of living" (Scoones 1998 in Göhler 2003). The livelihood approach is about human beings, with all of their possibilities and strengths in their local situations. In the livelihood concept, the subsistence economy is of major importance.

- ***Task:*** *Contrast in a few words the different positions taken by developing and undeveloped compared to more-developed countries in sustainability discourse.*
- ***Question:*** *What role did the Stockholm Conference play in the north–south conflict? What activities followed this conference?*

1.2.3 The Brundtland Commission

In 1983, the United Nations appointed a World Commission on Environment and Development (WCED) chaired by the Norwegian Minister, President Gro Harlem Brundtland. The Brundtland Commission, as it came to be known, published its final report *Our Common Future* (WCED 1987), providing what came to be the best-known definition of the concept of sustainable development.

The WCED report built on the findings of the first environmental conference in Stockholm and the insight that the environment, the economy, and the society are mutually dependent and interrelated. Three basic principles were important for the Brundtland Commission in its problem analysis and recommendations for action: the global perspective, the linking of the environment and development, and the pursuit of justice. The report distinguished between two different perspectives on justice:

- The intergenerational perspective, in regard to responsibility for future generations

- The intragenerational perspective, in the sense of responsibility for different peoples living today, with a duty for wealthy countries to compensate poor countries

Faces of Sustainability
Gro Harlem Brundtland

- Born 1939
- Minister President of Norway (three terms)
- 1983–1987 Chair of the World Commission on Environment and Development (WCED)
- 1998–2003 Director General of the UN World Health Organization (WHO)
- Since 2007 a Special UN Envoy on Climate Change (Fig. 2.2)

Fig. 2.2 Gro Harlem Brundtland (Nett 2008)

The Brundtland Commission's most cited definition of sustainable development was: "*To make development sustainable – to ensure that it meets the needs of the present without compromising the ability of future generations to meet their own needs*" (WCED 1987: 8). Sustainable development is a process that aims at achieving a state of sustainability. The Brundtland Commission report called for the international community of nations to take urgent action. This demand was extensively discussed at the 1992 United Nations Conference on Environment and Development (UNCED) in Rio de Janeiro and subsequently implemented in a number of documents (cf. UN 1992a), most importantly in the *Agenda 21* (UN 1992b).

- ***Task:*** *Research and discuss the importance of the activities following the Rio Conference on Environment and Development* (UNCED) 1992.
- ***Question:*** *How do you personally evaluate the compatibility of economic growth and the tenets of sustainable development?*

2 Sustainable Development: Theoretical Concepts

Specifying the concept of sustainability and developing strategies for its implementation is an enormous challenge. And there are a number of different approaches that can be found in the literature about the goals, strategies, and instruments of sustainable development. The German Advisory Council on the Environment (SRU), an important commission set up by the German government, critically comments that the discussion on sustainable development is marked by an inflationary use of the term, which is partly due to the influence of interest groups, and that in general, there is a lack of precision concerning the concept and its definition (SRU 2002).

2.1 Ethical Implications

The concept of sustainable development is not the result of scientific research; instead, it is an ethically grounded concept. The ethical norms upon which this concept is based, one of the most common being the principle of fairness or justice, are not subject to critical examination in most works on sustainability, nor are reasons given for it. The German Advisory Council did attempt to ground the concept ethically in its 1994 Annual Report (cf. SRU 1994). In an ethics of responsibility, which it defines as "the unity of wisdom and duty" (SRU 1994: 51), the Council distinguishes three ethical elements of sustainable development:

- The responsibility of humanity for its natural environment
- The responsibility of humanity for its social world
- The responsibility of humanity for itself

Against a background of continuing ecological crises, the Council underscores the growing urgency of addressing the issue of environmental ethics. In its own attempts to deal with this issue, the Council follows an anthropocentric approach based on the principle of personality, that is, "the moral autonomy of a human being, i.e., his or her dignity as an individual person." It is on the basis of a human being's individuality and rationality that the Council derives the responsibility of humankind for the natural world. The core of this comprehensive set of environmental ethics is the interrelatedness of all social systems with nature, for which they coined the concept of "retinity":

> If human beings would like to preserve their personal dignity as rational creatures, both regarding the self and others, they can only fulfil their implied responsibility for nature if they make the interrelatedness of all their civilizational activities and products with nature, the basis of life, the principle of their actions. (SRU 1994: 54)

This responsibility of human beings for nature refers on one hand to securing the existence of nature and understanding that nature has its own importance and on the other to securing the natural basis for human life.

In addition to environmental friendliness, the Council identifies social acceptance or social appropriateness as a further criterion for sustainable human activity. The responsibility for the social world extends both to one's own social group or society and to present and future generations. The most important ethical principle is, according to the Council, "the demand for universal solidarity as a condition for the creation of social justice" (SRU 1994: 56).

Furthermore, the Council refers to the responsibility of individual human beings for themselves and for the success of their own lives, which is their essential purpose as free human beings. This means that the state is obligated to secure the right of the individual citizen to autonomy and the free development of their personality, as well as a just and equitable coexistence and the preservation of the natural basis of human existence. The real ethical challenge is in developing an ethical stance that sees freedom as freedom *in* responsibility for our natural and social world. The Council points out that, for the development of such an ethical stance, it is necessary to have a nuanced awareness of values, ethical sensitivity, and the capacity for judgment. These must be learned in social processes that develop ethical capacities.

Action that can be ethically justified and that is oriented to the idea that sustainability can be grounded, according the Council, on the principles of personality and retinity, as well as its compatibility with the environment, the society, and the individual person. Sustainability thus does not describe a scientifically observed fact. Instead, it is an ethical concept that conveys an idea of "how the world should be" (UBA 2002c: 16; Renn et al. 1999). It is about how people would like to live today and tomorrow, as well as about what kind of future is desirable (Coenen and Grunwald 2003). This discourse is related to environmental ethics and the relationship between human beings and their natural and artificial world, which is largely influenced by the interests, values, and ethical attitudes of social actors.

The ethical component of sustainability development becomes especially apparent when issues of the national or global distribution of exploitation and pollution rights are at stake, whether those resources are natural or socioeconomic. It is hardly surprising that, given the variety of cultures, political systems, and interests in the world, there are at times strongly divergent ideas about what is a fair distribution of these rights. There are also different views in science, politics, and social interest groups within countries as to how the concept of sustainable development should be defined and implemented (cf. Fig. 2.3).

Sustainability is also interpreted as a "regulative" idea, a concept that originated with the German philosopher Immanuel Kant. Regulative ideas are not concepts that specify something we have experienced but instead are practical regulating principles. Similar to the concepts of freedom and justice, sustainability should be understood as an open-ended and positive concept that is only a provisional specification of something. This open-endedness is due to the fact that social understandings of sustainable development are dependent on time and situation, as well as on culture and knowledge (cf. Enquete-Kommission 1998).

At this point, it should be emphasized again that ethical questions cannot be decided scientifically. Questions with a normative content can only be decided in social decision-making processes (cf. Kopfmüller et al. 2001). Sustainability research must always remain aware that it is part of social perceptual and evaluation processes. A scientific discussion of the concept of sustainability can provide and critically reflect on knowledge that helps orient social decision-making, but it cannot make normative principles themselves. "Scientific statements thus have, theoretically speaking, the structure of if-then statements" (Kopfmüller et al. 2001: 348).

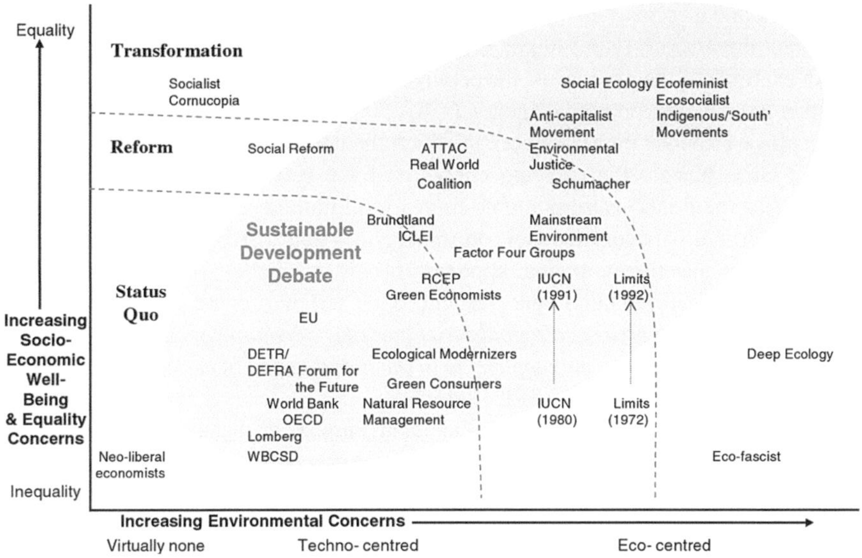

Fig. 2.3 Mapping different views of sustainable development (Hopwood et al. 2005: 41). The linking of justice criteria and environmental concerns illustrates the wide range of approaches regarding their normative priorities and so provides a useful basis for the critical evaluation of different interpretations and constructs in sustainability discourse

- ***Task:*** *Discuss the challenges related to the ethical implications of the concept of sustainable development and its "translation" into political action.*
- ***Question:*** *What problems do you see with the demand for intergenerational justice and what solutions would you propose?*

2.2 Dimensions of Sustainability

In the literature on the concept of sustainability, there is general agreement that sustainability can only be achieved by integrating the different dimensions of social development. However, there are different views as to the relative importance of these dimensions. Konrad Ott, a philosopher at the University of Kiel, Germany, points out that it is often assumed that the three dimensions, or pillars, of sustainability are equally important without the question of their equality ever being argued (Ott 2009). Some approaches give, for example, a primary role to the ecological dimension.

There are different ideas about the number and importance of the dimensions of sustainability, but in general, they can be divided into the "unidimensional" and "multidimensional" approaches (Tremmel 2003). In the unidimensional model, one dimension, for example, the ecological dimension is given a fundamental priority in case of a conflict between dimensions. Economic and social aspects are seen then as the causes and effects of environmental degradation but are not considered equally important dimensions (Kopfmüller et al. 2001).

An example of the primacy of the ecological dimension is given by the UBA, the German federal environmental agency (UBA 2002a, b). In their sustainability studies, ecology provides the framework in which the development of the economy and society takes place: "the carrying capacity of the ecosystem must therefore be accepted as the final, insurmountable limit for all human activities" (UBA 2002b: 2). The German Advisory Council on the Environment (SRU) in its 2002 Report also proposes that the ecological approach be given primacy, in particular, in the integration of environmental concerns into other policy sectors: "*This thought* [*of sustainability*] *has a clear ecological focus and thus accounts for the fact that environmental protection has, in comparison to the achievement of economic and social goals, the farthest still to go*" (SRU 2002: 68).

Multidimensional models, on the other hand, emphasize the equal importance of, typically, two to eight dimensions; however, the most common is the model with the three dimensions of ecology, society, and the economy. This model was introduced to the German debate on sustainability by the Enquete Commission of the German Federal Parliament on the Protection of Humanity and the Environment. Sustainability policy should be understood as a social policy in which the three dimensions of ecology, society, and the economy are of equal importance (Deutscher Bundestag 1998): "The main goal of sustainability is the maintenance and improvement of ecological, economic and social capabilities. These are mutually interdependent and cannot be optimized separately without endangering development processes as a whole" (Deutscher Bundestag 1998: 33).

As a result, on the one hand, "economic development and social well-being are only possible to the extent that nature as the basis of life is not endangered" (ibid.). On the other hand, ecological objectives are hard to reach if socioeconomic problems predominate in society or for the individual: "An ecologically dominant sustainability policy will always lose out in social decision-making processes when other problems prove to be more immediate, more tangible or more virulent and so more urgent and attractive for political action. Even if they can be successful they will be ineffective, since in the end only a policy integrating the three dimensions will be able to overcome the conceptual weakness of a discussion of the environment separated from economic and social issues" (Deutscher Bundestag 1998: 31 f.).

Two levels of argumentation are advanced in favor of the three-dimensional approach. First of all, together with natural resources, economic, social, and cultural values are seen as resources that, in their totality, provide the basis for satisfying human needs. Secondly, society can be endangered by ecological as well as economic or social risks. The carrying capacity of natural as well as social systems thus limits the scope for action of sustainable development. The environment, society, and the economy should be understood as independent but interrelated subsystems "whose functionality and resistance to breakdown should be preserved in the interests of future generations" (Kopfmüller et al. 2001: 49).

Economic dimension	Ecological dimension
caring economy; recycling economy; material flow management; environmental management system; environmentally friendly, innovative technologies; eco-design (operating life, disposability, aesthetics); prices reflecting ecological and social costs; polluter-pays principle; regional and local marketing networks; fair trade	efficient use of resources; nature's rhythms (regeneration, "propertime"); biodiversity; ecological lifecycle systems; regenerative energy; precautionary principle; avoiding ecosystem degradation (reducing pollutants, emissions, waste)
Social dimension	**Cultural dimension**
promoting human health; equal rights to the use of natural resources and to development; intrasocial justice; accounting for the interests of future generations; democratization; participation of all population groups in all areas of life, networks, livelihood through work	Ethical verification; sustainable lifestyle; holistic perception of nature; aesthetic perception of sustainable development; local cultural diversity of paths to sustainable development; traditional knowledge; experience of time; material culture; consumer awareness; local community; international exchange; global responsibility; cosmopolitan culture

Fig. 2.4 The four-dimensional model of sustainability (Following Stoltenberg 2010)

The goal of sustainable development is in this sense the avoidance of irreversible damage in all three dimensions.

The controversies in this discussion are found on two levels. On one hand, there are the arguments between advocates of the unidimensional and the multidimensional models, as discussed above. There is, however, a further controversy among those who endorse the unidimensional model but have different ideas about which dimension should be given priority. At an international level, developing and undeveloped countries have so far clearly given priority to social and economic development perspectives (including the issue of the global distribution of resources), which leads to their demands that the more-developed countries take the first step and shoulder the main burden. In contrast, countries in the northern hemisphere put ecological issues in the foreground (not least because they can afford to) and demand that the countries of the southern hemisphere take the initiative in solving these problems, where they believe progress can often be made at lower costs (Fig. 2.4).

In addition to the dimensions specified in the *Brundtland Report* of nature, society, and the economy, the additional dimensions most often discussed are the cultural, institutional, and, in developing and undeveloped countries, the political. Culture is defined (e.g., Meyer's Universal Dictionary 1983) as what human beings have created in a given period of time and in a defined region in their interaction with the environment. This includes, for example, language, religion, ethics, law, technology, science, art, and music, but also the processes of creating culture and cultural models, including individual and social lifestyles and types of behavior. Culture can therefore be understood as consisting of cultural values, world views, norms, and traditions which shape human beings' use of nature, their social interaction, and their economic means of production and consumption. This is a pragmatic understanding of culture which "enquires about the systems of knowledge that structure individual and social practice" (Holz and Stoltenberg 2011). Culture is understood less as a theoretical concept and more as an operative one. "A process of reflecting on sustainable, ethical values is primarily a cultural task. Sustainable development requires a change to a sustainable lifestyle" (Teller and Ax 2003: 89f). Calls for a culture of sustainability can be located at this level (Stoltenberg and

Michelsen 1999; Reisch 2002; Stoltenberg 2010). In this understanding, culture plays an important role in attaining a sustainable society and should be viewed as a separate dimension "since due to the concept of 'sustainability', our lifestyles, value systems, our education and economic systems or our way of developing technology as cultural background" of the other dimensions must be "critically evaluated and, if necessary, changed" (Stoltenberg 2000: 12).

- ***Task:*** *Find aspects of a problem of global unsustainable development, determine which sustainability dimension they are part of, and suggest appropriate proposals for solutions.*
- ***Question:*** *How do you evaluate the self-evident way that the ecological dimension is considered to have priority over the other dimensions?*

2.3 Weak and Strong Sustainability

A distinction is made in the scientific literature between *weak* and *strong sustainability* (see von Hauff 2014; Ott 2009; Meyer-Abich 2001; Scherhorn and Wilts 2001; SRU 2002; Ott and Döring 2008; Egan-Krieger et al. 2009). The chief distinction is based on what should be preserved over the long term and, closely related to this, whether existing types of capital are substitutable (see Table 2.1). Capital here is generally defined as a stock "whose yield is available and of use to *homo economicus*" (SRU 2002: 65).

A problematic aspect in the history of economics was the reduction of natural factors of production to "land" and "resources." Land and nonrenewable resources are now seen to be only components of "natural capital," the complex of which is increasingly recognized in more recent economic theory (Held and Nutzinger 2001). However, it is difficult to be more precise about what natural capital is, as its components are interrelated. Lists of different types of natural capital accordingly suffer from overlapping items. In fact, it is not possible to create a list of differentiated, unambiguous, distinct elements of natural capital. Instead, natural capital is characterized by all-encompassing terms such as "natural resource base," "natural basis of life," "ecosystem capacity," "stability of ecological systems," "biodiversity," etc. (SRU 2002: 64).

There are a number of different types of capital (cf. Ott 2009; SRU 2002):

- Natural capital (natural resources, such as water and air)
- Manufactured capital (machines, factories, equipment, infrastructure)
- Cultivated natural capital (forests, plantations, domesticated animals)
- Social capital (moral concepts, institutions)
- Human capital (person-specific knowledge, such as education and skills)
- Knowledge capital (nonperson specific, stored, and retrievable knowledge)

The concept of weak sustainability "is based on the understanding that the substitutability of different kinds of capital is, in principle, to a large extent unlimited" (Ott 2009). This implies that natural capital can be replaced or substituted by other

Table 2.1 Concepts of sustainability

	Very weak sustainability	Weak sustainability	Strong sustainability	Very strong sustainability
What should be preserved?	Total capital (human-made and natural)	Essential natural capital	Nonrenewable natural capital	Nature has its own value
Why?	Human welfare	Human welfare	Human welfare and obligations to nature	Obligations to nature
Management strategy?	Maximization of economic growth	Sustainable economic growth	Zero growth; sustainable growth if environment is not endangered	Zero growth, sometimes reduction of economic values
Substitutability between human-made and natural capital?	Unlimited in principle	Not always possible between man-made and natural capital	Not always possible between man-made and nonrenewable natural capital	Rejects substitutability debate
Ethics?	Instrumental value of nature	Instrumental value of nature	Priority: value of the ecosystem	Intrinsic value of nature

Eblinghaus and Stickler 1998; Dobson 2002; Rieckmann 2004; Steurer 2001

types of capital (natural capital), for example, forests by parks or natural lakes by swimming pools. The assumption is that it does not matter what the physical composition of the capital stock is, that is, passed on to the next generation. What is important is that the total capital and total utilization, in effect, the total level of welfare remains constant. Weak sustainability is thus related to neoclassical utility theory, in which it is irrelevant how the utility is created.

The paradigm of weak sustainability emerged as a reaction to the first report by the Club of Rome "The Limits to Growth" (Meadows 1972). In 1974, at the "Symposium on the Economics of Exhaustible Resources", the topic of economic growth with finite resources was discussed. Joseph Stiglitz proposed three factors that were not addressed in the Club of Rome study, which called into question the findings of the report "The Limits to Growth." These three factors recognize the central importance of technological progress: "There are at least three economic forces offsetting the limitations imposed by natural resources: technical change, the substitution of man-made factors of production (capital) for natural resources, and returns to scale" (Stiglitz 1974: 123). These three factors, according to the view of the economists who attended the symposium, make it possible for all people living with a constant per capita consumption in the future to have at least the same level as the people living today. Commenting on the 1987 Brundtland Report, Robert Solow defined sustainability as follows: "I could say this about that: it is an obligation to conduct ourselves so that we leave to the future the option or the capacity to be as well off as we are" (Solow 1993: 181). In the final analysis this means that there is no imperative to preserve certain elements of nature.

The capital stock left to subsequent generations is composed of cumulative real capital and natural capital (i.e., the combined state of the environment). Both kinds of capital are generally interchangeable over time, although the substitution of natural by real capital is the dominant form. It is crucial here that consumer goods can be made available in the same volume at all times and the level of consumption is preserved for the individual, that is, our children and their children (v. Hauff 2014: 47). In this context, it is all about how to evaluate the future costs of exploiting nature today or the ongoing reduction of natural capital. The questions arise as to intertemporal justice, which examines how, for example, pollution and the use of finite resources can be projected to future periods.

This is the position taken by the neoclassical theorists with regard to sustainable growth, which led to the paradigm of weak sustainability and remains, to this day, the dominant neoclassical position on sustainability.

Steurer (2001) sees the "quantitative growth paradigm" in weak sustainability. Finally, it may be said of the weak sustainability paradigm that the proponents of solutions to problems – similar to many growth theorists – focus mainly on technological progress or technical solutions. A key assumption according to the Hartwick Rule is that technical advances lead to the substitution of natural resources. This is the context of the term technical optimism. However, whether these solutions always arrive at the required time is an issue that is often neglected (Table 2.2).

Advocates of strong sustainability, on the contrary, believe that human-made capital and natural capital can only be complementary and are thus only interchangeable to a very limited extent (cf. Daly 1999a; Ott 2009; Ott and Döring 2008).

Faces of Sustainability
Herman E. Daly

- Born in 1938.
- Professor emeritus at the University of Maryland.
- 1988–1994 Senior Economist in the Environment Department of the World Bank.
- Daly was one of the first to warn of the ecological limits to economic growth. He was the originator of the management rules for sustainable development (Fig. 2.5).

Fig. 2.5 Herman E. Daly (The European 2011)

Table 2.2 The level model of sustainability

Level	Theoretical status (cf. Stegmüller 1980; Ott and Döring 2008: 345 ff.)
1. Idea (theory of inter- and intragenerational justice)	Core theory
2. Conception (strong or weak sustainability, mediating conceptions)	
3. Constant natural capital rule, management rules	
4. Guidelines (resilience, sufficiency, efficiency)	Bridge principle
5. Action dimension (nature conservation, agriculture and forestry, fishery, climate change, etc.)	Application cases
6. Target systems, special concepts and models, and indicators	
7. Implementation, institutionalization, instrumentation	

Following Döring 2009

Some scholars would require that the volume of the individual elements of natural capital (such as climate factors, landscapes, biodiversity, etc.) should be kept as constant as possible. The assumption is that human beings are dependent on the ecological functions of nature, and so, these are not substitutable (cf. SRU 2002). However, a certain degree of substitution is possible within specific types of capital. For example, the loss of a forested area can be replaced by reforesting in another area, or the use of oil can be compensated by investments in renewable energies. The "limits to growth" paradigm can be seen in the concept of strong sustainability (Steurer 2001). The environmental space concept is an attempt to operationalize strong sustainability (cf. Table 2.1) by defining "the resource base and sink functions that people use in their natural environment without irreversibly damaging it" (SRU 2002: 65).

Strong sustainability is the opposite of the neoclassical sustainability concept and was developed by the proponents of ecological economics. They soundly reject the substitution rule. One of the most important advocates of ecological economics is Herman E. Daly. In many publications over the past decades, he advanced the idea of a "steady-state economy". This aims at a stationary state. Thoughts on the stationary state have been introduced by other economists too. For example, Adam Smith wrote about a stationary state back in the 1700s (Smith 1776: 99). However, he concluded that this state leads to poverty, and from his reasoning, he deduced that only growth can guarantee prosperity. Other economists in contrast to Smith assumed the existence of a stationary state and thought of it as desirable. These include economists such as Malthus, Marx, Mill, Schumpeter, and Keynes.

In more recent times, the steady-state approach has been driven primarily by Daly, who was inspired by John Stuart Mill. He justifies the limits of quantitative growth with two laws of thermodynamics. In doing so, he borrowed from the writings of Georgescu-Roegen, who among others, called for a greater involvement of scientific principles in economics (Georgescu-Roegen 1971). In contrast to the neoclassical economists, he came to the conclusion that quantitative growth not only reaches a limit at a certain point but even becomes uneconomical. His considerations are based on the realization that on a microeconomic level, a state can occur that can be labeled as uneconomical.

A business or a household aims for a level of activities that is optimal. If this level is exceeded by other activities, it may be that the additional costs (marginal costs) are greater than the marginal utility. Daly describes this case as "uneconomic". Considering this from a macroeconomic perspective, the microeconomic variables mentioned are to be aggregated to the macroeconomic level. Consequently, ever more natural resources (green flow) are used to produce tangible goods (brown flow). "As we expand the brown flow, we reduce the green flow" (Daly 1999b: 5). This case, according to Daly, represents "uneconomic growth".

To specify this point in more detail, it is necessary to examine the costs and benefits of increasing growth separately from one another. The economic limit of growth is reached when the marginal costs of growth are equal to the marginal utility, that is, when an economy has reached its optimum size. This assumes capital stock is kept at a constant level. This can be achieved if the employment figures remain constant, which, in turn, requires that the population also remain constant. This implies that the birthrate and immigration is balanced with the death rate and emigration.

The steady-state economy can be thought of as an economic system that is designed to ensure a constant supply of tangible goods that is sufficient to provide the "good life" for the population. However, the specifics of this design have not been adequately demonstrated. It is also not clear what happens if such a state is not achieved for all people. The neoclassical economists, in particular, are quick to critique other points. Primarily, this concerns the question of the macroeconomic effects, which have not been sufficiently analyzed by Daly. In the context of an economy with no growth: effects on the labor markets, wealth distribution, poverty, the financial sector, commerce, and the tax system which, in turn, affects the state budgets.

Conclusion: The neoclassical position stands in irreconcilable opposition to the ideas of environmental economics even up to this day.

- ***Task:*** *Choose an example of non-sustainable development, go through the possibilities and limits of applying the concept of strong sustainability, and come up with ideas for possible solutions.*
- ***Question:*** *In which areas of international environmental policy can elements of the strong sustainability concept be found?*

2.4 The Integrative Concept of Sustainable Development

From 1998 to 2002, a number of different members of the Helmholtz Association of German Research Centres (HGF) carried out a research project to define and implement the concept of sustainable development. They developed an integrative concept of sustainable development, defined by its constitutive elements, its goals, and its rules (cf. Kopfmüller et al. 2001; Coenen and Grunwald 2003), and applied it to the situation in Germany. It has to be understood that apart from this German discourse, there is extensive relevant work from other communities and countries (among others, the triple bottom line (Elkington 1997), sustainability assessment (Gibson 2006; Hardi and Zdan 1997), or the development of indicators for sustainable developments (Bossel 1999)).

Constitutive elements of the integrative HGF concept are those that, based on a review of the research literature, were found to be essential in shaping the concept of sustainable development. A central constitutive element is intra- and intergenerational justice. The Brundtland Commission famously defined sustainable development as a state "that meets the needs of the present without compromising the ability of future generations to meet their own needs" (WCED 1987). Intragenerational justice refers then to the needs of the present generation and has as its goal that all human beings on Earth are able to enjoy a decent life – one that at least satisfies its fundamental needs. Intergenerational justice, on the contrary, strives to ensure that future generations will also be able to satisfy their needs. This will only be possible if the present generation passes on to the next generation the conditions that will allow them to choose their own lifestyle. As this critically involves the fair distribution of natural resources, economic goods and basic social goods, distributive justice plays an important role. In the literature, there are a number of different perspectives on the relationship between intra- and intergenerational justice. The HGF approach is based on the definition of sustainable development discussed above and sees both principles as equally important.

Another constitutive element is sustainable development's global orientation. The global validity of sustainability is based on three levels. The first is an ethical one. In a global ethic, all human beings have a moral right to satisfy their basic needs, to fulfil their desire for a better life, to preserve the ecosystem functions essential to life, and to have equal access to global resources. Secondly, there is a problem-oriented justification. Many of the known sustainability problems, such as the anthropogenic greenhouse effect, the destruction of the ozone level, the loss of biodiversity, population growth, and unemployment, are global problems and differ only in their regional characteristics. And thirdly, there is a strategic justification for sustainable development. The problems of non-sustainable development are global and the strategies for their solutions must be identified, developed, and implemented on a global level.

A further constitutive element is the anthropocentric approach. The concept of sustainable development is above all a concept that focuses on human needs. Human beings are conceded rights to use nature, but these rights are related to duties. The thoughtful and careful use of natural resources is an example of such a duty, and it is in the best interests of human beings. It is crucial to preserve the functions nature fulfils for humankind over the long term. The use of nature does not only consist of exploiting raw materials or disposing of waste materials but also of cultural uses, such as the "aesthetic use" of landscapes. This is referred to as an "enlightened anthropocentric approach". Missing from this discourse is the concept of nature or nonhuman creatures having their own rights (e.g., animal rights).

Part of the HGF research project was to first develop "general goals of sustainable development" (cf. Coenen & Grunwald) so as to then specify the constitutive elements. These goals are the necessary conditions for a sustainable development that would fulfil the constitutive elements described above.

General goals include:

- *Safeguarding human life:* It is of utmost importance that the present generation be prevented from destroying the conditions for the life of future generations. That means first of all that the functions of nature necessary for human life must be maintained. A further consequence is that all human beings in the world are ensured the possibility to lead a decent life.
- *Maintaining society's productive potential*: Coming generations must have similar opportunities to meet their needs, which may differ from those of the present generation. This gives us another general goal of sustainable development, namely, that the productive capacity of (global) society be preserved in a very general sense over time. In addition to natural (renewable and nonrenewable) resources, society's productive potential includes human knowledge.
- *Preserving the scope for development:* The requirement not to endanger the possibility for future generations to satisfy their needs must include both material and immaterial needs. Today's generation thus should not restrict the scope for coming generations to take different paths of development. This also means that the possibilities for individuals to develop themselves must be preserved, both today and in the future.

In order to operationalize these goals, so-called sustainability rules have been developed. A distinction is made between:

- *Substantive rules of sustainability*: These rules are considered the minimum conditions for reaching the goals of sustainability. They are also referred to as the "what rules" of sustainability.
- *Instrumental rules of sustainability*: These rules describe the institutional, economic, and political conditions for sustainable development. They are about how to fulfil the minimum conditions for sustainable development and so, are also referred to as the "how rules" (cf. Kopfmüller et al. 2001).

The substantive rules of sustainability can be ordered according to the general goals of sustainable development described above:

1. *Preserving human life:* Dangers and unacceptable risks for human health from anthropogenic environmental degradation are to be avoided. Basic human needs (housing, food, clothing, health) must be satisfied, and major risks to life (illness, invalidity) must be minimized. All members of society must be guaranteed the

opportunity to secure a livelihood (including raising children and being cared for in old age) by work freely taken up. This rule goes beyond satisfying basic human needs to ensuring an autonomous life. The use of the natural environment is to be justly distributed by means of the fair participation of all. Extreme differences in income or wealth are to be reduced. Poverty that makes it impossible to take an active role in social life must also be eliminated.

2. *Maintaining society's productive potential*: The rate at which renewable resources are used must not exceed their regenerative capacity nor endanger the productivity or functionality of the ecosystem. The availability of resources known to be nonrenewable is to be preserved over time. This involves reducing their consumption (sufficiency), increasing their productivity (efficiency), or replacing them with renewable resources (consistency). Releasing materials into the environment must not exceed the capacity of environmental media and ecosystems to absorb them. Technological risks with potentially catastrophic effects for human beings and the environment are to be avoided. Manufactured, human, and knowledge capitals are to be developed so that the economic productivity can be maintained or enhanced.
3. *Preserving the scope for development:* All members of society must have the same opportunities regarding access to education, information, professions and occupations, public office, and social, political, and economic positions. All members of society must be able to participate in socially relevant decision-making processes. This includes, for example, maintaining or enhancing democratic types of decision-making or conflict regulation. The cultural heritage of humanity and cultural diversity are to be preserved. Cultural and natural landscapes or landscapes with special characteristics and beauty are to be preserved. In order to guarantee the social integrity of society, a sense of law and justice, tolerance, and solidarity and an orientation to the common good, as well as the potential to deal with conflicts nonviolently, are to be strengthened.

The instrumental rules of sustainability are the so-called how rules. These are about economic and institutional aspects of sustainable development (cf. Coenen and Grunwald 2003).

1. *Internalization of social and ecological costs*: Prices must reflect the ecological (e.g., resource scarcity, degraded ecosystems) and social costs (e.g., child labor, health risks, unemployment) that are created in economic processes.
2. *Appropriate discount rate*: Discounting must not discriminate against present or future generations.
3. *Responsible debt-making*: New debts should be limited to investments that serve to satisfy future needs.
4. *Fair global economic conditions*: Fair participation in economic processes, especially access to the market for developing and undeveloped countries.
5. *Promoting international cooperation*: Countries, NGOs, companies.
6. *Increasing social awareness of relevant problems*: Increasing perception and awareness of problems, consciousness of problems, and consciousness of the

possibilities for all social actors to take action through institutional innovation.
7. *Development of institutional conditions*: For the analysis and evaluation of the effects of social actions.
8. *Increasing the capacity for governance*: New types of social governance are necessary for sustainable development.
9. *Promoting the self-organization potential of social actors*: New types of cooperative and participative decision-making need to be developed that will contribute to the strengthening of civil society while still functioning alongside established institutions.
10. *Strengthen the balance of power*: Opinion-building processes, negotiation, and decision-making processes should be constructed so that all actors have the possibility to articulate their interests and demands. This process should be transparent. All those involved should have the same opportunities to gain acceptance for their position.

These rules provide a normative base for sustainable development and serve as a means to achieve its goals. In order for these rules to be practically relevant, they need to be guided by indicators, which are a further step to operationalizing the integrative concept of sustainable development.

- ***Task:*** *Compare the integrative concept of sustainable development with the concept of strong sustainability and find the similarities and differences of both approaches.*
- ***Question:*** *Which strengths and weaknesses do you see in the integrative concept of sustainable development?*

The normative concept of sustainable development has quite a long history. Since the Brundtland Report and the Rio Conference "Environment and Development" in 1992, a variety of different implementation strategies are discussed. Thus, a distinction is made between strong and weak sustainability, and different dimensions of sustainability – ecological, economic, social, and cultural – are referred to. Sustainable development has to be understood as a process that ought to include as many people as possible and is confined within well-defined corridors (upper and lower limits) and is limited. Scientific research on non-sustainable development and possible solutions is usually inter- and transdisciplinary, directly involving relevant social actors. Sustainable development is understood as the search, thinking, and design process that focuses on the idea of a just and sustainable society.

References

Ayres RU (1978) Resources, environment, and economics: applications of the materials/energy balance principle. Wiley-Interscience, New York

Ayres RU, Kneese AV (1969) Production, consumption, and externalities. Am Econ Rev 59:282–297

Bossel H (1999) Indicators for sustainable development: theory, method, applications. International Institute for Sustainable Development, Manitoba. Available at https://www.iisd.org/pdf/balatonreport.pdf. Accessed 1 Dec 2014

Boulding KE (1966) The economics of the coming spaceship Earth. In: Jarrett HE (ed) Environmental quality in a growing economy. Johns Hopkins University Press, Baltimore

Coenen R, Grunwald A (eds) (2003) Nachhaltigkeitsprobleme in Deutschland. Analyse & Lösungsstrategien. Ed. Sigma, Berlin

Daly HE (1973) Toward a steady-state economy. W.H. Freeman, San Francisco

Daly HE (1977) Steady state economics: the economics of biophysical equilibrium and moral growth. W.H. Freeman, San Francisco

Daly HE (1999a) Wirtschaft jenseits von Wachstum: die Volkswirtschaftslehre nachhaltiger Entwicklung. Pustet, Salzburg

Daly HE (1999b) Uneconomic growth in the theory and in fact. In: The first annual Feasta Lecture, Trinity College, Dublin, 26th April, 1999. Available at http://www.feasta.org/documents/feastareview/daly.htm. Accessed 5 Aug 2015

Deutscher Bundestag (1998) Konzept Nachhaltigkeit: vom Leitbild zur Umsetzung. Abschlussbericht der Enquete-Kommission "Schutz des Menschen und der Umwelt – Ziele und Rahmenbedingungen einer nachhaltig zukunftsverträglichen Entwicklung" des 13. Deutschen Bundestages. Bonn: Dt. Bundestag, Referat Öffentlichkeitsarbeit

Di Giulio A (2004) Die Idee der Nachhaltigkeit im Verständnis der Vereinten Nationen – Anspruch, Bedeutung und Schwierigkeiten. LIT, Münster

Dobson A (2002) Fairness and futurity: essays on environmental sustainability and social justice. Oxford University Press, Oxford

Döring R (2009) Sustainability, natural capital and nature conservation – introduction. In: Döring R (ed) Sustainability, natural capital and nature conservation. Metropolis, Marburg, pp 5–15

Eblinghaus H, Stickler A (1998) Nachhaltigkeit und Macht: Zur Kritik von Sustainable Development. 3. Auflage. IKO – Verl. für Interkulturelle Kommunikation, Frankfurt am Main

Elkington J (1997) Cannibals with forks: the triple bottom line of twenty-first century business. Capstone, Oxford

Enquete-Kommission "Schutz des Menschen und der Umwelt" des Deutschen Bundestages (1998) Konzept Nachhaltigkeit. Vom Leitbild zur Umsetzung. Deutscher Bundestag, Bonn

Georgescu-Roegen N (1971) The entropy law and the economic process. Harvard University Press, Cambridge, MA

Gibson RB (2006) Sustainability assessment: basic components of a practical approach. Impact Assess Proj Apprais 24:170–182

Göhler D (2003) Livelihood Strategien unter besonderer Berücksichtigung der Waldressourcen. Dargestellt am Beispiel der Fokontany Tsilakanina im Nordwesten Madagaskars. Deutsche Gesellschaft für Technische Zusammenarbeit (GTZ) im Auftrag des Bundesministeriums für wirtschaftliche Zusammenarbeit, Eschborn

Grober U (2012) Sustainability: a cultural history. Green Books, Cambridge

Harborth H-J (1991) Dauerhafte Entwicklung statt globaler Selbstzerstörung. Eine Einführung in das Konzept des Sustainable Development. Edition Sigma, Berlin

Hardi P, Zdan T (eds) (1997) Assessing sustainable development: principles in practice. International Institute for Sustainable Development, Manitoba. Available at https://www.iisd.org/pdf/bellagio.pdf. Accessed 1 Dec 2014

Held M, Nutzinger HG (eds) (2001) Nachhaltiges Naturkapital – Perspektive für die Ökonomik. Campus, Frankfurt/Main et al

Herrera AO, Scolnik HD, Menke-Glückert P, Janic O (1977) Grenzen des Elends. Das Bariloche-Modell: So kann die Menschheit überleben. S. Fischer, Frankfurt am Main

Holz V, Stoltenberg U (2011) Mit dem kulturellen Blick auf den Weg zu einer nachhaltigen Entwicklung. In: Sorgo G (ed) Die unsichtbare Dimension. Bildung für nachhaltige Entwicklung im kulturellen Prozess. Forum Umweltbildung, Wien, pp 15–34

Hopwood B, Mellor M, O'Brien G (2005) Sustainable Development: Mapping Different Approaches. Sustain Dev 13:38–52
Kopfmüller J, Brandl V, Jörissen J, Paetau M, Banse G, Coenen R, Grunwald A (2001) Nachhaltige Entwicklung integrativ betrachtet. Konstitutive Elemente, Regeln, Indikatoren. Ed. Sigma, Berlin
Meadows DH, Meadows DL, Randers J, Behrens WW III (1972) Limits to growth. New American Library, New York
Meadows DH, Randers J, Meadows DL (2005) The limits to growth: the 30-year update. Earthscan, London. Available at http://www.mnforsustain.org/meadows_limits_to_growth_30_year_update_2004.htm. Accessed 30 Apr 2014
Meyer-Abich KM (2001) Nachhaltigkeit – ein kulturelles, bisher aber chancenloses Wirtschaftsziel. Zeitschrift für Wirtschafts- und Unternehmensethik 2:291–310
Ott K (2009) On substantiating the conception of strong sustainability. In: Döring R (ed) Sustainability, natural capital and nature conservation. Metropolis, Marburg, pp 49–72
Ott K, Döring R (2008) Theorie und Praxis starker Nachhaltigkeit. 2nd edition, Metropolis, Marburg
Peters W (1984) Die Nachhaltigkeit als Grundsatz der Forstwirtschaft, ihre Verankerung in der Gesetzgebung und ihre Bedeutung in der Praxis – die Verhältnisse in der Bundesrepublik Deutschland im Vergleich mit einigen Industrie- und Entwicklungsländern. Dissertation, Universität Hamburg
Reisch L (2002) Kultivierung der Nachhaltigkeit – Nachhaltigkeit als Kultivierung? GAIA 11:113–118
Renn O, Knaus A, Kastenholz H (1999) Wege in eine nachhaltige Zukunft. In: Breuel B (Hrsg.) Agenda 21. Vision: nachhaltige Entwicklung. Campus, Frankfurt am Main, pp 17–74
Rieckmann M (2004) Lokale Agenda 21 in Chile. Eine Studie zur Implementation eines lokalen Agenda 21-Prozesses in der Cuenca del Lago Llanquihue. oekom, München
Schanz H (1996) Forstliche Nachhaltigkeit. Sozialwissenschaftliche Analyse der Begriffsinhalte und Funktionen. Universität Freiburg, Institut für Forstökonomie, Freiburg im Breisgau
Scherhorn G, Wilts CH (2001) Schwach nachhaltig wird die Erde zerstört. Gaia 10:249–255
Smith A (1776) An inquiry into the nature and causes of the Wealth of Nations. Strahan, London
Solow RM (1993) Sustainability – an economist's perspective. In: Dorfman R, Dorfman NS (Hrsg.) Economics of the environment, selected readings. Norton & Company, New York/London, pp 179–187
SRU – The German Advisory Council on the Environment (1994) Umweltgutachten 1994. *Für* eine *dauerhaft-umweltgerechte Entwicklung*. Metzler-Poeschel, Stuttgart
SRU – The German Advisory Council on the Environment (2002) Environmental report 2002. Towards a new leading role. Available at http://www.umweltrat.de/SharedDocs/Downloads/EN/01_Environmental_Reports/2002_Environmental_Report_summary.pdf?__blob=publicationFile. Accessed 30 Apr 2014
Stegmüller W (1980) Neue Wege der Wissenschaftsphilosophie. Springer, Berlin/Heidelberg/New York
Steurer R (2001) Paradigmen der Nachhaltigkeit. Zeitschrift für Umweltpolitik und Umweltrecht 24:537–566
Stiglitz JE (1974) Growth with exhaustible natural resources – efficient and optimal growth paths. Rev Econ Stud 41. Jg.:123–137
Stoltenberg U (2000) Umweltkommunikation in Lokalen Agenda 21-Prozessen. In: Stoltenberg U, Nora E (eds) Lokale Agenda 21. Akteure und Aktionen in Deutschland und Italien. – Agenda 21 Locale. Attori ed Azioni in Germania ed in Italia. VAS, Frankfurt am Main, pp 11–14
Stoltenberg U (2010) Kultur als Dimension eines Bildungskonzepts für eine nachhaltige Entwicklung. In: Parodi O, Banse G, Schaffer A (eds) Wechselspiele: Kultur und Nachhaltigkeit. edition sigma, Berlin, pp 293–311
Stoltenberg U, Michelsen G (1999) Lernen nach der Agenda 21. Überlegungen zu einem Bildungskonzept für eine nachhaltige Entwicklung. NNA-Berichte 12(1):45–54

Teller M, Ax C (2003) Nachhaltigkeit gilt als Schlüsselbegriff für eine zukunftsfähige Welt. Wechselwirkung & Zukünfte 25(1):83–92

The European (2011) Herman Daly. Available at: http://c0964762.cdn.cloudfiles.rackspacecloud.com/images/6693/insight/herman_daly_interview.jpg?1313489087. Accessed 30 Apr 2014

Tremmel J (2003) Nachhaltigkeit als politische und analytische Kategorie. Der *deutsche Diskurs um nachhaltige Entwicklung* im *Spiegel* der *Interessen* der *Akteure*. oekom, München

UBA – The Umweltbundesamt (2002a) Sustainability in Germany: creating a lasting environmentally compatible future. Erich Schmidt, Berlin

UBA – The Umweltbundesamt (2002b) Kommunale Agenda 21 – Ziele und Indikatoren einer nachhaltigen Mobilität. Texte 8/02. Erich Schmidt, Berlin

UBA – The Umweltbundesamt (2002c) Nachhaltige Entwicklung in Deutschland. Die Zukunft dauerhaft umweltgerecht gestalten. Erich Schmidt, Berlin

UN – United Nations (1992a) Rio declaration on environment and development. Available at http://www.un.org/documents/ga/conf151/aconf15126-1annex1.htm. Accessed 30 Apr 2014

UN – United Nations (1992b) United Nations conference on environment & development Rio de Janeiro, Brazil, 3–14 June 1992 – Agenda 21. Available at: http://sustainabledevelopment.un.org/content/documents/Agenda21.pdf. Accessed 30 Apr 2014

Nett VG (2008) Fulgte hjertet og hjalp Gro. VG Nyheter from 8 Jan 2008. Available at http://www.vg.no/nyheter/innenriks/helse-og-medisin/fulgte-hjertet-og-hjalp-gro/a/515142/. Accessed 7 Aug 2015

von Egan-Krieger T, Schultz J, Thapa PP, Voget L (eds) (2009) Die Greifswalder Theorie starker Nachhaltigkeit. Beiträge zur Theorie und Praxis starker Nachhaltigkeit, 2. Metropolis, Marburg

von Hauff M (2014) Nachhaltige Entwicklung. Oldenbourg, München

WCED – World Commission on Environment and Development (1987) Our common future. Oxford University Press, New York. Available at http://www.un-documents.net/our-common-future.pdf. Accessed 30 Apr 2014

Wichterich C (2002) Sichere Lebensgrundlagen statt effizienterer Naturbeherrschung – Das Konzept nachhaltige Entwicklung aus feministischer Sicht. In: Görg C, Brand U (eds) Mythen globalen Umweltmanagements: “Rio+10” und die Sackgassen “nachhaltiger Entwicklung”. Westfälisches Dampfboot, Münster, pp 72–91

Chapter 3
Transformational Sustainability Research Methodology

Arnim Wiek and Daniel J. Lang

Abstract Sustainability science can roughly be differentiated into two distinct research streams – a "descriptive-analytical" and a "transformational" one. While the former is primarily concerned with describing and analyzing sustainability problems, the latter aims at developing evidence-supported solution options to solve these problems. This chapter presents relevant methodological guidelines and requirements as well as five exemplary research frameworks for transformational sustainability research. The frameworks are for (1) complex problem-handling, (2) transition management and governance, (3) backcasting, (4) integrated planning research, and (5) the transformational sustainability research (TRANSFORM framework). The TRANSFORM framework aims at synthesizing key components of the other frameworks. The frameworks provide guidelines for transformational sustainability research; yet, willingness and capacity of academic, governmental, private, and nonprofit organizations to use them for knowledge-generating operations are still fairly low. To truly support sustainability transformations, much more of this solution-oriented sustainability research is needed.

Keywords Descriptive-analytical sustainability research • Transformational sustainability research • Knowledge types • Methodological frameworks • Evidence-supported solution options

---

A. Wiek (✉)
School of Sustainability, Arizona State University,
PO Box 875502, Tempe, AZ 85287-5502, USA
e-mail: arnim.wiek@asu.edu

D.J. Lang
Faculty of Sustainability, Institute for Ethics and Transdisciplinary Research,
Leuphana University of Lüneburg, Scharnhorststr. 1, 21335 Lüneburg, Germany
e-mail: daniel.lang@leuphana.de

H. Heinrichs et al. (eds.), *Sustainability Science*, DOI 10.1007/978-94-017-7242-6_3

1 Descriptive-Analytical vs. Transformational Sustainability Research

Sustainability research addresses problems that pose major threats to the viability and integrity of societies around the world (Kates et al. 2001; Clark and Dickson 2003; Jerneck et al. 2011; Sarewitz et al. 2012; Miller et al. 2014). Yet, the term "addresses" is ambiguous, which is why the field of sustainability science has mainly developed in two distinctive streams (Wiek et al. 2012).

The first one is primarily concerned with addressing sustainability problems through describing and analyzing them—their complexity, dynamics, and cause-effect relations (Turner et al. 2003; Ostrom 2009; Collins et al. 2011; De Vries 2013). The dominant methodological approach here is systems thinking and modeling, applied to past, current, and future sustainability problems. According to its main features, this stream has been called the "descriptive-analytical."

The second stream addresses sustainability problems by developing evidence-supported solution options for them (Sarewitz et al. 2012; Miller et al. 2014; Wiek et al. 2015). In this context, solutions are real-world changes that depend on actions executed by stakeholders other than researchers. Solution options, by contrast, are evidence-supported, actionable *knowledge* that, if applied, can *lead* to such real-world changes toward sustainability. Solutions to sustainability problems are generally *not* simple technical fixes or command-control procedures; they are often as complex as the problems themselves and require long-term processes that involve real-world experimentation, collective learning, and continuous adaptation. The second stream is therefore primarily concerned with providing evidence for how successfully to intervene in sustainability problems in order to resolve or at least mitigate them. For this, a sufficient problem understanding is advantageous; yet, gaining this understanding is here undertaken pragmatically, without losing sight of the ultimate objective to develop evidence-supported solution options (Sarewitz et al. 2012). With its intention to transform problems toward solutions, this stream has been called the "transformational."

The fact that a solution-oriented perspective is distinctly different from a problem-focused one has been acknowledged in several fields over the past decade. As Robinson and Sirard (2005, p. 196) point out for the field of public health research, "Knowing a cause of a problem, while sometimes a helpful first step, does not directly translate into knowing how to intervene to solve that problem." Let's illustrate the difference between descriptive-analytical and transformational sustainability research with examples from climate change research. A great deal of research in this area addresses emission sources, pathways, atmospheric CO_2 concentrations, temperature changes, and effects such as sea-level rise, as well as impacts on societies, for example, migration from coastal regions. This research enhances our understanding of the complex cause-effect relations in the human-climate system. However, it does not provide any knowledge as to what we can *do* in order to mitigate or adapt to climate change effectively. The latter is being pursued in transformational climate change research. Here, researchers develop and test different strategies that can *change* the current emission sources, pathways, atmospheric CO_2 concentrations, temperature changes, effects, and impacts toward a sustainable vision.

For *transformational* sustainability research, which is the focus of this chapter, it is important to develop clear methodological guidelines (as it is important for any other field). Such guidelines provide researchers with instructions and quality criteria on how to conduct transformational sustainability research. They enable researchers to select, combine, and apply methods in pursuit of designing and testing solution options. While such guidelines might be informed by existing methodologies, we cannot simply carry over methodologies of established disciplines and hope to accomplish transformational results with approaches that were not built for this purpose. If transformational solutions are the ultimate goal, we need to develop and adopt research methodologies that are capable of reaching this goal (Miller et al. 2014).

Three general methodological requirements apply to transformational sustainability research: first, transformational research needs to apply *suitable methods*; such methods are transparent, structured, and replicable sequences of steps that generate knowledge *as ingredients of* solution options. Such solution options should be composed of different types of knowledge (Grunwald 2007): they should (1) be based on, at least, a sufficient understanding of the problem (descriptive-analytical/system knowledge); (2) be guided by a coherent and sustainability-inspired vision (normative/target knowledge); and (3) outline concrete transition and intervention strategies, i.e., action plans that detail how to resolve the problem and reach the vision (instructional/transformation knowledge). Thus, second, transformational sustainability research needs to employ *methodological frameworks* that combine different types of methods to generate such multifaceted actionable knowledge. And, third, transformational sustainability research is concerned with real-world problems and aims at actionable knowledge that stakeholders are willing and able to implement. Therefore, there is broad agreement that such research has to be carried out in *close collaboration* between scientists and nonacademic stakeholders from business, government, and civil society (Clark and Dickson 2003; Talwar et al. 2011; Lang et al. 2012). As recent reviews have addressed the third requirement (e.g., Spangenberg 2011; Lang et al. 2012), this chapter focuses on the first two requirements.

The terms "research" and "research methodology" often refer to advanced academic research. Yet, we use these terms here in a much broader sense, referring to a particular type of knowledge generation that also includes student research and research conducted by professionals. The key condition is that the respective research activity adheres to quality criteria, including validity, reliability, saliency, and so forth; these criteria need to be adapted to the specific objectives of transformational sustainability research. While transformational sustainability research can draw on a spectrum of suitable methodological frameworks (e.g., Kajikawa 2008; Jerneck et al. 2011; Wiek et al. 2012), it is important to understand the similarities and differences among them. This enables students, researchers, and professionals to tailor the transformational research methodology to the specific objectives and needs of their respective projects.

- **Task**: *Illustrate for a research area related to sustainability challenges other than climate change research (e.g., urbanization, food provision, public health) the difference between the descriptive-analytical and the transformational stream in sustainability science. Try to formulate research questions for both streams.*

2 Methodological Frameworks for Transformational Sustainability Research

Several suitable methodological frameworks have been developed and applied that combine different methods in a meaningful sequence in order to generate actionable knowledge or, in other words, evidence-supported solution options for sustainability challenges. For pragmatic reasons, we present only *basic framework types* here and neglect all frameworks that are based on minor variations of framework elements.[1]

We first describe four prominent methodological frameworks that have been widely applied in sustainability projects. They all fulfill, to varying degrees, the following requirements of methodological frameworks for transformational sustainability research (a subset of the first two requirements outlined above): (1) they allow for addressing "wicked" problems similar to sustainability problems; (2) they combine methods in a way that generates solution options; and (3) they provide empirical evidence for the effectiveness of the solution options generated. In the descriptions below, we refer back to these requirements. The sequence of steps the respective framework proposes is indicated in Table 3.1. For this, we differentiate three families of procedures and methods, corresponding to the three knowledge types mentioned above. First, procedures and methods that produce descriptive-analytical or system knowledge offer insights on the past, current, or future state of the problem addressed. This *descriptive-analytical family* includes, for example, methods for systems modeling and scenario analysis (Ostrom 2009). Second, procedures and methods that produce normative or target knowledge offer insights on the (un)sustainability of past, current, or future states of the problem. This *normative family* includes, for example, methods for assessment and visioning (Swart et al. 2004; Wiek and Iwaniec 2014). Third, procedures and methods that produce instructional or transformation knowledge offer insights on how to resolve the problem and achieve the sustainable vision. This *instructional family* includes, for example, intervention research methods (Fraser et al. 2009). Evaluating the impact of interventions draws from methods in the descriptive-analytical and the normative method families.

[1] The literature uses the terms "framework", "method", "approach", and "tool" sometimes interchangeably, sometimes as distinctly different (not consistently). There is no need to differentiate between these terms here. We focus on frameworks as defined above, irrespective of the fact that some of the frameworks are labeled, for instance, as "methods" (e.g., complex problem-handling) or "approaches" (e.g., backcasting).

Table 3.1 Overview of four methodological frameworks for transformational sustainability research.

	Complex Problem Handling	*Transition Management and Governance*	*Backcasting*	*Transdisciplinary Case Study*
Step 1	Problem analysis	Problem analysis	Envisioning normative scenarios	System analysis
Step 2	Goal setting	Constructing sustainable vision	Current state analysis & appraisal	Scenario construction
Step 3	Intervention design, analysis, selection	Transition strategy design	Backcasting pathways	Multi-criteria assessment (current state and scenarios)
Step 4	Implementation	Transition experiments		Strategy derivation
Step 5	Intervention evaluation	Evaluation		
Step 6		Multiplication		

The text of those steps that are dominated by activities other than research is shaded in gray; the steps mainly using methods of the descriptive-analytical family are shaded in light gray, those using mainly methods of the normative family in dark gray, and those mainly using methods of the instructional family in black

The first framework is the *complex problem-handling* framework developed beginning in the early 1990s by Dorien DeTombe (2001). This framework addresses complex societal problems that are similar to sustainability problems, as they display dynamic features, include many phenomena, involve many actors, and have severe impacts on society. The approach is solution oriented and encompasses all phases of problem handling, from building awareness of the problem to evaluating interventions. The complex problem-handling framework has been applied to a variety of complex societal problems ranging from climate change to children born of war (DeTombe 2008; Mochmann and DeTombe 2010). What types of methods are being adopted and how they are sequentially combined in the complex problem-handling framework is indicated in Table 3.1. The complex problem-handling framework focuses on the problem to be resolved. While goals are recognized as additional points of reference, the main emphasis is put on the problem analysis and the intervention analysis, each with several sub-steps. However, there is a lack of actual provision of solution options and subsequent implementations (with impact on the real world). The framework's focus on the problem and the intervention is shared, for instance, with the *intervention research* framework (Fraser et al. 2009), with even more emphasis put on the elaboration of intervention options. While the complex problem-handling framework lacks the step of empirically following through to the evaluation stage (no evaluative impact studies have been conducted so far, to our knowledge), the intervention research framework has a strong track record in intervention testing. However, intervention research does not always tackle problems as complex as sustainability problems.

The second framework is the *transition management and governance* approach developed beginning in the early 2000s by Jan Rotmans, Derk Loorbach, and other researchers (Rotmans et al. 2001; Loorbach 2010). The transition management and governance framework addresses complex, unstructured, persistent problems of a specific type that "cannot be solved with simple, short-term solutions" (Loorbach 2010, p. 164). The framework includes a process model that comprises policy design and is intended to develop transition strategies toward sustainability (Loorbach 2010). The framework has been applied and evaluated in several "transition experiments," including transition projects on regions, industry, and business, as well as societal sectors (health and energy sector), originally mainly in Belgium and the Netherlands (Loorbach and Rotmans 2010). However, formal impact evaluations are still missing, in part due to the long-term approach of the transition experiments conducted. What types of methods are being adopted and how they are sequentially combined in the transition management and governance research framework are indicated in Table 3.1.[2] The transition management and governance research framework focuses on both the problem and, to an even greater degree, the vision. From these reference points, transition strategies are being developed and tested in transition experiments. The main emphasis is put on the developing, testing, and multiplying of these transition strategies, which are elaborated in several sub-steps. Variations of the transition management and governance framework incorporate, among others, the backcasting approach (Voß et al. 2009), which is, however, a complete framework by itself and therefore discussed separately below.

The third framework is the *backcasting* approach developed beginning in the early 1980s by John Robinson (2003). Others have further developed the backcasting framework or developed alternative backcasting frameworks (e.g., Holmberg 1998; Quist and Vergragt 2006). The backcasting framework has been developed to address "complex societal problems such as sustainability challenges" (Robinson 2003, p. 842). The framework leads from "articulating the nature of the desired end-point conditions" to "analysing how those may be achieved" (Robinson 2003, p. 848 f.). In more recent projects, the framework has been used for fostering social learning and building collective capacity for sustainability (Robinson 2003). The backcasting framework has been applied in various research projects on energy, regional development, and climate change (e.g., Robinson 2003; Quist and Vergragt 2006). Reflexive impact studies provide initial evidence of the impacts of backcasting studies (Robinson et al. 2011; Talwar et al. 2011). What types of methods are being adopted and how they are sequentially combined in the backcasting research framework are indicated in Table 3.1. The rationale behind the backcasting research framework is best understood through the intention of building an alternative to the forecasting approaches predominant in energy and resource studies in the 1970s and 1980s. In response to the challenges of prediction and guidance for action, the backcasting framework employs an explicitly *normative* scenario approach (versus predictive or exploratory future studies) combined with methods that construct *pathways* of "how desirable futures can be attained" (Robinson 2003, p. 842). The approach puts strong emphasis on the construction of desirable and sustainable

[2] While the methods are structured sequentially in this process model, Loorbach (2010) emphasizes the flexible character of the model: "In reality, there is no fixed sequence of the steps in transition management. The cycle only visualizes the need to connect activities and presents some possible logical connections but does not suggest a sequential order of activities" (p. 172). This position supports the general concept employed in this chapter that there is no single right way of creating solution options for sustainability problems (there are multiple). Yet, most of the empirical transition research projects follow the outlined sequence.

future states. The title of the framework indicates the intention of "working backwards from a particular desired future end-point or set of goals to the present, in order to determine […] the policy measures that would be required to reach that point" (ibid.). With its more recent turn toward capacity building and social learning as main objectives, the framework and its applications tend to put even more emphasis on the creation and construction of sustainable future visions than on the actual backcasting part. The backcasting framework is, for instance, similar to the sequence of sustainability science components suggested by Kajikawa (2008).

The fourth research framework is the *integrated planning research* approach developed beginning in the 1990s by Roland Scholz and other researchers (Scholz and Tietje 2002; Scholz et al. 2006; Wiek and Walter 2009). The integrated planning research framework addresses a new kind of complex systemic and ill-defined problem that requires a new type of problem solving (Scholz et al. 2006). The framework intends to contribute to societal problem-solving efforts through methodologically sound research that yields strategies toward sustainability (Wiek and Walter 2009). It has been applied in numerous empirical studies addressing the sustainability challenges of a railroad company, a regional economy, a national nuclear waste disposal program, and so forth (Scholz et al. 2006; Krütli et al. 2010). Evaluative studies provide first evidence of the impacts of some integrated planning research projects (Walter et al. 2007). What types of methods are being adopted and how they are sequentially combined in the integrated planning research framework are indicated in Table 3.1. The rationale behind the integrated planning research framework is the recognition of ill-defined problems combined with the conviction that the current status bears the potential for its transformation in itself. Ill-defined complex problems require extra effort to understand the systems in which they are positioned. A thorough understanding of the current state, its inertia, and future path dependencies (foresight) allows for revealing the current and near-future opportunities to change this path. The integrated planning research framework shares basic assumptions with similar planning research and integrated assessment frameworks (e.g., Ravetz 2000).

Based on the review presented in this section, we can summarize that there are at least four distinct ways, or *frameworks*, to create solution options for sustainability problems (with several sub-variations). These frameworks are differentiated through the specific sequence of methods. While all four frameworks comply with the requirement to arrange and combine methods from all essential families of methods, they put relative emphasis on the different steps/methods within each framework. For simplification purposes, one might summarize that the complex problem-handling framework puts emphasis on the problem analysis (what is the structure of the problem?), the transition management and governance framework on strategy building (what is a promising transition/intervention strategy?), the backcasting framework on visioning (what is a sustainability vision?), and the integrated planning research framework on foresight and sustainability assessment (how might the problem develop in the future and how sustainable are different future states?). All these questions are legitimate questions, and there is no universally "right" way to develop solution options for sustainability problems. Selecting the most appropriate framework depends on several factors, including the specific context of the problem, the capacity of the research team, and so forth.

A recently developed framework, called TRANSFORM, synthesizes key features of the aforementioned frameworks and integrates foresight, backcasting, and intervention research (Fig. 3.1) (Wiek et al. 2011, 2012, 2013; Lang and Wiek 2012; Wiek 2014). The TRANSFORM framework, similar to the other ones, has been designed for developing solution options for sustainability problems and eventually to *transform* the status quo toward sustainability. It entails two corresponding, yet reverse and complementary, research streams: the first is *foresight*, in which researchers analyze and assess past and current states of the problem, as well as project the problem into the future to depict the diversity of plausible future states (I and IIa); the second stream is *backcasting*, in which researchers construct and assess sustainable future visions, as well as trace these visions back to the current state of the problem (pathways) (IIb and I). As indicated in the figure, scenarios and visions inform and complement one another. Finally, researchers design and test *transition and intervention strategies* (III) that contribute to mitigating the current state of the problem, achieving the sustainable visions, and actively avoiding undesirable scenarios. In order to use a broad evidence base, build capacity, and develop shared ownership for the intervention strategies, this framework calls for a close collaboration of researchers from different disciplines and stakeholders in government, businesses, and civil society. The TRANSFORM framework has been applied to projects on urban sustainability, including land-use planning, mitigating urban sprawl effects, water governance, mitigating childhood obesity, and transit-oriented development (Wiek et al. 2012; Kay 2012; Xiong et al. 2012; Bernstein et al. in press; Wiek et al. in press).

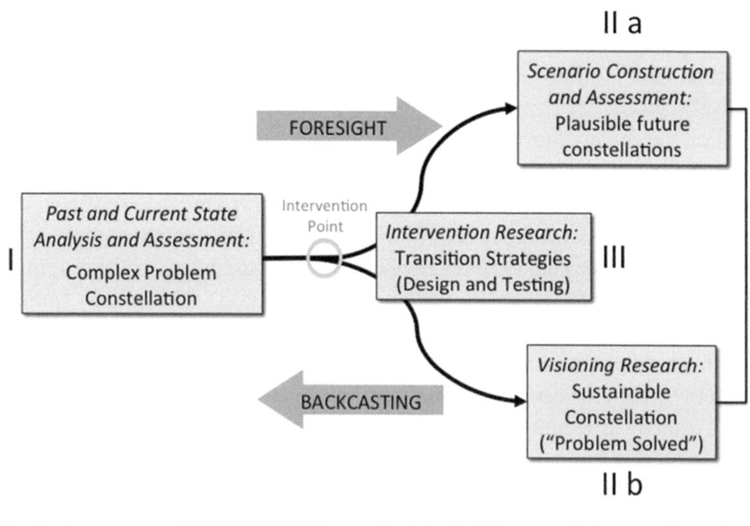

Fig. 3.1 Basic structure of TRANSFORM, a methodological framework for transformational sustainability research integrating foresight, backcasting, and intervention research (Adapted from Wiek et al. (2011))

- **Tasks**:
  1. *Given the characteristics of the methodological frameworks introduced, elaborate on potential strengths and weaknesses with regard to fostering sustainability transitions. For specific sustainability problems, argue why one of the frameworks might fit better than the other ones.*
  2. *Imagine a specific sustainability challenge. Describe this challenge in a few sentences, and develop a rough outline for a transformational sustainability research project, using the TRANSFORM framework that integrates foresight, backcasting, and intervention research (Fig.* 3.1*). Outline for each step (module): (1) what is being done, (2) who is involved, and (3) what are the expected outcomes?*

3 Outlook

Transformational sustainability research develops evidence-supported solution options for sustainability problems. It shares with descriptive-analytical research the intention to provide credible knowledge, i.e., sufficient evidence for the effectiveness of the interventions. Yet, it differs with respect to the ultimate objective of fostering transformation. There is not one but several methodological frameworks that can guide students, researchers, and professionals in their transformational sustainability research pursuits. All these frameworks have their particular focal points, be it the problem, the projected trajectories of the problem, the goals (or visions), or the intervention strategies themselves. More recently, efforts have been undertaken to synthesize the key features of those frameworks in order to avoid blind spots and fully utilize their strengths.

While progress is being made, there are several challenges ahead. Minor ones relate to a more careful comparison of the different frameworks in order to create a structured pool of frameworks. Such a pool would allow for choosing the most suitable framework for a given project and adopting it to the specific objectives and needs. More challenging is to advance the provision of evidence. While each framework provides a good initial structure, the ultimate objective is to generate evidence-supported solution options. The majority of research projects currently undertaken can still benefit from enhancing their efforts to demonstrate empirically that the designed solution options actually work in practice, as well as to learn from what has not worked. This anticipates the third and major challenge, which pertains to the issue of urgency. The reviewed frameworks offer robust guidelines for transformational sustainability research. Yet, willingness and capacity of academic, governmental, private, and nonprofit organizations to embrace these frameworks fully and use them for their knowledge-generating operations is still fairly low. The mismatch between the call for transformational results and the inertia of business-as-usual operations prevails. Much more needs to be done individually and collectively in transforming societies, governments, and companies around the world toward sustainable trajectories.

References[3]

Bernstein MJ, Wiek A, Brundiers K, Pearson K, Minowitz A, Kay B, Golub B (2013, in press) Mitigating urban sprawl effects – a collaborative tree and shade intervention in Phoenix, Arizona, USA. Local Environ. DOI:10.1080/13549839.2014.965672

Clark WC, Dickson NM (2003) Sustainability science: the emerging research program. PNAS 100:8059–8061

Collins SL, Carpenter SR, Swinton SM, Orenstein DE, Childers DL et al (2011) An integrated conceptual framework for long-term social-ecological research. Front Ecol Environ 9:351–357

De Vries B (2013) Sustainability science. Cambridge University Press, New York

DeTombe DJ (2008) Climate change: a complex societal process – analyzing a problem according to the COMPRAM methodology. J Organ Transform Soc Change 5(3):235–266

*DeTombe DJ (2001) COMPRAM, a method for handling complex societal problems. Eur J Oper Res 128:266–281

Fraser MW, Richman JM, Galinsky MJ, Day SH (2009) Intervention research: developing social programs. Oxford University Press, New York

Grunwald A (2007) Working towards sustainable development in the face of uncertainty and incomplete knowledge. J Environ Policy Plan 9:245–262

Holmberg J (1998) Backcasting: a natural step in operationalising sustainable development. Greener Manag Int 23:30–51

Jerneck A, Olsson L, Ness B, Anderberg S, Baier M et al (2011) Structuring sustainability science. Sustain Sci 6:69–82

Kajikawa Y (2008) Research core and framework of sustainability science. Sustain Sci 3:215–239

Kates RW, Clark WC, Corell R, Hall JM, Jaeger CC et al (2001) Sustainability science. Science 291:641–642

Kay B (2012) Developing and testing transition strategies for urban sustainability – case studies in transition research in Phoenix, Arizona. PhD thesis, School of Sustainability, Arizona State University, Tempe, AZ

Krütli P et al (2010) Technical safety vs. public involvement? A case study on the unrealized project for the disposal of nuclear waste at Wellenberg (Switzerland). J Integr Environ Sci 7:229–244

Lang DJ, Wiek A (2012) The role of universities in fostering urban and regional sustainability. In: Mieg HA, Töpfer K (eds) Institutional and social innovation for sustainable urban development. Earthscan, London, pp 393–411

Lang DJ, Wiek A, Bergmann M, Stauffacher M, Martens P, Moll P, Swilling M, Thomas C (2012) Transdisciplinary research in sustainability science – practice, principles and challenges. Sustain Sci 7(Supplement 1):25–43

*Loorbach D (2010) Transition management for sustainable development: a prescriptive, complexity-based governance framework. Gov Int J Policy Adm Inst 23:161–183

Loorbach D, Rotmans J (2010) The practice of transition management: examples and lessons from four distinct cases. Futures 42:237–246

*Miller TR, Wiek A, Sarewitz D, Robinson J, Olsson L, Kriebel D, Loorbach D (2014) The future of sustainability science: a solutions-oriented research agenda. Sustain Sci 9:239–246

Mochmann IC, DeTombe DJ (2010) The COMPRAM methodology and complex societal problems – an analysis of the case of children born of war. Organizacija 43(3):113–124

Ostrom E (2009) A general framework for analyzing sustainability of social-ecological systems. Science 325:419–422

Quist J, Vergragt P (2006) Past and future of backcasting: the shift to stakeholder participation and a proposal for a methodological framework. Futures 38:1027–1045

[3] *The sources highlighted with an asterisk (*) are recommended as further readings.*

Ravetz J (2000) Integrated assessment for sustainability appraisal in cities and regions. Environ Impact Assess Rev 20:31–64

*Robinson JB (2003) Future subjunctive: backcasting as social learning. Futures 35:839–856

*Robinson TN, Sirard JR (2005) Preventing childhood obesity: a solution-oriented research paradigm. Am J Prev Med 28:194–201

Robinson J, Burch S, Talwar S, O'Shea M, Walsh M (2011) Envisioning sustainability: recent progress in the use of participatory backcasting approaches for sustainability research. Technol Forecast Soc Change 78:756–768

Rotmans J, Kemp R, Van Asselt M (2001) More evolution than revolution: transition management in public policy. Foresight 3(1):17

Sarewitz D, Kriebel D, Clapp R, Crumbley C, Hoppin P, Jacobs M, Tickner J (2012) The sustainability solutions agenda. New Solut 22(2):139–151

Scholz RW, Tietje O (2002) Embedded case study methods: integrating quantitative and qualitative knowledge. Sage, Thousand Oaks

Scholz RW, Lang D, Wiek A, Walter A, Stauffacher M (2006) Transdisciplinary case studies as a means of sustainability learning: historical framework and theory. Int J Sustain High Educ 7:226–251

Spangenberg JH (2011) Sustainability science: a review, an analysis, and some empirical lessons. Environ Conserv 38:275–287

Swart RJ, Raskin P, Robinson J (2004) The problem of the future: sustainability science and scenario analysis. Glob Environ Chang 14(2):137–146

Talwar S, Wiek A, Robinson J (2011) User engagement in sustainability research. Sci Public Policy 38:379–390

Turner BL, Kasperson RE, Matson PA, McCarthy JJ, Corell RW et al (2003) A framework for vulnerability analysis in sustainability science. PNAS 100:8074–8079

Voß J-P, Smith A, Grin J (2009) Designing long-term policy: rethinking transition management. Policy Sci 42:275–302

Walter A, Helgenberger S, Wiek A, Scholz RW (2007) Measuring social effects of transdisciplinary research – design and application of an evaluation method. Eval Progr Plan 30:325–338

Wiek A (2014) TRANSFORM – a framework for transformational sustainability research. Working paper. Sustainability Transition and Intervention Research Lab, School of Sustainability, Arizona State University, Tempe, AZ

Wiek A, Iwaniec D (2014) Quality criteria for visions and visioning in sustainability science. Sustain Sci 9:497-512

Wiek A, Walter A (2009) A transdisciplinary approach for formalized integrated planning and decision-making in complex systems. Eur J Oper Res 197:360–370

Wiek A, Kay B, Forrest N (in press) Worth the trouble?! An evaluative scheme for Urban Sustainability Transition Labs (USTL) and an application to the USTL in Phoenix, Arizona. In: Frantzeskaki N, Coenen L, Broto C, Loorbach D (eds) Urban sustainability transitions. Routledge series on sustainability transitions. Routledge, London

Wiek A, Withycombe L, Redman CL (2011) Key competencies in sustainability – a reference framework for academic program development. Sustain Sci 6:203–218

Wiek A, Ness B, Brand FS, Schweizer-Ries P, Farioli F (2012) From complex systems analysis to transformational change: a comparative appraisal of sustainability science projects. Sustain Sci 7(Supplement 1):5–24

Wiek A, Guston DH, van der Leeuw S, Selin C, Shapira P (2013) Nanotechnology in the city: sustainability challenges and anticipatory governance. J Urban Technol 20:45–62

Wiek A, Harlow J, Melnick R, van der Leeuw S, Fukushi K, Takeuchi K, Farioli F, Yamba F, Blake A, Geiger C, Kutter R (2015) Sustainability science in action – a review of the state of the field through case studies on disaster recovery, bioenergy, and precautionary purchasing. Sustain Sci 10:17–31

Xiong A, Talbot K, Wiek A, Kay B (2012) Integrated health care for communities – participatory visioning and strategy building for a New Mountain Park Health Center Clinic in Phoenix. Project report. Sustainability Transition and Intervention Research Lab, School of Sustainability, Arizona State University, Tempe, AZ

Chapter 4
Green and Sustainable Chemistry

Klaus Kümmerer and James Clark

Abstract The products of the chemical and pharmaceutical industries are indispensable for our high standard of living and health. Estimations say that about 100,000 chemicals are available on the market, mostly used in combination with other chemicals. Consumers may be unaware that the products of chemical industries provide the functionality they expect when buying or using a certain product. Often the contribution of chemistry is not clear to the consumer, as chemicals are used to improve or enable certain production processes, to improve the efficacy or the lifetime of a product or to generate a specific colour or taste (e.g. food additives, preservatives). In other words, the benefit of modern chemistry and pharmacy cannot be overestimated.

Contrary to current perception, which is dominated by the legacies of the past, chemistry can and will contribute in many ways to sustainability through its products and processes. However, it is important that chemistry itself becomes more sustainable. Sustainable chemistry encompasses green chemistry but is much more than that. An overview of green and sustainable chemistry and its important achievements are presented, and some possible future contributions are outlined.

Keywords Chemistry • Sustainability • Design • Resource • End of life • Biorefinery

1 Introduction

The products of the chemical and pharmaceutical industries are an indispensable basis of our high standard of living and health. Estimations say that about 100,000 chemicals are available on the market, most of them used in combination with other chemicals and often constituting complex products. Sometimes consumers are not even aware that the products of chemical industries provide the functionality they

K. Kümmerer (✉)
Sustainable Chemistry and Resources, Institute of Sustainable and Environmental Chemistry, Faculty of Sustainability, Leuphana University of Lüneburg, Lüneburg, Germany
e-mail: klaus.kuemmerer@uni.leuphana.de

J. Clark
York Green Chemistry Centre of Excellence, University of York, York, UK

H. Heinrichs et al. (eds.), *Sustainability Science*, DOI 10.1007/978-94-017-7242-6_4

expect when they buy or use a certain product, the contribution of chemistry often going unnoticed as chemicals are used to improve or enable certain production processes, to improve the efficiency or the lifetime of a product and to generate a specific colour or taste (e.g. food additives, preservatives). In other words, the benefit of modern chemistry and pharmacy can hardly be overestimated. In many areas, chemistry and pharmacy make up the backbone for sustainable development. This includes among others a pivotal role in the so called megatrends:

- Natural resources and environment
- Demographics- Globalization
- Technology and Innovation
- Consumption patterns

Chemistry is fundamental for challenges related to these megatrends such as alternative feedstock, environmental technology, nutrition and health, clean air and water, intelligent and efficient materials, renewable energy to mention just a few. Thereby chemistry can contribute much to sustainability. However, at the same time chemistry itself has to become sustainable.

According to the OECD (2008), the value of chemical production will be roughly $4000 billion (US) in 2015 and rise to $5500 billion by 2030. Most of this increase is expected for non-OECD countries. However, there are also challenges and a backside to the coin. Population growth and climate change will place great pressure on resources in the future. Increasing income and health will result in an increase in products and wastes.

Nowadays, most western countries have measures for proper and effective treatment and the prevention of emissions into air, water and soil stemming directly from production and manufacturing in place (Kümmerer 2010a, 2011; Schwarzenbach et al. 2006). That is often not the case in less developed countries where the products used in developed countries are synthesised and manufactured (Larsson et al. 2007). Interestingly, the introduction of chemicals into the environment is often unavoidably connected to the proper use of certain products of the chemical and pharmaceutical industries, such as pharmaceuticals, disinfectants, contrast media, laundry detergents, surfactants, anticorrosives used in dishwashers, personal care products and pesticides, to name just a few. It has been learned in recent years that even if advanced effluent treatment were to be applied, a significant portion of these chemicals would still remain in the wastewater. Incomplete removal of the chemicals leads to introduction into the aquatic cycle, where they can undergo further distribution and transformation (Fig. 4.1). Follow-up problems of such an end-of-the-pipe measure are increased energy demand and formation of unwanted reaction products that can even be more toxic and persistent in the environment than the parent compounds. Additionally, such advanced technologies often cannot be applied in developing countries.

Other chemicals, such as flame retardants or textile chemicals, are washed out during laundering, and still others, again stemming from, e.g., furniture, carpets, computers and other items, enter the indoor air and the environment because of their volatility. In the air, chemicals may be distributed globally if their lifetime is higher than approximately 10 days.

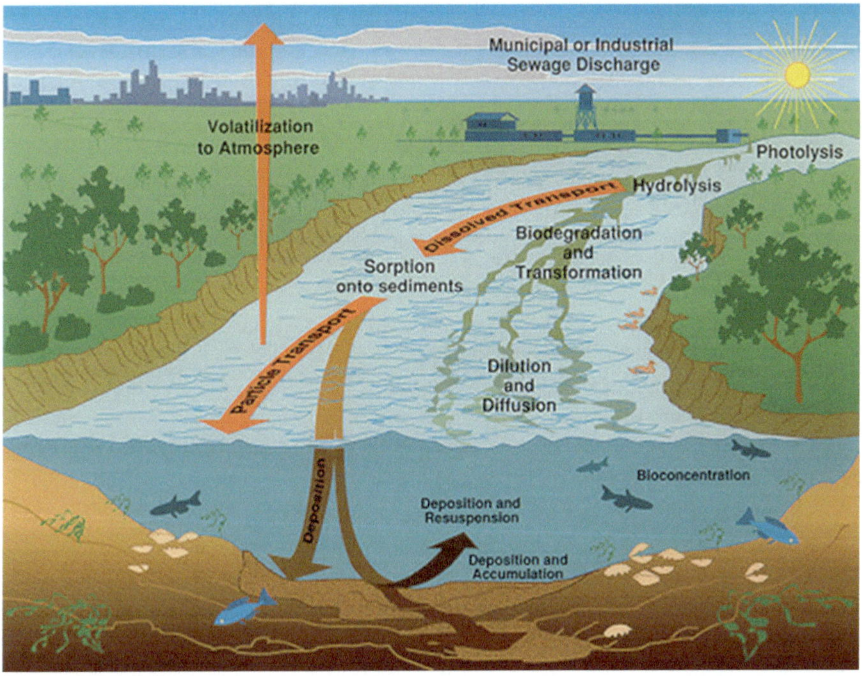

Fig. 4.1 Fate of pollutants in the aquatic environment (Source: U.S. Geological Survey, http://toxics.usgs.gov/regional/emc/transport_fate.html)

In some instances, it is not just the individual molecules but also the products themselves that pose a risk to the environment. An example is the pollution of the sea by plastics stemming from packaging such as bottles or bags, as well as from other plastic products such as rope. They are present as tiny particles (Andrady 2011) that adsorb other toxic chemicals and can cause the death of animals after ingestion by mechanically injuring them as well as by poisoning them through release of the formerly adsorbed chemicals. This pollution has economic consequences too.

2 Green and Sustainable Chemistry

Both "green" and "sustainable" chemistry embrace the full life cycle of chemicals and not just one stage of that cycle:

- Raw materials
- Synthesis
- Production
- Use
- Fate after use ("end of life")

Sustainable chemistry includes economical, social and other aspects related to manufacturing and application of chemicals and products. It aims not only at green synthesis or manufacturing of chemical products but also includes the contribution of such products to sustainability itself. In the Rio Declaration within Agenda 21 adopted in Rio de Janeiro in 1992, it was stated that it is important for research to intensify for the development of safe substitutes for chemicals with long life cycles (Agenda 21, # 19.21). Principles that address a more integrative view were subsequently established in the European Union in 1996 by a council directive (EC 1996). In general, use of the best available techniques, efficient energy use and prevention of accidents and limitations of their consequences were addressed. In Annex IV of the directive, specific measures were specified:

1. The use of low-waste technology;
2. The use of less hazardous substances;
3. The furthering of recovery and recycling of substances generated and used in the process, and of waste, where appropriate;
4. Comparable processes, facilities or methods of operation which have been tried with success on an industrial scale.
5. Technological advances and changes in scientific knowledge and understanding.
6. The nature, effects and volume of the emissions concerned.
7. The commissioning dates for new or existing installations.
8. The length of time needed to introduce the best available technique.
9. The consumption and nature of raw materials (including water) used in the process and their energy efficiency.
10. The need to prevent or reduce to a minimum the overall impact of the emissions on the environment and the risks to it.
11. The need to prevent accidents and to minimize the consequences for the environment.

An amendment came into force in 2010 as 2010/75/EU (ABl. EG L 334, p. 17–119).

Anastas and Warner (1998) published some similar simple rules of thumb addressing more or less the same points. These rules (later called the "12 principles") had their roots in the United States' Pollution Prevention Act of 1990 (http://www2.epa.gov/green-chemistry/basics-green-chemistry#definition):

1. Prevent waste: Design chemical syntheses to prevent waste. Leave no waste to treat or clean up.
2. Maximize atom economy: Design syntheses so that the final product contains the maximum proportion of the starting materials. Waste few or no atoms.
3. Design less hazardous chemical syntheses: Design syntheses to use and generate substances with little or no toxicity to either humans or the environment.
4. Design safer chemicals and products: Design chemical products that are fully effective yet have little or no toxicity.
5. Use safer solvents and reaction conditions: Avoid using solvents, separation agents, or other auxiliary chemicals. If you must use these chemicals, use safer ones.

6. Increase energy efficiency: Run chemical reactions at room temperature and pressure whenever possible.
7. Use renewable feedstocks: Use starting materials (also known as feedstocks) that are renewable rather than depletable. The source of renewable feedstocks is often agricultural products or the wastes of other processes; the source of depletable feedstocks is often fossil fuels (petroleum, natural gas, or coal) or mining operations.
8. Avoid chemical derivatives: Avoid using blocking or protecting groups or any temporary modifications if possible. Derivatives use additional reagents and generate waste.
9. Use catalysts, not stoichiometric reagents: Minimize waste by using catalytic reactions. Catalysts are effective in small amounts and can carry out a single reaction many times. They are preferable to stoichiometric reagents, which are used in excess and carry out a reaction only once.
10. Design chemicals and products to degrade after use: Design chemical products to break down to innocuous substances after use so that they do not accumulate in the environment.
11. Analyze in real time to prevent pollution: Include in-process, real-time monitoring and control during syntheses to minimize or eliminate the formation of by-products.
12. Minimize the potential for accidents: Design chemicals and their physical forms (solid, liquid, or gas) to minimize the potential for chemical accidents, including explosions, fires, and releases into the environment.

At the Johannesburg World Summit in 2002, as part of the millennium goals set up, it was agreed upon to substitute dangerous compounds, to increase resource efficiency and to cooperate for the development of a better management of chemicals globally, including education and training. This resulted in the establishment of a Strategic Approach to International Chemicals Management (SAICM; http://www.saicm.org).

There are estimates that green chemicals will save industry $65.5 billion by 2020 (http://www.navigantresearch.com/newsroom/green-chemicals-will-save-industry--65-5-billion-by-2020). However, it was not clearly defined in this context what "green chemicals" would exactly mean – the ones that fulfil one or a few of the above rules of thumb or the ones that fulfil all of them.

In general, only rarely are aspects that go beyond the chemicals themselves and their technical issues addressed by green chemistry, whereas sustainable chemistry generally includes all aspects of a product related to sustainability, e.g. social and economic aspects related to the use of resources, the shareholders, the stakeholders and the consumers (Fig. 4.2).

Integrating the principles of green and sustainable chemistry into synthesis of chemicals as well as the manufacturing of new materiala and complex porducts requires the chemist doing his work to think in an open-minded interdisciplinary manner and to take into consideration the world outside the laboratory from the very

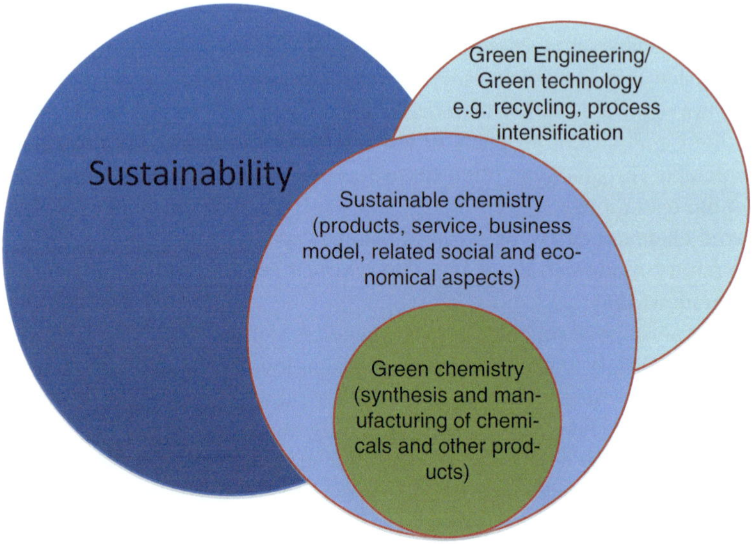

Fig. 4.2 The relationship of sustainability, sustainable chemistry green engineering, green technology, and green chemistry

beginning. This includes accounting for not only the functionalities of a molecule that are necessary for its application but also their impact and significance at the different stages of its life cycle.

3 Green Chemistry Metrics

It is important to be able to quantify the change when changes are made to chemical processes (Constable et al. 2007, Lapkin and Constable 2008). This enables us to quantify the benefit from the new technology introduced (if there are benefits). This can aid in in-house communication (to demonstrate the value to the workforce) as well as external communication. For yield improvements and selectivity increases, simple percentages are suitable, but this simplistic approach may not always be appropriate. For example, if a toxic reagent is replaced by a less toxic one, the benefit may not be captured by conventional methods of measuring reaction efficiency. Equally, these do not capture the mass efficiency of the process – a high-yielding process may consume large amounts of auxiliaries such as solvents and reagents, as well as those used in product separation and purification. Ideally, we also need to find a way to include energy and water, both of which have been commonly used in a rather cavalier way but which are now subject to considerable interest that can vary depending on the location of the manufacturing site.

Numerous metrics have been formulated over time and their suitability discussed at great length. The problem observed is that the more accurate and universally applicable the metric devised, the more complex and unemployable it becomes. A good metric must be clearly defined, simple, measurable, and objective rather than subjective and must ultimately drive the desired behaviour. Some of the most popular metrics are:

- E factor (which effectively measures the amount of product compared to the amount of waste – the larger the E factor, the less product-specific the process; the fine and pharmaceutical manufacturing sectors tend to have the highest E factors)
- Effective mass yield (the percentage of the mass of the desired product relative to the mass of all non-benign materials used in its synthesis – this includes an attempt to recognise that "not all chemicals are equal" – important and very real, but very difficult to quantify)
- Atom efficiency/economy (measures the efficiency in terms of all the atoms involved and is measured as the molecular weight of the desired product divided by the molecular weight of all of the reagents; this is especially valuable in the design "paper chemistry" stage, when low atom efficiency reactions can be easily spotted and discarded)
- Reaction mass efficiency (essentially the inverse of the E factor)

Of course, the ultimate metric is life cycle assessment (LCA), but this is a demanding exercise that requires a lot of input data, making it inappropriate for most decisions made in a process environment. However, some companies do include LCA impacts such as greenhouse gas production in their in-house assessment, for example, to rank solvents in terms of their greenness. It's also essential that we adopt a "life cycle thinking" approach to decision making so that we don't make matters worse when greening one stage in a manufacturing process without appreciating the effects of that change on the full process, including further up and down the supply. An integrated zero waste biorefinery that sequentially exploits an extraction, followed by biochemical and thermal processing, with internal recycling of energy and waste gases, is viewed as a model system. Extraction of secondary metabolites prior to their destruction in subsequent processes can significantly increase the overall financial returns.

4 Natural Resources and Chemistry

4.1 The Fossil Age

The resources of chemistry are inorganic materials, such as metals and minerals, that are gained by mining. Mining is often connected to severe environmental pollution and also has a social impact. For organic chemistry, oil is still by far the most important resource. It is also used as a resource for energy in chemical industry. This resource is limited, as are gas and coal. It is a hot bed of discussion nowadays

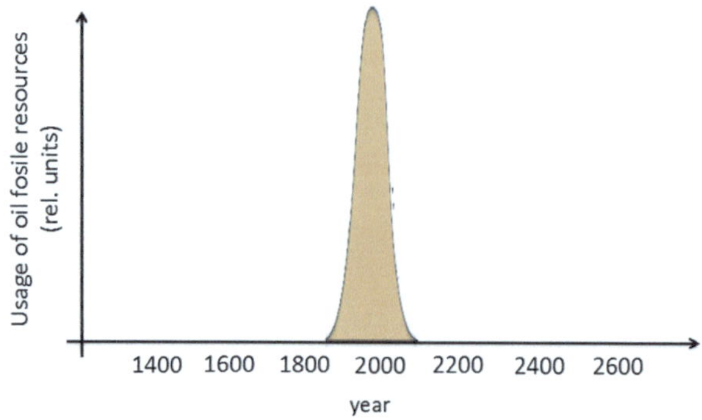

Fig. 4.3 The fossil age

as to how long these fossil resources will be available (Aleklett et al. 2012). However, even in the best case, it will be a matter of only decades or a century at the most (Fig. 4.3).

4.2 *The Biorefinery and Its Potential for Replacing the Petrochemical Industry*

The twentieth century saw a boom in the chemical industry with the emergence of an organic chemical manufacturing industry based on a cheap carbon feedstock, oil. This revolutionised the main energy source away from bio-resources, thereby creating the basis of the petroleum refinery we know today. It also helped create the chemical industry that has dominated the world for over 50 years.

Environmental and political concerns over the impact of continued fossil fuel use, their depletion and security of supply, combined with a growing population, have created a need for renewable sources of carbon. Over the last two decades, there has been a global policy shift back towards the use of biomass as a local, renewable and low-carbon feedstock. The "biorefinery" concept is a key tool in utilising biomass in a clean, efficient and holistic manner, whilst maximising value and minimising impact. However, the use of biomass as a source of energy, chemicals and materials is not new and has been taking place for millennia. The biorefinery concept is analogous to today's petroleum refineries. Biorefineries are ideally integrated facilities for conversion of biomass into multiple value-added products, including energy, chemicals and materials (Figs. 4.4 and 4.5). It is important that biorefineries utilise a range of low-value, locally sourced feedstocks, which don't compete with the food sector, including low-value plants such as trees, grasses and heathers, energy crop and food crop by-products (wheat straw), marine resource wastes, seaweeds and food wastes.

The main transformations available to the biorefinery can be classified as extraction, biochemical and thermochemical processes. The application of green chemical technologies (including supercritical fluid extraction, microwave processing, bio-

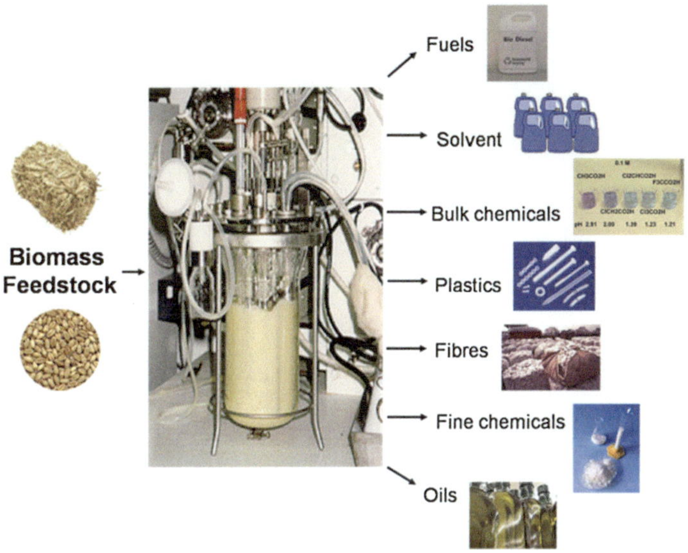

Fig. 4.4 Products from biomass

conversion, catalytic and clean synthesis methods) are all utilised with the aim of developing new, genuinely sustainable, low environmental impact routes to important chemical products, materials and bioenergy. These methodologies are usually studied independently of one another; however, the integration and blending of technologies and feedstocks is a way to increase the diversity of products and the socio-economic and environmental benefits of the biorefinery.

4.3 Valorisation of Waste

Modern society is based on a linear model of extraction of resources (oil, minerals, etc.), processing (e.g. chemical manufacturing, then formulation), use and disposal. This can only be sustainable if the disposal returns the resource to us in a useful form and on a reasonable timescale, typically measured within a human lifetime (<100 years). But with most types of resource, we have not been doing this. With metals, for example, we have been dispersing original virgin ores in the form of waste into landfills where recovery is difficult (Graedel 2011; Dodson et al. 2012). With organic materials, the situation is somewhat more complex – some are recycled by nature through biodegradation but many are not (e.g. non-biodegradable plastics). Our efforts at recycling are woefully inadequate (e.g. we recycle only about 1 % of the 260 million MT of plastics produced each year). Also, by stepping outside of the natural quick cycles of fast rotation bio-resources (plants, trees, etc.) and using a large proportion of bio-resources with very long cycles measured in millions of years (oil, coal, etc.), we have created an unsustainable economic model.

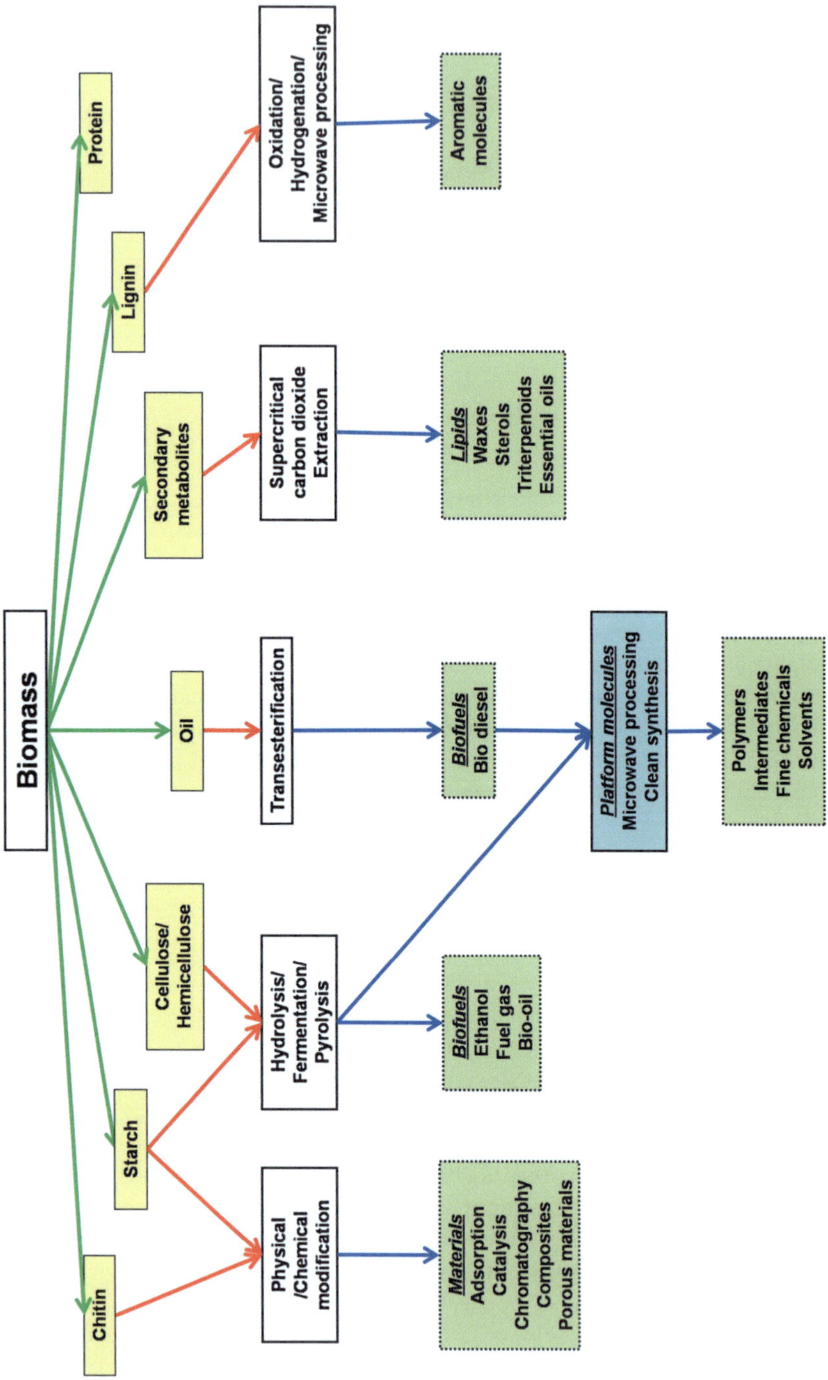

Fig. 4.5 Biomasses, chemical processes and products

We need to change our consumption pattern through adoption of a circular economy (Ellen MacArthur Foundation 2014) whereby we only consume resources in a way that can return them in a useful form on a sensible timescale. This is being "resource intelligent"! With a circular economy in place, our rate of consumption will then be limited by our efficiency – the closer we get to 100 % efficiency, the more we can enjoy the benefits of our planet!

5 Synthesis and Manufacturing

5.1 Solvent Selection

A significant number of organic solvents are strictly regulated and their use restricted (e.g. hexane, dichloromethane). This is set to increase, with new regulatory constraints being developed (REACH, VOC directives). It is likely that under REACH and other chemical-related legislation, many of the commonly used solvents in chemical and pharmaceutical manufacturing, as well as in other sectors, will be subject to authorisation or restrictions in use. As we seek alternatives, we should also be aware of incentives to encourage the use of bio-based chemicals, such as the EU prioritising some groups of chemicals, including solvents, for example, in the production of new standards.

In order to justify a replacement solvent, reaction efficiency must not be compromised simply to reduce the burden on the environment. Inferior reaction performance is not appealing on the grounds of increased waste and energy consumption, not to mention any economic implications. The search for greener solvent replacements can be systemised, thereby also providing additional justification for any proposed substitution.

Pharmaceutical manufacturing is one very important area in which new solvent restrictions could have dramatic effects. Typically, for the production of one kilogram of final API (active pharmaceutical ingredient), about 25–100 kg of waste are produced, meaning that 96–99 % of the overall process mass is discarded and requires, therefore, appropriate disposal. According to a report published by GlaxoSmithKline Pharmaceuticals in 2007, solvent usage accounts for 85–90 % of the total input material of a process. Besides, these figures do not include water usage, which is extensive in workup steps and as a reaction medium (Constable et al. 2007). The large volume of solvents used in drug manufacture, and the nature thereof, has become a matter of major concern in recent years. There is now a major search for greener solvents for use in industries like pharmaceuticals (Kerton 2013). This includes new bio-based solvents such as limonene (used in cleaning, but also a possible reaction solvent), cymene, cyrene (dihydrolevoglucosenone) and organic carbonates, as well as nonconventional solvents like water and liquid or supercritical carbon dioxide. There was a lot of research into ionic liquids as non-volatile, powerful solvents, but these have proven to be severely limited by factors including cost, some evidence for toxicity and difficulty in purification.

An example of the development of a new route of synthesis with fewer steps is given in Figs. 4.6 and 4.7.

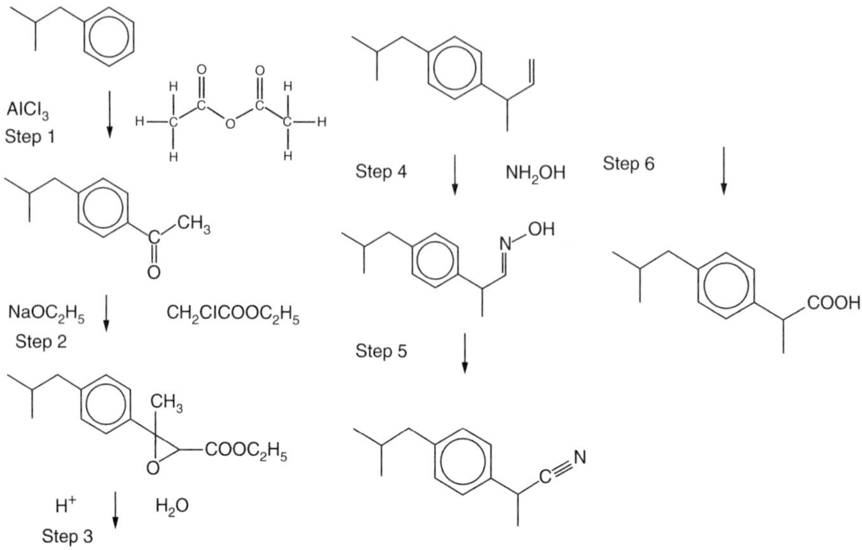

Fig. 4.6 Old route of synthesis of the pain killer ibuprofen (Source: http://www.rsc.org/learn-chemistry/resources/chemistry-in-your-cupboard/nurofen/5)

Fig. 4.7 New route of synthesis of the pain killer ibuprofen (Source: http://www.rsc.org/learn-chemistry/resources/chemistry-in-your-cupboard/nurofen/6)

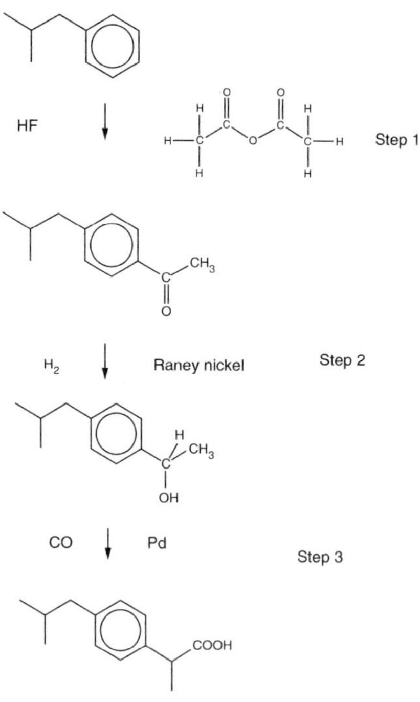

6 Products and End of Life

6.1 Benign by Design

Many chemical products end up in the environment not because of improper use but because of proper use (see above). One of the biggest challenges nowadays related to chemicals is persistence in the environment (see, e.g. http://chm.pops.int/default.aspx). The correlation of structure and composition of a chemical, encoded in a formula, and its properties is at the core of chemistry and chemical language. A change in the structure of a chemical will result in different properties. The consideration of the functionalities of a molecule and the properties that are correlated with them and their significance and impact along their entire life cycle brings into the foreground the targeted design of new chemicals at the very earliest stage of their conceptualisation (Fig. 4.8) including end of life. This approach is called "benign by design" (Kümmerer 2007).

Substructures and functional groups are already known that may improve degradability by chemical processes such as hydrolysis, photolysis and biodegradation under environmental conditions (see Table 4.1). With the help of computer-based models such as (quantitative) structure activity relationships ((Q)SARs), a more systematic assessment can be done, e.g. of biodegradability (Rücker and Kümmerer 2012) or toxicity and physical chemical properties of molecules (Cronin and Madden 2010; Ekins 2007; Boethling and Mackay 2000). A big advantage then is that molecules can be assessed even before synthesis. This not only saves money and time, it also gives guidance as to which molecules may possess the desired low toxicity and fast and complete mineralization when they are introduced into the environment at the end of their life. The first steps on the road to greener pharmaceuticals are already done (Rastogi et al. 2014).

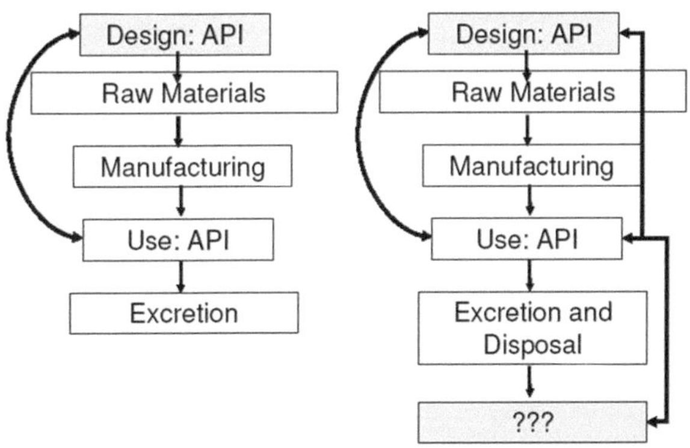

Fig. 4.8 Conventional (*left*) and sustainable approaches (*right*) for the design of new chemicals and active pharmaceutical ingredients (APIs): the end of life of the molecule is already taken into account at the very beginning (Source: Kümmerer 2010a)

Table 4.1 Examples of chemical functionalities and their impact on biodegradability in the environment

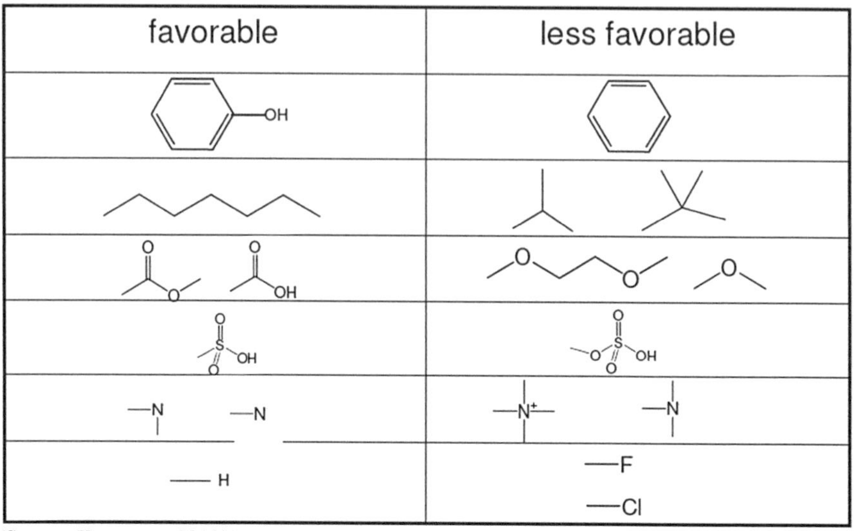

(Source: Kümmerer 2010b)

6.2 *Limits of Recycling and Material Flows*

Inorganic molecules, metals and complex products are different from the above-described "small" organic molecules. Chemical products such as plastics often contain a mix of molecules, such as the polymer itself, and other molecules for the modification of their properties, such as softness or resistance against light. For them, as well as for inorganic materials and products, recycling is an option for recollecting the constituents. However, if the products are not designed for recycling, it may be difficult or even impossible to recycle the products themselves or extract components for further use. Furthermore, recycling needs additional energy and good logistics. Most often, the so-called recycling, in fact, is down cycling, that is, the quality of the regained products/and or constituents is lower than the original one. The laws of thermodynamics tell us that there is always a loss of material and quality through recycling.

In general, independent of recycling, the more complex the materials themselves, the higher diversity of products and chemicals that constitute a given material's flow, and the bigger those flows are, the bigger the loss. Even if the synthesis of a chemical or the manufacturing of products can be called green or sustainable, in the end, non-sustainability may result if the related material flows are too diverse and too big.

6.3 New Business Models

Thinking in terms of functionality or offering a service can avoid some of these pitfalls. For example, if the functionality "wood preservation" is needed, a wood preservative can deliver it. However, a wise construction may avoid giving water the access that makes the preservative necessary.

Another example would be a company that does not buy solvents, but instead leases them and returns them to the deliverer after usage. This has the advantage for the deliverer that, in taking back the solvent, they can use all their solvent-related knowledge and experience to make them most effective (e.g. solvent selection). Now, they have a specific interest in having the most efficient use of solvents. The leasing company has the same interest. This is a win-win situation – just selling a solvent is a win-lose situation: the provider wants to sell as much solvent as possible, the customer to buy as little as possible.

Another example is the use of disinfectants (Schülke 2011): A provider wants to sell as much disinfectant as possible. However, the goal behind the application of disinfectants is to safeguard a proper standard of hygiene. If the provider of the disinfectant is responsible for providing the necessary standard of hygiene, they will aim to use as little disinfectant as possible. As the manufacturer of a disinfectant has lots of knowledge about disinfectants and regulation on hygiene, they can provide training and education on the right use of disinfectants and application of other measures to maintain the necessary hygienic standards. In fact, they can save money by spending less money for the raw materials and synthesis/manufacturing of the disinfectant and earn money by selling a service – the maintenance of the appropriate standard of hygiene. Furthermore, fewer physical resources were needed; less energy for synthesis and delivery, less packaging material and less introduction of chemicals into the aquatic environment will result.

7 Conclusions

Chemistry has been – in modern words – a success story starting with the advent of chemical science working in close interaction with the beginning of chemical industries. Chemistry has been, is and will be a fundamental part of the modern way of life, contributing, for instance, to health benefits. But along the way, the other side of the coin has become apparent related to the intoxication of humans and pollution of the environment. Green and sustainable chemistry takes this into account through life cycle thinking and measures to prevent such negative impact from the very beginning and through all stages. It provides various goods and services that are needed for a successful transition to sustainable societies. Both the chemistry of the past and the present and even more so that of the future, that is, sustainable chemistry, are examples for successful interaction between disciplines (interdisciplinarity), basic science and everyday life (transdisciplinarity). There will be no future without chemistry. As to the challenges we face nowadays, however, we assert that chemistry needs to significantly change its ways in order to be part of a sustainable future.

Questions

1. What are the typical stages of life of a chemical product and what connection does each one have with sustainability?
2. What is a biorefinery and what would be its role in the future?
3. Where and how does chemistry contribute to sustainability?
4. Is there a difference between sustainable chemical products (molecules and materials) and sustainable chemistry?
5. What are typical environmental problems related to chemistry? What are solutions to these provided by sustainable chemistry?
6. What is peak oil and what is its significance for chemistry?

References

Aleklett K, Qvennerstedt O, Lardelli M (2012) Peaking at peak oil. Springer, New York/Heidelberg/Dordrecht/London

Anastas PT, Warner JC (1998) Green chemistry: theory and practice. Oxford University Press, New York/Oxford

Andrady AL (2011) Microplastics in the marine environment. Mar Pollut Bull 62:1596–1605

Boethling RS, Mackay D (2000) Handbook of property estimation methods for chemicals – environmental and health sciences. Lewis Publisher, Boca Raton

Constable CJ, Jimenez-Gonzalez C, Henderson RK (2007) Perspective on solvent use in the pharmaceutical industry. Org Process Res Dev 11:133–137

Cronin MTD, Madden JC (eds) (2010) In silico toxicology: principles and applications. RSC Publishing, Cambridge

Dodson JR, Hunt AJ, Parker HL, Yang Y, Clark JH (2012) Elemental sustainability: towards the total recovery of scarce metals. Chem Eng Process 51:69–78

EC Council Directive 96/61/EC of 24 September 1996 concerning integrated pollution prevention and control Official Journal L 257, 10/10/1996 P. 0026–0040

Ekins S (ed) (2007) Computational toxicology. Wiley, Hoboken

Ellen MacArthur Foundation (ed) (2014) A new dynamic: effective business in a circular economy. Ellen MacArthur Foundation Publishing, New York. Available online at: http://www.ellenmacarthurfoundation.org

Graedel T (2011) On the future availability of the energy metals. Annu Rev Mater Res 41:323

Jang YC, Hong S, Lee J, Lee MJ, Shim WJ (2014) Estimation of lost tourism revenue in Geoje Island from the 2011 marine debris pollution event in South Korea. Mar Pollut Bull 81:49–54

Kerton MF (2013) Alternative solvents for green chemistry. Royal Society of Chemistry, Cambridge

Kümmerer K (2007) Sustainable from the very beginning: rational design of molecules by life cycle engineering as an important approach for green pharmacy and green chemistry. Green Chem 9:899–907

Kümmerer K (2010a) Pharmaceuticals in the environment. Ann Rev Environ Resour 35:57–75

Kümmerer K (2010b) Rational design of molecules by life cycle engineering. In: Kümmerer K, Hempel M (eds) Green and sustainable pharmacy. Springer, Berlin/Dordrecht/London/New York

Kümmerer K (2011) Emerging contaminants. In: Wilderer P (ed) Treatise on water science, vol 3. Academic, Oxford, pp 69–88
Lapkin A, Constable D (eds) (2008) Green chemistry metrics: measuring and monitoring sustainable processes. Wiley-Blackwell, Oxford
Larsson DG, de Pedro C, Paxeus N (2007) Effluent from drug manufactures contains extremely high levels of pharmaceuticals. J Hazard Mater 148:751–755
OECD (2008) Environmental outlook to 2030. OECD, Paris
Rastogi T, Leder C, Kümmerer K (2014) Designing green derivatives of β-blocker Metoprolol: a tiered approach for green and sustainable chemistry. Chemosphere 111:493–499
Rücker C, Kümmerer K (2012) Modeling and predicting aquatic aerobic biodegradation – a review from a user's perspective. Green Chem 14:875–887
Schülke (2011) Chemikalenleasing – effiziente und nachhaltige Krankenhaushygiene. (Chemical leasing – efficient and sustainable hygiene in hospitals). Schülke, Noderstedt
Schwarzenbach R, Escher BI, Fenner K, Hofstetter TB, Johnson AC, von Gunten U, Wehrli B (2006) The challenge of micropollutants in aquatic systems. Science 313:1072–1077

Chapter 5
Sustainability and Ecosystems

Henrik von Wehrden, Goddert von Oheimb, David J. Abson, and Werner Härdtle

Abstract Maintenance of human well-being is highly dependent on nature. The natural environment provides a source of both directly used goods and services that support human livelihoods and an intrinsic value that contributes to human flourishing. Today, much of the planet is influenced or even transformed by human activity, and natural ecosystems are increasingly under threat. Ecology and conservation biology are crucial for understanding and quantifying changes in ecological systems. Moreover, ecology in conjunction with other branches of science provides key insights to enable management options for supporting a sustainable future for our planet. Here, we give an overview of the relationship between biodiversity, ecosystems, and sustainability. First, we introduce the notion of biodiversity, then we present the links between biodiversity, ecosystem functions, and services, in which ecosystem services are the benefits people derive from ecosystems. Finally, we outline the current threats to ecological integrity and provide a brief overview of the links between ecology and other disciplines within sustainability science.

Keywords Biodiversity • Ecosystem function • Ecosystem services

---

H. von Wehrden (✉)
Centre for Methods, Leuphana University, Lüneburg, Germany

Institute of Ecology, Faculty of Sustainability, Leuphana University, Lüneburg, Germany

FuturES, Leuphana University, Scharnhorststr. 1, 21335 Lüneburg, Germany
e-mail: henrik.von_wehrden@leuphana.de

G. von Oheimb
Institute of General Ecology and Environmental Protection, Technische Universität Dresden, Tharandt, Germany

D.J. Abson
Faculty of Sustainability, Leuphana University, Lüneburg, Germany

FuturES, Leuphana University, Scharnhorststr. 1, 21335 Lüneburg, Germany

W. Härdtle
Institute of Ecology, Faculty of Sustainability, Leuphana University, Scharnhorststr. 1, 21335 Lüneburg, Germany

H. Heinrichs et al. (eds.), *Sustainability Science*, DOI 10.1007/978-94-017-7242-6_5

1 Introduction

The notion of sustainability is a largely anthropocentric concept (regarding the world in terms of human values and experiences) concerned with sustaining human well-being (Richardson 1997). However, the maintenance of human well-being is highly dependent on nature. The natural environment, here conceptualized via the concept of biodiversity, is a source of both directly used goods and services that support human livelihoods and an intrinsic value that contributes to human flourishing (Randall 1991). Today, much of the planet is influenced or even transformed by human activity, and natural ecosystems are increasingly under threat (Rockström et al. 2009). Ecology and conservation biology are crucial for understanding and quantifying changes in ecological systems. Moreover, ecology in conjunction with other branches of science provides key insights to enable management options for supporting a sustainable future for our planet (Cardinale et al. 2012). This chapter gives an overview of the relationship between biodiversity, ecosystems, and sustainability. First, we introduce the notion of biodiversity, then we present the links between biodiversity, ecosystem functions, and services, in which ecosystem services are the benefits people derive from ecosystems. Finally, we outline the current threats to ecological integrity and provide a brief overview of the links between ecology and other disciplines within sustainability science.

2 Biodiversity

Biodiversity (the contraction of "biological diversity") is a term used to describe the totality and variety of life on Earth. Formally, the Convention on Biological Diversity defines biodiversity as "the variability among living organisms from all sources including terrestrial, marine and other aquatic ecosystems, and the ecological complexes of which they are part; this includes diversity within species, between species, and of ecosystems" (http://www.cbd.int). This view is now generally accepted, and we can consider three levels of biodiversity: genetic diversity, species diversity, and ecosystem diversity.

Genetic diversity is the heritable variation within and between populations of organisms. It finds its expression in both externally visible features (e.g., the color of a dog's fur) and at the level of molecules (e.g., the blood groups of humans). Ultimately, genetic diversity is based on the variation in the sequence of the four base pairs which, as components of nucleic acids, constitute the genetic code. For the survival of species, genetic diversity is of major importance, since it allows species to adapt to environmental change. From a human perspective, the enormous variability of numerous cultivated plants and domestic animals is a crucial aspect of biodiversity, both in terms of species diversity and regarding the genetic diversity of these species. For example, more than 35 mammalian and bird species have been domesticated for use in agriculture and food production, and there are more than

8,000 recognized breeds. Similarly, rice originates from one species, while the International Rice Database holds records of about 100,000 rice cultivars. The genetic diversity in a given species provides the basis for adaptations to future environmental changes, and maintaining genetic diversity of on-farm plants and animals is, thus, a key issue in sustaining agricultural production (e.g., Zhu et al. 2000; Tilman et al. 2006).

Species richness, that is, the number of species in a given area, represents the most commonly used metric for characterizing the diversity of life. Species diversity is an abundantly used measure of biodiversity within the scientific literature, so much so that it is often used as a synonym for biodiversity. This is because species are well-known and distinct units of diversity, and they are often relatively easy to identify in the field. Worldwide, about 1.8 million species have been described to date, and estimates for the total number of species existing on Earth range from 5 to 30 million (Millennium Ecosystem Assessment 2005). In terms of species numbers alone, much of the biodiversity on Earth appears to consist essentially of insects and microorganisms.

Ecosystem diversity is the diversity of ecosystems (i.e., the embedding of species or communities interacting with the non-biotic components of the environment), natural communities, and habitats in a given place. Because an ecosystem is a community of organisms and their physical environmental interactions, ecosystem diversity is, in essence, the variety of ways that species interact with each other and their environment. While there are many well-defined measures of genetic and species diversity, it is difficult to assess ecosystem diversity precisely due to the complexity of the interactions and the lack of sharp boundaries between ecosystems. It has to be kept in mind, however, that in order to fully capture biodiversity, it has to be integrated with other metrics.

Biodiversity is not evenly distributed across the Earth (Kleidon and Mooney 2000). For many taxonomic groups, there is a latitudinal gradient in species richness, with high numbers of species occurring in the tropics. Moving from the low to high latitudes, the species richness decreases. Determining why these gradients occur has long been a core issue in ecological research (Gaston 2000). A few environmental variables can explain a substantial proportion of variation in the spatial patterns of biodiversity, such as temperature, water availability, net primary productivity (i.e., the amount of biomass produced in a given area), or evapotranspiration (i.e., the sum of evaporation and plant transpiration from the land surface to the atmosphere). There is, however, no pattern without variations and exceptions, and we are far from a universal theory or model for predicting the spatial distribution of biodiversity. Nevertheless, it is possible to locate the areas that are the most immediately important for conserving biodiversity. These so-called biodiversity hotspots hold especially high numbers of endemic species, i.e., species that are prevalent in, or peculiar to, a specific locality (Myers et al. 2000). In very recent scientific studies, priority areas worldwide have been mapped out that suggest schemes for protecting vulnerable species and focusing conservation efforts (Jenkins et al. 2013; Le Saout et al. 2013).

More information on biodiversity can be found in Gaston and Spicer (2009).

- ***Questions***

  1. *How is biodiversity defined?*
  2. *Which group of species has the highest number of species?*
  3. *What drives the distribution of biodiversity and individual species, and what limits it?*

3 Ecosystem Functioning and Services

The different biotic (living) abiotic (non-living physical and chemical elements) components of the ecosystems found on the planet show strong interactions and relations, which taken together are coined as "ecosystem processes" (Daily 1997). For instance, carbon fluxes, pollination networks, and herbivory rates are all examples of ecosystem processes, and the joined effects of all these different processes are usually referred to as "ecosystem functioning." Biodiversity is often positively related to ecosystem functioning, in which higher rates of biodiversity are linked to higher rates of ecosystem functioning (Cardinale et al. 2006). Due to the alarming rates at which biodiversity is decreasing and ecosystems are being degraded on a global scale (see below), considerable research efforts in ecology have investigated the biodiversity-ecosystem function linkages. Many experimental approaches focusing on the relation between biodiversity and ecosystem functioning have been established within the last decades in different environmental settings (e.g., Tilman et al. 2006). Controlled environments were created through manipulation of ecosystems, enabling the study of important ecosystem processes and their relationships to biodiversity. Experimental designs frequently involve a manipulation of the variance of different species combinations within given levels of biodiversity. These different species combinations are suggested to account for the variance found within real ecosystems.

Specific species are grouped in relation to specific ecosystem processes based on their key characteristics, which are called traits; several databases and large research projects are currently generating coherent standards on these species' information and characteristics, thereby reducing pattern complexity within ecosystem functioning research while often increasing the explanatory power of ecological models. In addition, phylogenetic databases are increasingly gaining importance in ecological research (Winter et al. 2013). Phylogenetics defines the systematic position of a species (e.g., its family or genus) into the wider systematic background and quantifies the relationships between different species (see below).

In order to understand the complexity and interrelations within ecosystems and their relations to human well-being, the concept of ecosystem services has gained momentum in the last couple of decades (Seppelt et al. 2011). The concept of ecosystem services attempts to model directly or indirectly appropriated ecosystem structures, functions, or processes that contribute to human well-being (Millennium Ecosystem Assessment 2005). The Millennium Ecosystem Services Assessment was the first large collaborative approach to generating pivotal knowledge on ecosystem services on a planetary scale (see Box 5.1). Several national and regional studies and assessments investigating the complexity of ecosystem services dynamics have followed since, and an increasing number of studies derive system knowledge on local scales. Ecosystem services are strongly related to biodiversity through complex indirect relations between ecological functions and human well-being.

The ecosystem services approach is rooted in the attempt to understand the main sources of human well-being in complex dynamic socio-ecological systems (Daily 1997). Defining the boundary of one system is often the first challenge, since the borders of most systems are not discrete, but instead show linkages and interactions across different scales and system components (Post et al. 2007). Land use within human-dominated landscapes is of primary importance in the context of understanding ecosystem services, since many ecosystem services are specifically linked to one or several land-use types, e.g., carbon sequestration (Foley et al. 2005). The variability in ecosystems service provision across space is driven by two key factors. Categorical phenomena such as land use (e.g., forests and agricultural land) drive broad scale dynamics of ecosystem services provision. However, within a given ecosystem, gradual changes in ecological structures can also alter the provisioning of ecosystem services. For example, primary productivity changes along climatic

Box 5.1: The Millennium Ecosystem Assessment

The Millennium Ecosystem Assessment (MA) was a groundbreaking, global interdisciplinary scientific endeavor undertaken between 2001 and 2005 and involved the collaboration of more than 1,360 experts worldwide, carried out under the auspices of the United Nations. The MA assessed the consequences of ecosystem change for human well-being and developed scenarios to consider how ecosystems and the provision of ecosystem services may change in the future. The MA produced five main reports (A Framework for Assessment, Current States & Trends, Scenarios, Policy Responses, and Multiscale Assessments), as well as a number of shorter synthesis reports (http://www.millenniumassessment.org). The main finding of the MA was that over the past 50 years, humans have rapidly, and often irreversibly, changed ecosystems, largely to meet rapidly growing demands for food, freshwater, timber, fiber, and fuel. This has resulted in a substantial and largely irreversible loss in the diversity of life on Earth.

The MA findings provide a state-of-the-art scientific appraisal of the condition and trends in the world's ecosystems and the services they provide, as well as the scientific basis for action to conserve and use them sustainably. The MA was crucial in the development of other major scientific endeavors related to ecosystem services, such as The Economics of Ecosystems and Biodiversity (www.teebweb.org/) and the recently set up Intergovernmental Panel on Biodiversity and Ecosystem Services (www.ipbes.net).

gradients, which, in turn, affects the amount of food (provisioning ecosystem services) that can be provided within a particular system.

Besides being able to generate descriptive knowledge (functional relationships between ecosystem functions and human well-being) about a specific system, normative evaluations (judgments of how the world should be) are a key component in understanding the value that humans ascribe to ecosystem services. While monetary values are often ascribed to ecosystems services, normative evaluations can go beyond economic realms and consider non-monetary values that humans ascribe to ecosystems. For example, sense of place, cultural identity, and the intrinsic value of nature are recognized ecosystem services that cannot easily be quantified in terms of monetary values. The ecosystem services approach is increasingly recognized and applied in policy and conservation planning and research. A global institution was recently found to foster and link biodiversity and ecosystem services research and its implementation with stakeholders (www.ipbes.org).

Current ecology and conservation approaches are working both bottom-up and top-down. While the ecosystem functioning research investigates fundamental relationships between parts of the ecosystem on a local scale (bottom-up), the ecosystem service concept usually focuses on a wider scale (which can be top-down) relating generalized ecosystem functions to human well-being. However, a holistic understanding of ecosystem services needs to acknowledge local-regional complexity, i.e., combining bottom-up and top-down approaches.

- ***Questions***
  1. *How are ecosystem services defined?(Give some examples.)*
  2. *How are values ascribed to ecosystem services?*
  3. *Does the concept demand a bottom-up or a top-down approach?*

4 Global Threats to Biodiversity and Ecosystem Functioning

Human activities have altered the world's biodiversity, the functioning of ecosystems, and related ecosystem services in multiple ways. Global biodiversity currently changes at an unprecedented rate, and there is evidence that biodiversity losses are strongly linked to both important ecosystem processes and society's use of natural resources (Sala et al. 2000).

Despite the multiples ways in which humans have affected and will affect the functioning of ecosystems in the course of this century, recent research has identified five drivers of global change that are considered the most important regarding their impact on biodiversity loss and shifts in ecosystem functions. Ranked according to their projected impact on (terrestrial) biodiversity loss, these drivers are "land-use changes," followed by "climate change," "nitrogen deposition," "biotic exchange," and elevated levels of "atmospheric carbon dioxide" (Fig. 5.1, Sala et al. 2000).

Land-use change: This driver is expected to have the largest global impact on biodiversity by the year 2100, mostly because of its devastating effects on habitat

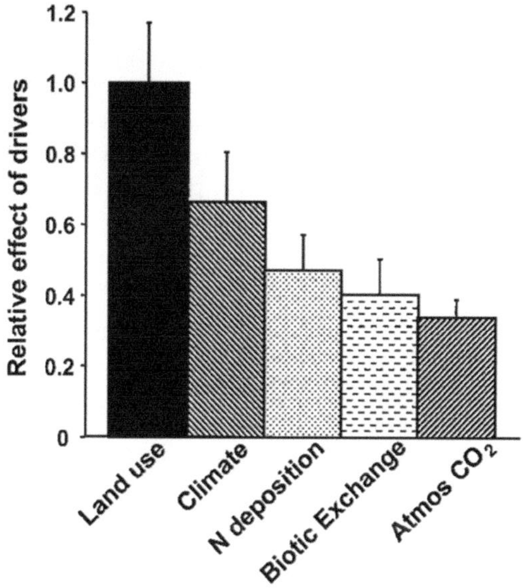

Fig. 5.1 Relative effects of major drivers of changes on biodiversity. *Thin bars* are standard errors and represent the variability among the different biomes analyzed (from Sala et al. 2000)

availability and related species extinctions. Land-use changes have affected tropical rain forests in particular, which are currently subject to extensive clear-cutting due to non-sustainable harvest of timber and their conversion to arable land or crop plantations. Since tropical rain forests host a huge proportion of the earth's biodiversity, they are considered the most important "biodiversity hotspots" worldwide (Myers et al. 2000).

Climate change: Climate change will be the second most important driver of biodiversity loss, mostly as a result of the expected warming at higher latitudes (but also at higher elevations, e.g., in the Alps). Recent analyses suggest that 15–37 % of a sample of 1,103 land plants and animals would eventually become extinct as a result of climate changes expected by 2050 (Ball 2012).

Nitrogen deposition: Anthropogenic nitrogen emissions (resulting from fuel combustion and agricultural activities) have tripled since the beginning of industrialization in the nineteenth century. As the productivity of plants in most ecosystems is limited by nitrogen, increasing nitrogen availability not only impacts plant productivity but strongly affects the competition between plant species (so-called interspecific competition) in different environments. As a consequence, many weak competitors that are unable to use an increasing nitrogen supply to enhance their biomass productivity have gone extinct in nitrogen fertilized environments. Nitrogen deposition will affect biodiversity hotspots during the course of this century due to increasing use of artificial fertilizers in previously low-input agricultural systems.

Biotic exchange: Biotic exchange describes the shift in the species composition and species assemblages in a particular area due to the encroachment of neobiota (non-native species introduced by humans). Depending on their competitiveness, introduced species with an aggressive spreading behavior are often classified as "invasive species" or "invaders." Introduced species are considered a particular problem in the tropics and subtropics, where neobiota often finds ideal conditions for growth and reproduction (in contrast to the temperate and boreal zone, where strong winters often limit or prevent the establishment of neobiota).

Atmospheric carbon dioxide: Increasing levels of atmospheric carbon dioxide (attributable to the combustion of fossil fuels) affect one of the most important biochemical processes typical of primary producers, photosynthesis. Since plants have different mechanisms in fixing atmospheric carbon dioxide, increasing levels will affect the interspecific competition of species characterized by a respective fixation mechanism. This will provide a competitive disadvantage for some species, which in turn may cause their extinction in the long term.

Despite an increasing knowledge of ecosystem responses to global change, our ability to anticipate global change impacts on ecosystem functions and services is still very limited (Loreau et al. 2001). Moreover, it is very likely that global change drivers will interact in a non-additive way, which in turn might cause antagonistic (canceling or damping) or synergistic (amplifying) effects on ecosystem responses. However, anticipating these responses to co-occurring global change drivers (such as climate change and increasing nitrogen deposition) will be crucial in the guidance of management, policy, and conservation efforts aimed at long-term mitigation of global change effects (Zavaleta et al. 2003).

- ***Questions***

  1. *What are the major threats to biodiversity?*
  2. *Is climate change a threat to biodiversity today?*
  3. *How can we mitigate the effects of global change?*

5 Developing Solutions to These Threats: Interdisciplinary and Transdisciplinary Efforts

Tackling the challenges our planet is facing regarding global environmental change calls for a fundamental change in societal and individual human behavior (Fischer et al. 2012). Many of these challenges, such as depletion of ecosystem services, loss of biodiversity, and changes in environmental entities and characteristics, have been primarily investigated by natural sciences. The natural sciences generate baseline information to identify critical changes to the environment, which in turn often lead to discussions both from policy-makers and civil society. Ecological research and conservation biology are therefore of primary importance for achieving a sustainable future and harmonizing people and nature, especially but not exclusively if they aid normative and transformative knowledge creation (e.g., for ecosystem services, see Abson et al. 2014). Many discussions that found their way into the broad societal discussion started as basic research, including climate change, waldsterben (forest loss due to acid rain), and ocean acidification. Basic natural science research contributes to many applied aspects of environmental problems that link other disciplines to sustainability. Topics rooted in ecology and conservation can help identify socio-ecological complexity and link to numerous other aspects of sustainability. However, ecology alone cannot tackle sustainability problems that are inherently linked to both natural and social systems.

The social sciences, stakeholder involvement, and mutual learning (both across scientific disciplines and, more broadly, in society) help to identify key drivers of ecosystem services and biodiversity loss. Many topics in ecology and conservation demand an exchange of knowledge about socio-ecological systems, such as governance and behavioral change, which are generated in other scientific disciplines. Consequently, socio-ecological research must link fundamental knowledge and system understanding between different disciplines and calls for strong interdisciplinary connections and exchange, including solution-orientated research. Ideally, this transdisciplinary process (engaging multiple scientific disciplines and the wider society) creates a transformative process based on scientific evidence that links back to, and helps change, societal dynamics and processes toward more sustainable human-environmental interactions.

- ***Questions***
  1. *How is ecological research linked to policymaking?*
  2. *How can ecological research benefit from stakeholder involvement?*

Further Reading

Chan KMA, Shaw MR, Cameron DR, Underwood EC, Daily GC (2006) Conservation planning for ecosystem services. PLoS Biol 4:2138–2152

Daily GC, Polasky S, Goldstein J, Kareiva PM, Mooney HA, Pejchar L, Ricketts TH, Salzman J, Shallenberger R (2009) Ecosystem services in decision making: time to deliver. Front Ecol Environ 7(1):21–28

Walker B, Salt D (2012) Resilience practice: building capacity to absorb disturbance and maintain function. Island Press, Washington, DC

References

Abson DJ, von Wehrden H, Baumgärtner S, Fischer J, Hanspach J, Härdtle W, Heinrichs H, Klein AM, Lang DJ, Martens P, Walmsley D (2014) Ecosystem services as a boundary object for sustainability. Ecol Econ 103:29–37

Ball P (2012) Feeling the heat. Nature 492:174–176

Cardinale BJ, Srivastava DS, Duffy JE, Wright JP, Downing AL, Sankaran M, Jouseau C (2006) Effects of biodiversity on the functioning of trophic groups and ecosystems. Nature 443:989–992

Cardinale BJ, Duffy JE, Gonzalez A, Hooper DU, Perrings C, Venail P, Narwani A, Mace GM, Tilman D, Wardle DA, Kinzig AP, Daily GC, Loreau M, Grace JB, Larigauderie A, Srivastava DS, Naeem S (2012) Biodiversity loss and its impact on humanity. Nature 486:59–67

Daily G (ed) (1997) Nature's services: societal dependence on natural ecosystems. Island Press, Washington, DC
Fischer J, Dyball R, Fazey I, Gross C, Dovers S, Ehrlich PR, Brulle RJ, Christensen C, Borden RJ (2012) Human behavior and sustainability. Front Ecol Environ 10:153–160
Foley JA, DeFries R, Asner GP, Barford C, Bonan G, Carpenter SR, Chapin FS, Coe MT, Daily GC, Gibbs HK (2005) Global consequences of land use. Science 309:570–574
Gaston KJ (2000) Global patterns in biodiversity. Nature 405:220–227
Gaston KJ, Spicer JI (2009) Biodiversity: an introduction. Wiley, Malden
Jenkins CN, Pimm SL, Joppa LN (2013) Global patterns of terrestrial vertebrate diversity and conservation. PNAS 110(28):E2602–E2610
Kleidon A, Mooney HA (2000) A global distribution of biodiversity inferred from climatic constraints: results from a process-based modelling study. Glob Chang Biol 6:507–523
Le Saout S, Hoffmann M, Shi Y, Hughes A, Bernard C, Brooks TM, Bertzky B, Butchart SHM, Stuart SN, Badman T, Rodrigues ASL (2013) Protected areas and effective biodiversity conservation. Science 342:803–805
Loreau M, Naeem S, Inchausti P, Bengtsson J, Grime J, Hector A, Hooper D, Huston M, Raffaelli D, Schmid B (2001) Biodiversity and ecosystem functioning: current knowledge and future challenges. Science 294:804–808
Millennium Ecosystem Assessment (2005) Ecosystems and human well-being. Island Press, Washington, DC
Myers N, Mittermeier RA, Mittermeier CG, da Fonseca GAB, Kent J (2000) Biodiversity hotspots for conservation priorities. Nature 403:853–858
Post DM, Doyle MW, Sabo JL, Finlay JC (2007) The problem of boundaries in defining ecosystems: a potential landmine for uniting geomorphology and ecology. Geomorphology 89:111–126
Randall A (1991) The value of biodiversity. Ambio 20:64–68
Richardson D (1997) The politics of sustainable development. In: Baker S, Kousis M, Richardson D, Young S (eds) The politics of sustainable development. Routledge, London, pp 43–60
Rockström J, Steffen W, Noone K, Persson Å, Chapin SI, Lambin E, Lenton TM, Scheffer M, Folke C, Schellnhuber HJ, Nykvist B, de Wit CA, Hughes T, van der Leeuw S, Rodhe H, Sörlin S, Snyder PK, Costanza R, Svedin U, Falkenmark M, Karlberg L, Corell RW, Fabry VJ, Hansen J, Walker B, Liverman D, Richardson K, Crutzen P, Foley J (2009) Planetary boundaries: exploring the safe operating space for humanity. Ecol Soc 14(2)
Sala OE, Chapin FS, Armesto JJ, Berlow E, Bloomfield J, Dirzo R, Huber-Sanwald E, Huenneke LF, Jackson RB, Kinzig A, Leemans R, Lodge DM, Mooney HA, Oesterheld M, Poff NL, Sykes MT, Walker BH, Walker M, Wall DH (2000) Biodiversity – global biodiversity scenarios for the year 2100. Science 287:1770–1774
Seppelt R, Dormann CF, Eppink FV, Lautenbach S, Schmidt S (2011) A quantitative review of ecosystem service studies: approaches, shortcomings and the road ahead. J Appl Ecol 48:630–636
Tilman D, Reich PB, Knops JM (2006) Biodiversity and ecosystem stability in a decade-long grassland experiment. Nature 441:629–632
Winter M, Devictor V, Schweiger O (2013) Phylogenetic diversity and nature conservation: where are we? Trends Ecol Evol 28(4):199–204
Zavaleta ES, Thomas BD, Chiariello NR, Asner GP, Shaw MR, Field CB (2003) Plants reverse warming effect on ecosystem water balance. PNAS 100:9892–9893
Zhu Y, Chen H, Fan J, Wang Y, Li Y, Chen J, Fan J, Yang S, Hu L, Leung H (2000) Genetic diversity and disease control in rice. Nature 406:718–722

Chapter 6
Sustainability Assessment of Technologies

Sjouke Beemsterboer and René Kemp

Abstract Sustainability has multiple dimensions. This chapter wants to stress that there is an inherent element of subjectivity in sustainable development that needs to be acknowledged even when sustainable development is at heart about improved states of the environment. Understanding of objectivity, subjectivity, and development can serve a more fruitful discussion about choices in sustainability. The aim of the chapter is to assess available methods for appraising the sustainability of innovation with regard to three key aspects for sustainability assessment: the ability to objectify impacts, the extent to which normative aspects are considered, and the coproduction of impacts between technology and environment.

Keywords Sustainability assessment • Assessment methods • Innovation • Perspectives • Coproduction

1 Introduction

Innovation and technical change are hailed by many proponents as solutions to the sustainability problems of modern society. Examples are electric battery and hydrogen cars, renewable energy technologies, smart grids, and smart housing, to name just a few. The enthusiasm for technological innovations as means toward sustainability is understandable, as they score positively in regard to certain environmental aspects. In this chapter, we want to examine the issue of the sustainability of green technologies and innovation. We will examine methods for environmental assessment and methods for dealing with the normative aspects of sustainable development. The aim of the chapter is to assess available methods for appraising the sustainability of innovation with regard to three key aspects for

S. Beemsterboer, M.A. (✉) • R. Kemp
International Centre for Integrated assessment and Sustainable development (ICIS), Maastricht University, P.O. Box 616, 6200 MD Maastricht, Netherlands
e-mail: sjouke.beemsterboer@maastrichtuniversity.nl; rene.kemp@maastrichtuniversity.nl

H. Heinrichs et al. (eds.), *Sustainability Science*, DOI 10.1007/978-94-017-7242-6_6

sustainability assessment: the ability to objectify impacts, the extent to which normative aspects are considered, and the coproduction[1] of impacts between technology and environment.

We will argue that sustainability should not be used as a label for technologies, as each and every technology has aspects that are problematic from an environmental point of view. It is better used as a yardstick to measure bigger or smaller contributions of technologies to sustainability criteria.

A second reason for not using the term sustainability as a label is that sustainable development involves normative choices about *what* we value (clean air, quietness), *how much* we value it, and issues of equity and justness having to do with normative views about whether it is right to eat meat, exploit nature in the way we do, and burn fossil fuels in the almost certain knowledge that this gives rise to potentially destructive climate change. A fundamental problem in sustainability assessments is how to *frame* sustainability (Bond and Morrison-Sounders 2013), as something environmental or normative and subjective. Our answer is that one should try to consider each of these aspects.

A starting point of this chapter is that science cannot determine what is sustainable; what it can do is offer evidence about the problems discussed in the name of sustainable development. It can also reveal the different perspectives on sustainability issues and make people mindful of the normative aspects in their own thinking and valuation and the implicit assumptions about progress. The radical implication of this is that sustainability goals cannot be determined in a fixed set of criteria, valid irrespective of time, place, and topic. They have to be agreed upon over time (Bond and Morrison-Sounders 2013; Gibson 2005). This does not mean that anything can go for sustainable development, but simply that sustainable development is neither something objective nor subjective. The chapter does not go into the philosophical aspects of this but examines methods for dealing with subjective and objective aspects and gives attention to the coproduction of impacts. It also draws attention to the political use of labels of sustainability by advocates of certain technologies, hiding problematic aspects of these technologies from public scrutiny (in terms of resource use, emissions, and waste), which enforces our conclusion that sustainability is to be used as a yardstick, not a label.

2 Methods for Assessing Technologies

There are different methods for assessing the sustainability contributions of technologies. To familiarize the reader with some of them, this chapter will first present a spectrum of methods that can be used to assess those contributions

[1] Depending on the scientific community, coproduction is also known as interaction effects.

Table 6.1 Methods for assessing sustainability contributions

Assessment method	Further reading
Cost–benefit analysis	Johansson (1993)
Dialogue methods	Cuppen (2010)
Ecological footprint (proxy methods)	Wackernagel and Rees (1996)
Life-cycle assessment	Baumann and Tillman (2004)
Material flow analysis	Brunner and Reichberger (2004)
Multi-criteria analysis	Figueira et al. (2005)
Scenario methods (incl. backcasting)	Swart et al. (2004), Holmberg (1998)
Procedural framework	
Environmental impact assessment	Glasson et al. (2012)
Integrated sustainability assessment	Weaver and Rotmans (2006)
Strategic environmental assessment	Therivel (2010)

Table 6.2 Classification of sustainability assessment methods

Styles		
Reductionist (indicator)	Holistic	
Monetary	Biophysical	Social
Product (micro)	Area/environment (macro)	

(Table 6.1). The readings included are intended to present an entry point for readers who seek more information.

Recently, efforts have been made to integrate various assessment methods into sustainability assessment. This has led to the publication of several overviews of sustainability assessment methods (Gasparatos and Scolobig 2012; Ness et al. 2007; Singh et al. 2012). Typically, they can be categorized on a number of levels. Such classifications provide insight into the research styles used in an assessment. It shows how the production of *objective information* is attempted (Table 6.2).

Generally, sustainability assessments benefit from recognizing the importance of context (Bond and Morrison-Sounders 2013; Gibson 2005). The overall context consists of the immediate physical context and the social context. The physical aspects are straightforward: a concentrated solar power plant in a sunny desert produces different results from one in a cloudy city. The social context refers to the social actors having views and value frames about the technology or practice which may lead them to reject certain options as inappropriate or fundamentally wrong. The sinking of the Brent spar oil platform serves as a useful example. According to Shell, it was an environmentally sound thing to do. But the general public educated in recycling saw this as an environmentally harmful activity. The sinking was seen as dumping and as setting a dangerous precedent. Shell was wrong to consider the sinking only on environmental and economic grounds. A *perspective-based* method would have revealed that other values were at stake (Cuppen 2010). Dialogue methods are a way to consider the different perspectives on problems and possible solutions and also to make people accept the outcomes of the assessment.

Acknowledging different perspectives and understanding the limits of methods for objectifying knowledge is a prerequisite for undertaking a useful sustainability assessment. They help the analyst to pick the right method and the user to grasp the qualities of an assessment.

A third complication is that the impacts of technologies are *coproduced* and dynamic. Dynamic elements in assessments are normally restrained to cause–effect chains of impacts on the environment. For example, the impact of increased CO_2 levels in the atmosphere on global temperatures, local precipitation levels, and biodiversity changes. These cause–effect chains are very complex and also include feedback effects. Yet, they only focus on the environment and disregard the impact of feedback effects on the technology.

Key technology-specific feedbacks derive from reflexivity, user practices, and rebound effects. The impact of risky technologies depends on the precautionary measures being undertaken to avoid risks and emergency strategies. For every product, the impacts depend on aspects of use and what is being done at the end of its lifetime. Better waste management systems help to reduce environmental impacts. Refrigerators have become more energy efficient, but they have also become bigger, encouraging people to store more food, and in so doing, they contribute to the practice of throwing away food. Impacts are thus tied up with practices, culture, economic frame conditions (prices), and systems of production and consumption. Most sustainability assessments do not include technology evolution (Karlström 2004; Sandén 2004) and do not consider scenarios of use that include rebound effects and interaction effects. In a dynamic sustainability assessment, coproduction of impacts between a technology and its environment is to be included.

To summarize, there are a number of conditions that improve the validity of sustainability claims:

1. Present *objectified* information on the impacts of a technology.
2. Be attentive to different *perspectives* on technology, impacts, and sustainability.
3. Include *coproduction* of impacts between a technology and its context.

In a utopian world, sustainability assessments would be flawless on all three criteria. In Fig. 6.1, this sweet spot [S] is depicted at the intersection of the three

Fig. 6.1 Key conditions in sustainability assessment

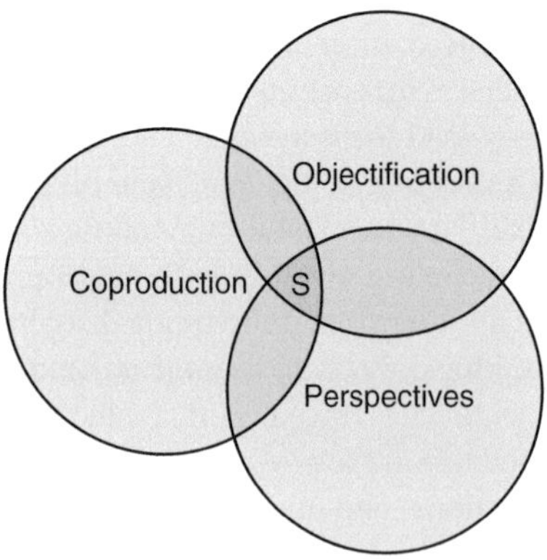

Table 6.3 Conformity of assessment methods with key conditions of sustainability assessment

Assessment method	Condition		
	Objectification	Perspectives	Coproduction
Backcasting		X	X
Cost–benefit analysis	XX		
Dialogue methods		XX	
Ecological footprint (proxy methods)	XX		
Life-cycle assessment	XX	X	
Material flow analysis	XX	X	
Multi-criteria analysis	XX	X	
Scenario analyses		XX	X
Assessment procedural framework			
Environmental impact assessment			
Integrated sustainability assessment			
Strategic environmental assessment			

conditions of sustainability assessment. But then again, sustainability assessment would not exist in a utopian world, as we would all live in perfect harmony. Each assessment method has its strengths and weaknesses. This requires a practitioner to select strategically from the vast quantity of methods that are available. For each key condition, this chapter will illustrate its application in a practical sustainability assessment (Table 6.3).

2.1 *Objectification of Assessments: Life-Cycle Assessment*

Many methods can be identified that aim to objectify sustainability contributions. They are predictive and grounded in the paradigm of technical measurement (Guba and Lincoln 1989). Some widely used methods include cost–benefit analysis (CBA), ecological footprint, life-cycle assessment (LCA), material flow assessment (MFA), and multi-criteria analysis (MCA). This section will focus on **LCA** as a way to illustrate objectification in knowledge produced on the environmental performance of a technology (Box 6.1).

Box 6.1: A Life-Cycle Assessment of a Coal-Based Power Plant with Carbon Capture and Storage Facility

Koornneef et al. (2008) have conducted an LCA of a coal-based power plant with carbon capture and storage (CCS) facility.[2] An LCA consists of four stages: goal and scope definition, life-cycle inventory analysis, life-cycle impact assessment, and interpretation of results. The goal and scope definition states the main question that the assessment tries to answer: what environmental trade-offs and benefits result from CCS? Greenhouse gas emissions are central in this assessment, as they determine the allocation of CO_2 credits upon which CCS is financed. The specific technologies assessed are a post-combustion coal power plant CCS facility with a monoethanolamine solvent and two reference cases without CCS. For each case, the life cycle of the facility is addressed from cradle-to-grave. The assessment uses a functional unit of 1 kWh of electricity generated at the power plant (Koornneef et al. 2008).

In the life-cycle inventory analysis, the assessment provides an overview of environmentally relevant flows for key air pollutants, emissions to water, resources, wastes, and byproducts in metric units per kWh. Examples of inventory results are 200 g CO_2 per kWh and 67.97 mg nitrate per kWh for the facility with CCS (Koornneef et al. 2008). In the life-cycle impact assessment stage, "raw results" from the inventory analysis are analyzed to make statements on the actual impacts of environmental loads. After having obtained the results, a number of tests are conducted to establish their robustness. In this case, a sensitivity analysis is performed for a number of parameters to estimate the impact of deviations in key data or assumptions – such as the impact of changes in CO_2 removal efficiency on other impact categories (Koornneef et al. 2008).

The assessment concludes that, due to CCS, greenhouse gas emissions are reduced to 243 g/kWh. This is 78 and 71 % lower compared to the two reference plants without CCS. However, the assessment shows that CCS does lead to increases in the other categories, i.e., human toxicity, ozone layer depletion, freshwater eco-toxicity potential, eutrophication, acidification, and photochemical oxidation potential (Koornneef et al. 2008).

[2] Carbon capture and storage (CCS) can be defined as "a process consisting of the separation of CO_2 from industrial and energy-related sources, transport to a storage location and long-term isolation from the atmosphere. […] an option in the portfolio of mitigation actions for stabilization of atmospheric greenhouse gas concentrations" (IPCC 2005).

Regardless of results, an assessment as described in Box 6.1 cannot by itself make a valid sustainability claim. To determine a sustainability contribution requires a judgment on the value of a reduction in greenhouse gas compared to increases in other pollutants. Such judgment cannot be based on scientific knowledge alone but depends as well on notions about what is valued in nature. In LCA, such subjective elements are to some extent incorporated by different impact assessment methods (Baumann and Tillman 2004).

Beyond valuing outcomes differently, there are other aspects to assessment that matter. Review studies on LCA of CCS show that results depend primarily on the boundaries of the technology studied and choices regarding other system boundaries, type of impacts, method of valuation, and weighting (Corsten et al. 2013; Marx et al. 2011). The importance of such methodological choices is well recognized within the LCA community (Baumann and Tillman 2004; Finnveden et al. 2009). However, they are not confined to LCA but have to be made when using any assessment method.

Different assessment communities address these decisions differently. Generally, choices are made based on the goal of the study, traditions within a research community, knowledge of the assessor, funding, and so on. Principally, they are a matter of competing perspectives and interests. The LCA community has responded to such normative flexibility by developing guidelines to harmonize practices, demanding critical reviews of comparative studies leading to statements disclosed to the public, and a general call for transparency (Baumann and Tillman 2004). Such procedural streamlining is one option but does not dissolve differences. For a method that aims to present objectified results, such variety presents a fundamental problem.

2.2 *Perspective-Based Assessment: Dialogue and Backcasting*

Some assessment methods address the problem of perspectives head on and focus on values, interests, and power. They include dialogue approaches and certain scenario methods. Examples of scenario-oriented methods include forecasting, backcasting, and sensitivity analysis. Dialogue methods contain the Delphi method, focus group, and consensus conference. Box 6.2 illustrates a perspective-based assessment that explores contributions of hydrogen to a future sustainable energy system. Methodologically, it focuses on *backcasting* and *dialogue groups*.

Box 6.2: Backcasting with Dialogue Groups for Contributions of Hydrogen to a Sustainable Energy System

Hisschemöller and Bode (2011) have conducted an assessment of possible sustainable uses of hydrogen in a future energy system in the Netherlands. In the project, 60 stakeholders were involved over a 4-year period. Different perspectives on the use of hydrogen were integrated. Specific visions were developed for transport, construction, and energy (natural gas) infrastructure. Two methods used in the assessment were a dialogue approach and backcasting (Hisschemöller and Bode 2011).

(continued)

Box 6.2: (continued)

Dialogue was identified as a methodology for enabling problem structuring, stimulating the expression of different viewpoints, and allowing for interaction between those viewpoints. This includes the selection of a broad range of stakeholders with divergent perspectives, articulation of perspectives, confrontation of views and knowledge claims, and synthesis (Hisschemöller and Bode 2011). Backcasting was used to stimulate the development of different visions. It considers boosting creativity by freeing stakeholders from current mental and institutional restrictions. The method stimulates participants to articulate different perspectives and build arguments to support those perspectives (Hisschemöller and Bode 2011).

One of the conclusions in the transport perspective was that hydrogen in combination with fuel cell vehicles has the potential to contribute to inner-city air quality. H_2 is to be produced from natural gas at large industrial plants and distributed on a dedicated H_2 grid. This perspective assumes that natural gas will be cheaper than oil and that battery electric transport will not pick up due to range issues and the environmental risks of batteries. Its use in the built environment is mainly driven by user wishes to increase autarky from the central electricity grid. Here, hydrogen offers a promising future as a decentralized generation option. Surplus energy from intermittent renewables is converted to H_2 through electrolysis. This perspective assumes that hydrogen and non-hydrogen options can be complementary and an institutional environment that favors small-scale over large-scale energy generation. It would also require the development of local heat grids and storage options. From the energy infrastructure perspective, climate change is undergirding hydrogen visions. H_2 is envisioned to be mixed with natural gas into Hythane and transported through the existing natural gas grid. This vision depends on the development of efficient extraction technologies to extract H_2 from the Hythane mix at the end user. A constant supply of hydrogen and gas is required to balance flows in the grid (Hisschemöller and Bode 2011, pp. 16–20).

It is concluded that H2 is "potentially promising for the future." A policy of diversification and niche development is proposed to stimulate different perspectives to develop in a protected environment (Hisschemöller and Bode 2011, p. 22).

Ignoring multiple perspectives makes the analysis one-sided by drawing on one single perspective that the assessor decides to adopt. Acknowledging different angles to a problem and solution increases the visibility of different perspectives and can stimulate mutual understanding. At the same time, it can also require the assessor to engage in the power struggle between incompatible perspectives. Depending on the need and desirability of integrating perspectives, the assessor will have to balance the number and types of perspectives to include. Including all perspectives would make the picture more realistic but also infinitely complex. Adopting too few perspectives prohibits valuable knowledge from entering the assessment. Also, the assessor will have to determine how to deal with the interaction between perspectives. It is possible to search explicitly for consensus but also to try to stimulate and manage conflict.

2.3 *Coproduction of Impacts: A Search for Methods*

Innovative technologies have effects that are beyond their direct impact. The impacts of technologies are coproduced by different actors through multiple and complex causal chains (Rip and Kemp 1997). Trying to assess the effect of a technological innovation in a system becomes complex very quickly. This can make it wise not to aim for prediction but for understanding and to base predictions on complex system dynamics. By including different feedback processes, the assessor can obtain rough estimates of impact (Sandén 2013). Well-known examples of feedback processes are economies of scale and increased user utility with diffusion (Andersson 2001; Rosenberg 1982; Utterback 1994).

Box 6.3 illustrates the *Socratic method*, an attempt to integrate the concept of coproduction with conditions of objectification and perspectives.

Box 6.3: A Socratic Method for Sustainability Policy Appraisal

Kemp and Weaver (Weaver and Kemp 2012) propose to consider the coproduction of effects in their assessment of policies or innovations that purposely intend to make a positive sustainability contribution. Instead of analyzing the question, "What are the likely impacts of this policy/innovation?," the following question is being asked: "Under what conditions could this policy/innovation contribute to sustainability and in which ways?"

With a Socratic dialogue, the authors propose a consideration of underlying values and assumptions together with causal linkages. This method requires the involvement of experts that have an understanding of various disciplines and are recognized as such. The Socratic method draws on expert knowledge about problems and causal links and considers value-based perspectives and coproduction aspects. The group of experts will engage in a series of discussions and dialogues aimed at getting a better understanding of a number of critical conditions for the sustainability contributions of a policy or an innovation in a system.

> The Socratic method draws on [1] foresight by examining technology evolution, [2] soft system methodology by considering causal links and different understandings of a problem situation, [3] environmental science in giving attention to material streams and environmental pressures, and [4] a sociotechnical perspectives on innovation, which views impacts of technologies as co-produced. (Weaver and Kemp 2012, p. 8)

(continued)

Box 6.3: (continued)

In an illustrative case study, Weaver and Kemp emphasize some interaction effects for the case of electric cars. A positive interaction effect occurs when electric cars are used involving park-and-ride and becomes the favored car of use in multicar households. It is also found that electric vehicles can stimulate the deployment of renewables by charging at off-peak times and providing storage for peak demand. A negative spillover effect of better batteries is the replacement of normal bikes by e-bikes in bike-using countries. In non-bike-using countries, the e-bike may substitute for motorized transport as a positive effect. There are also interaction effects with the grid (Weaver and Kemp 2012, p. 19). A consideration of these interaction effects may help to maximize positive benefits and reduce negative effects.

The Socratic method shows that including conditions of objectification, perspectives, and coproduction in assessments can be done by a *combination of methodologies*. From an assessment perspective, it can be interesting to see which other combinations can assess the sustainability effects of technologies.

3 Conclusion and Outlook

Sustainability has multiple dimensions. This chapter wants to stress that there is an inherent element of subjectivity in sustainable development that needs to be acknowledged even when that development is, at heart, about improved states of the environment. Understanding of objectivity, subjectivity, and development can serve a more fruitful discussion about choices in sustainability. One way of doing so is by integrating the three identified conditions – *objectification*, *perspectives* and *coproduction*, – in assessments.

First steps in this direction have already been taken. For example, stakeholder-based LCA calls for the inclusion of multi-stakeholder groups in LCA practices (Thabrew et al. 2009). Such practice is an example of the integration of perspectives into objectification methods. Alternatively, hybrid LCA extends assessment practices to include environmentally extended input–output tables (Finnveden et al. 2009; Hawkins et al. 2006). This can constitute a first step in combining the condition of objectification with notions of coproduction.

Also theoretically, relevant contributions have been made on aspects of coevolution and micro–macro links. Dijk (2010) has created a coevolutionary framework to study the dynamics of technological innovations which combines elements that can be objectified, such as sales levels and efficiency rates, with subjective perspectives from individuals and organizations. A number of researchers have developed insights by combining system dynamics with environmental assessments of technological development (Hillman 2008; Karlström 2004; Kushnir 2012; Sandén 2004). Such contributions underline the importance of integrating conditions of objectification, perspectives, and coproduction. They provide a useful starting point for developing assessment practices for the sustainability contributions of technologies.

Questions

1. What are the three key conditions for sustainability assessment of technological innovations?
2. In which ways do these three key conditions differentiate from each other? You are encouraged to use the examples of assessments in Boxes 6.1, 6.2, and 6.3 to illustrate your answer.
3. What condition for sustainability assessment do you have most experience with in your daily routine? Can you explain in which ways the other two conditions are/can be of importance?

Further Reading

Bond A, Morrison-Sounders A (2013) Challenges in determining the effectiveness of sustainability assessment. In: Bond A, Morrison-Sounders A, Howitt R (eds) Sustainability assessment: pluralism, practice and progress. Routledge, London/New York, pp 37–50

Dijk M (2010) Innovation in car mobility: co-evolution of demand and supply under sustainability pressures. PhD, Maastricht University – ICIS, Maastricht

Gibson R (2005) Sustainability assessment: criteria and processes. Earthscan, London

Kemp R (2010) Sustainable technologies do not exist! Ekonomiaz 75(3):11–13

Kushnir D (2012) Foresight and feedback: monitoring and assessing the environmental implications of emerging technologies. Chalmers University of Technology, Göteborg

Sandén B (2004) Technology path assessment for sustainable technology development. Innov Manag Policy Pract 6(2):316–330. doi:10.5172/impp.2004.6.2.316

Weblinks

A sustainability-oriented assessment of electromobility from different systems perspectives (safety, LCA, …) for different sociotechnical configurations can be found at http://www.chalmers.se/en/areas-of-advance/energy/cei/Pages/Systems-Perspectives-on-Electromobility.aspx

Database with short overviews of different sustainability assessment methods developed in a collaboration between universities led by Free University Amsterdam (VU). http://www.ivm.vu.nl/en/projects/Archive/SustainabailityA-test/index.asp

References

Andersson BA (2001) Material constraints on technology evolution. PhD, Chalmers University of Technology, Göteborg University, Göteborg (Sandén)

Baumann H, Tillman A (2004) The hitch hiker's guide to LCA. Studentlitteratur, Lund

Brunner P, Rechberger H (2004) Practical handbook of material flow analysis. CRC Press, Boca Raton
Corsten M, Ramírez A, Shen L, Koornneef J, Faaij A (2013) Environmental impact assessment of CCS chains – lessons learned and limitations from LCA literature. Int J Greenhouse Gas Control 13(0):59–71. doi:http://dx.doi.org/10.1016/j.ijggc.2012.12.003
Cuppen E (2010) Putting perspectives into participation: constructive conflict methodology for problem structuring in stakeholder dialogues. PhD, Vrije Universiteit, Amsterdam
Figueira J, Greco S, Ehrgott M (2005) Multiple criteria decision analysis: state of the art surveys, vol 78. Springer, New York
Finnveden G, Hauschild MZ, Ekvall T, Guinee J, Heijungs R, Hellweg S, … Suh S (2009) Recent developments in life cycle assessment. J Environ Manag 91(1):1–21
Gasparatos A, Scolobig A (2012) Choosing the most appropriate sustainability assessment tool. Ecol Econ 80:1–7
Glasson J, Therivel R, Chadwick A (2012) Introduction to environmental impact assessment. Routledge, London/New York
Guba EG, Lincoln YS (1989) Fourth generation evaluation. Sage, Newbury Park
Hawkins T, Hendrickson C, Higgins C, Matthews HS, Suh S (2006) A mixed-unit input–output model for environmental life-cycle assessment and material flow analysis. Environ Sci Technol 41(3):1024–1031. doi:10.1021/es060871u
Hillman K (2008) Environmental assessment and strategic technology choice. The case of renewable transport fuels. Doctor, Chalmers University of Technology, Göteborg
Hisschemöller M, Bode R (2011) Institutionalized knowledge conflict in assessing the possible contributions of H2 to a sustainable energy system for the Netherlands. Int J Hydrog Energy 36(1):14–24. doi:http://dx.doi.org/10.1016/j.ijhydene.2010.09.024
Holmberg J (1998) Backcasting: a natural step in operationalising sustainable development. Greener Manag Int 23:30–52
IPCC (2005) Special report on carbon dioxide capture and storage. In: Metz B, Davidson O, De Coninck H, Loos M, Meyer L (eds) IPCC special report on carbon dioxide capture and storage. Cambridge University Press, New York
Johansson P-O (1993) Cost-benefit analysis of environmental change. Cambridge University Press, Cambridge/New York
Karlström M (2004) Environmental assessment of polymer electrolyte membrane fuel cell systems. Consequences of an evolutionary perspective on technology development. Chalmers University of Technology, Goteborg
Koornneef J, van Keulen T, Faaij A, Turkenburg W (2008) Life cycle assessment of a pulverized coal power plant with post-combustion capture, transport and storage of CO2. Int J Greenhouse Gas Control 2(4):448–467. doi:http://dx.doi.org/10.1016/j.ijggc.2008.06.008
Marx J, Schreiber A, Zapp P, Haines M, Hake JF, Gale J (2011) Environmental evaluation of CCS using life cycle assessment – a synthesis report. Energy Procedia 4(0):2448–2456. doi:http://dx.doi.org/10.1016/j.egypro.2011.02.139
Ness B, Urbel-Piirsalu E, Anderberg S, Olsson L (2007) Categorising tools for sustainability assessment. Ecol Econ 60(3):498–508. doi:10.1016/j.ecolecon.2006.07.023
Rip A, Kemp R (1997) Technological change. In: Rayner S, Malone EL (eds) Human choice and climate change, vol 2. Battelle, Columbus, pp 327–399
Rosenberg N (1982) Inside the black box: technology and economics. Cambridge University Press, Cambridge
Sandén B (2013) Technical change and environmental assessment. Seminar on environment and energy systems analysis, 26–27 March 2013. Chalmers University of Technology, Göteborg
Singh RK, Murty HR, Gupta SK, Dikshit AK (2012) An overview of sustainability assessment methodologies. Ecol Indic 15(1):281–299. doi:http://dx.doi.org/10.1016/j.ecolind.2011.01.007
Swart RJ, Raskin P, Robinson J (2004) The problem of the future: sustainability science and scenario analysis. Glob Environ Chang 14(2):137–146

Thabrew L, Wiek A, Ries R (2009) Environmental decision making in multi-stakeholder contexts: applicability of life cycle thinking in development planning and implementation. J Clean Prod 17(1):67–76. doi:http://dx.doi.org/10.1016/j.jclepro.2008.03.008
Therivel R (2010) Strategic environmental assessment in action. Earthscan, London/Washington, DC
Utterback J (1994) Mastering the dynamics of innovation. Harvard Business School Press, Boston
Wackernagel M, Rees WE (1996) Our ecological footprint: reducing human impact on earth. New Society Publications, Gabriola Island
Weaver P, Kemp R (2012) A Socratic method for sustainability policy appraisal. Paper presented at the International conference on sustainability transitions 2012. Lyngby, Denmark
Weaver P, Rotmans J (2006) Integrated sustainability assessment: What? Why? How? In: Jäger J, Weaver P (eds) MATISSE working paper, vol 1. European Commission, Brussels

Chapter 7
Corporate Sustainability Management

Stefan Schaltegger, Erik G. Hansen, and Heiko Spitzeck

Abstract This chapter provides a practice-oriented introduction to corporate sustainability management for students. It first introduces a brief history of corporate sustainability in businesses with reference to selected critical incidents (e.g., accidents environmental pollution) and pioneering firms and (sustainable) entrepreneurs which have developed proactive sustainability strategies and practices already several decades ago. Then the sustainability triangle, a conceptual framework for corporate sustainability, is presented. It aims at economic, eco- and socio-effectiveness by integrating and linking the economic, environmental, and social dimensions through the concepts of eco-efficiency, socio-efficiency, and eco-justice. This heuristic helps to classify sustainability management practices, relate them to each other, check for their synergies, and discuss how to overcome potential trade-offs. For the implementation of sustainability management in practice, collaboration between various actors is necessary. The reminder of the chapter therefore introduces various forms of collaboration available to businesses in order to engage with value chain partners and broader societal partners. Ultimately, the collaboration forms are classified into interdisciplinary and transdisciplinary approaches. Each text part contains various questions addressing the readers for their own reflection of current practices concerning the introduced concepts. The chapter ends with a brief conclusion.

Keywords Sustainability management • Corporate sustainability • Stakeholder collaboration • Transdisciplinarity research • Corporate social responsibility

S. Schaltegger (✉) • E.G. Hansen
Centre for Sustainability Management, Leuphana University of Lüneburg, Scharnhorststr. 1, 21335 Lüneburg, Germany
e-mail: schaltegger@uni.leuphana.de; erik.hansen@uni.leuphana.de

H. Spitzeck
Sustainability Research Center, Fundação Dom Cabral, Av. Dr. Cardoso de Melo 1184, 15° andar, Vila Olímpia, São Paulo, SP 04548-004, Brazil
e-mail: heiko@fdc.org.br

H. Heinrichs et al. (eds.), *Sustainability Science*, DOI 10.1007/978-94-017-7242-6_7

1 Basic Questions

Let us start this chapter by looking at the basic questions. We invite you to sit down, take a piece of paper, and note down your answers to the following questions before entering into the content of this chapter:

- Which company would you consider to be a leader in sustainability?
- Why do you consider the company a leader in sustainability? To what degree does the company think and act differently from others?
- Why do you think the company engages in sustainability? What competitive advantages and other benefits might the company gain by being a leader in sustainability? Who else benefits from these actions?
- When and why did companies start to deal explicitly with sustainability?

To achieve improved sustainability performance, it is necessary to reduce negative environmental and social impacts, to create positive social and environmental impacts, and to do this in a way which supports the economic success of the company.

2 When Did Companies Start to Deal Explicitly with Sustainability? A Brief History of Sustainability Management

The *roots of sustainability management* go back quite far. Early exemplars are the anthroposophical and life reform movements with their related business activities. Anthroposophical businesses, often attached to Waldorf schools, biodynamic farms and gardens, anthroposophically extended medical practices, therapists, artists, scientists, colleges, and adult education centers (see the Directory of Initiatives at http://www.anthroposophy.org/DoI/intro.html), following Rudolf Steiner's (b.1861–d.1925) theories, focused on overcoming social and environmental problems long before sustainability was coined as a term and sustainability management became a consideration for mainstream organizations. Many decades later, environmental management was promoted strongly in the 1970s and 1980s as a consequence of the oil crises and the Club of Rome report. Committed people in business following their personal convictions, like Gottlieb Duttweiler, Heinz Hess, Claus Hipp, or Georg Winter, began to strive for economic success in their businesses through a combination of social, ecological, and market principles. Alternative business approaches, as discussed under the notion of social and sustainable entrepreneurship, have been invented and established new organizations looking beyond profit seeking (or maximization) alone. Apart from such sustainable entrepreneurs, various owners of large established (often family owned) companies (e.g., Henkel building community houses for their workers) had already begun applying practices more than 100 years ago and were philanthropically active, particularly to improve the social situation of their employees (for a general positioning of philanthropy as a company activity see e.g. Carroll 1979, 1991).

Environmental issues became a particular focus in the 1980/1990s as the result of various accidents in the chemical and oil industry (e.g., the Icmesa accident in Seveso, the ICI accident in Bhopal, the Sandoz accident in Schweizerhalle, the Exxon Valdez accident in Alaska; see e.g. Schaltegger et al. 2003). The perspective on environmental technologies for pollution prevention, clean production, and environmental services became the main focus in the USA, Europe, and Japan in the 1980s, and managers of large conventional companies started ever more to consider eco-efficient production, product development and innovation, business model innovation, the creation of business cases for sustainability, and the design of sustainable supply chains. In the 1990s, business received further encouragement to be aware of environmental and later also social impacts through the introduction of environmental and social management systems and standards for organizations (i.e., ISO 14000 series, EMAS, and BS 7750) and labels for certified products and services (e.g., EU-certified organic produce, fair trade-certified products). Social, environmental, and sustainability management systems and performance can now be certified and communicated in a standardized way. Efforts to improve the sustainability of companies are furthermore supported by awards (e.g., the German sustainability management award), rankings (e.g., the Dow Jones Sustainability Index), and media coverage on specific sustainability innovations and performances.

Exposed to concerns about the social and environmental impacts of their organizations and products, management has realized that social acceptance and legitimacy are part of sustainability management and that the consideration of stakeholder expectations is key to good strategic management (Freeman 1984; Stead and Stead 2008). Furthermore, operational management has not only understood that reducing energy consumption and pollution can go along with cost reductions but that sustainability considerations can be a basis for innovation (Hansen et al. 2009). Sustainability has thus become an argument in product and organizational development as well as in sustainable entrepreneurship and strategic management (Schaltegger et al. 2003).

The International Corporate Sustainability Barometer, a survey of the largest companies in 11 developed countries worldwide (Schaltegger et al. 2013c), shows that the main reasons for dealing with sustainability issues are to secure legitimacy (i.e., social acceptance) and to improve internal organization of the company, whereas the market forces are of clearly lower relevance than societal issues. Among the main stakeholders who are considered to be supportive of corporate sustainability engagement are NGOs, governments, and media, whereas banks, insurance companies, and suppliers are not perceived as supportive (Fig. 7.1).

A longitudinal study of the Corporate Sustainability Barometer for German companies, however, shows that market drivers seem to be of increasing (though still not paramount) importance in dealing with sustainability. This is also supported by a recent study by BCG and the MIT Sloan Management Review, who determined the primary reasons that companies embrace sustainability in their business models (Haanaes et al. 2012), as depicted in Fig. 7.2.

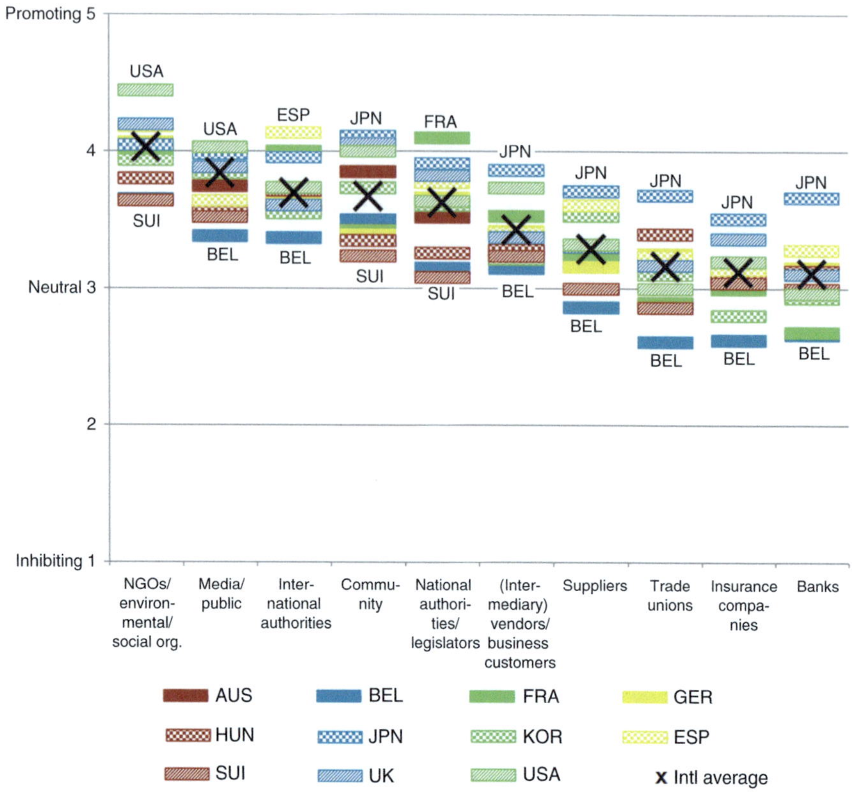

Fig. 7.1 Who motivates large companies to deal with sustainability issues? (Source: Schaltegger et al. 2013a, b, c, 19)

Overall, this shows that corporate management may be motivated by various reasons, including intrinsic, personal reasons (as is the case among the pioneers of sustainable entrepreneurship), seeking social acceptance and legitimacy, reducing risks and costs in production processes, or striving for market success (see also Schaltegger 2010; Hansen 2010, pp. 28ff.; Schaltegger et al. 2012).

3 What Do Sustainability-Oriented Companies Do Differently? Conceptualizing Sustainability Management

Sustainability management has been informed by the vision of integrating environmental, social, and economic perspectives into corporate management. The "Sustainability Triangle" (Fig. 7.3) represents this concept of sustainability management (similar to BMU et al. 2002; Dyllick and Hockerts 2002; Schaltegger and Burritt 2005).

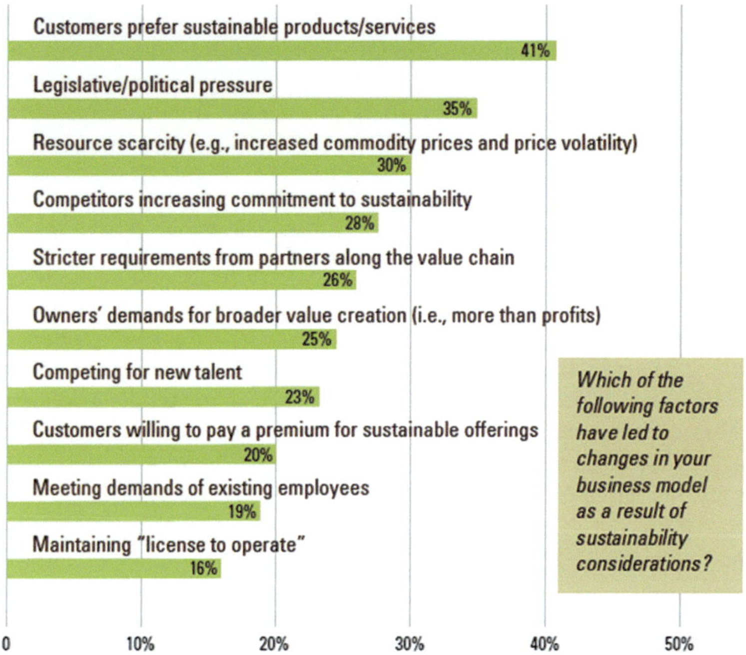

Fig. 7.2 Factors motivating companies to consider sustainability in their business model (Source: Haanaes et al. 2012, p. 7)

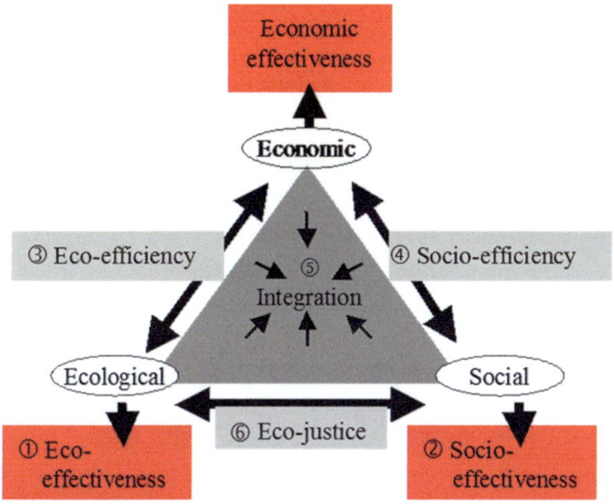

Fig. 7.3 The sustainability triangle of perspectives of corporate sustainability management (Source: Schaltegger and Burritt 2005, 189)

The Sustainability Triangle is designed both to help explain the three generally recognized components of corporate sustainability – the social, environmental, and economic perspectives, as well as the interrelationships between them – and to identify where a contribution is needed from sustainability management in order to support management with relevant information as to where performance contributions can and should be made. In the Triangle, each component is represented by one of the three corners, whereas the interrelationships are represented by the lines which connect each corner together. The corners therefore represent the company's effectiveness in achieving each component individually and are measured in absolute terms, whereas the lines represent different ways in which eco- and socio-efficiency, as well as eco-justice, can be defined by taking different combinations of the three perspectives and expressing the results as relative indicators. Effectiveness, measured in absolute terms, is the goal whenever management strives for the improvement of a single dimension (e.g., tons of waste avoided, additional income in poor regions), whereas efficiency, measured in relative terms, describes the relationship between different dimensions, e.g., socio-efficiency for the relationship between the social and economic dimensions (e.g., additional income in poor regions per unit of additional turnover).

The conventional aim of business management is economic effectiveness. The challenge for sustainability management is to support business leaders on the other aspects. A company's eco-effectiveness (i.e., ecological effectiveness) reflects how successful environmental management as part of sustainability management has been in reducing its impacts on the natural environment. This is usually expressed in terms of the absolute amounts of physical quantities, such as CO_2 emissions. Socio-effectiveness reflects how a company has performed with regard to social and cultural demands and to legitimate its activities. Therefore, topics such as stakeholder management (Freeman 1984) and how to respond to societal demands (Spitzeck 2009a) become important and can be measured, e.g., by reputation indexes, positive and negative media reporting, and the capacity of companies to create trusting relationships with nonmarket stakeholders. The economic challenge of business is to maximize its financial returns (profits) relative to financial resources such as capital invested. The equivalent challenge to sustainability management is to achieve the maximum environmental and social performance as economically as possible (see the "triple bottom line" approach by Elkington 2004), and these are measured by eco-efficiency and socio-efficiency, respectively. Eco-efficiency is defined as the relative proportions of an economic (monetary) measure and a physical (ecological) measure (Schaltegger and Burritt 2005; Schmidheiny and BCSD 1992; von Weizsäcker et al. 1997, 2009). It can be defined as the ratio of value added to environmental impact added per unit (e.g., 300 Euro sales per 1.2 kg of CO_2 impacts for one mobile phone). Environmental impact added is defined as the sum of all environmental impacts which are generated directly or indirectly by a product or activity, e.g., value added per tonne of CO_2 emitted. Similarly, socio-efficiency can be defined as the economic value added relative to social measures, such as the number of staff accidents. Eco-justice reflects the ratio between environmental and social objectives or indicators, e.g., environmental impacts relative to poverty.

The main challenge for sustainability management is to integrate all these different aspects. A good example is C&A's attempt to introduce organic textiles to the mass market (see Box 7.1). This integration challenge goes along with the observation and development that CSR and corporate sustainability have become ever more similar, although they were originally different approaches with similar goals related to sustainable development (see, e.g., Marrewijk 2003). Integration requires the consideration of all sustainability aspects and the links between them. Sustainability management can therefore also be seen as the art of overcoming trade-offs between social, environmental, and economic perspectives and the search for ways to integrate improvements in all dimensions. With regard to the links to the economic dimension, this also addresses business cases for sustainability (Schaltegger and Synnestvedt 2002). Several studies point out that sustainability leaders have a better reputation, counting on brand value, client fidelity, and preference, and better stakeholder relations, and therefore fewer risks, as well as positive and free media reporting (e.g., Hansen et al. 2010). In purely financial terms, sustainability management pays off in cases in which clients are prepared to pay a higher price (think of organic products), costs are reduced (e.g., with lower energy consumption), and attractiveness as an employer is improved, as well as access to capital markets being facilitated. The degree to which companies investigate and develop the business case for sustainability differs and is conceptualized as stages of organizational learning on a continuum from defensive, compliance, managerial, and strategic to civil (Zadek 2004).

To support sustainability management, a multitude of tools have been developed in theory and corporate practice, some addressing single aspects of sustainability management, others addressing two or more aspects. So far, the application of fully integrative sustainability management tools (see 5 in Fig. 7.3) is rare, but on an increasing trend. The future challenges of sustainability management may thus relate to the broad integration of sustainability aspects to overcome trade-offs and create multi-win solutions without compromising the effectiveness in one dimension. This challenge may seem unachievable, but as a vision, it provides direction and can serve to express ambition and enhance innovation for sustainable development.

Question Read the article by Simon Zadek (2004) which treats the issue of child labor at Nike. Now, imagine a large energy provider for electricity in your country. Analyze their corporate reports and website and identify their top three sustainability-related challenges. Position them into the organizational learning matrix and see if they represent a risk or opportunity for the company.

Box 7.1 Example: Clothing Made from Organic Cotton in the German Retail Industry (e.g., Hansen and Schaltegger 2013)

A very good example of voluntary sustainability management practices can be seen in the German clothing retail industry. Some of the large players have embarked on a transformation journey in which they change their product portfolio towards products with improved environmental and social performance. For example, C&A, Germany's largest textile retailer, engages strongly in the development of organic cotton practices, which has both environmental and social benefits compared to conventional cotton. C&A has increased its share of organic textiles – sold under the brand name "Bio Cotton" – to more than 20 % annually. By systematically developing alternative supplier structures together with CottonConnect (see http://www.cottonconnect.org), a nongovernmental organization (NGO), they are able to increase their organic cotton products year by year and thereby increase pressure to peers in the industry to invest in changes themselves.

Mostly, large industry players are not the first ones to develop more sustainable practices. Often, they adopt new processes, organizational structures, and products from smaller entrepreneurial companies – usually operating in niche markets. In the case of the textile industry, Hess Nature (http://www.hessnatur.com) was the first company to develop strictly environmental textiles, having already begun doing so in the 1980s. For example, they were one of the first to develop production standards to assure chemical-free end products, introduce organic cotton production in pilot projects, and adopt strict labor practices in textile production plants in developing nations.

Last but not least, large companies in industry often do not fully voluntarily transform their businesses – in the case of the textile industry, Greenpeace, particularly through its Detox campaign (http://www.greenpeace.de/detox), has played a major role in pushing the industry towards more eco-friendly production.

4 Collaboration as Key to Advance Sustainability Management

As the Nike example and the case of organic cotton demonstrate, sustainability management usually cannot be done successfully by individual actors alone. Developing solutions to sustainability-related problems, whether environmental or social, requires not only interdisciplinary (e.g., environmental experts and engineers working together on a solution) but usually transdisciplinary efforts, i.e., collaboration within firms, between firms, and across sectors (e.g., Schaltegger et al. 2013a).

4.1 *Collaboration Within the Firm: Integrative Approach to Sustainability*

In order to achieve this, sustainability management examines production processes and, also, all support functions, such as planning and control, human resources, financing, and organization. Sustainability management is therefore a multifunctional activity stretching right across the company. In fact, its success often depends on cooperation between different functions. To reduce waste at a construction site, a company needs to engage its architects responsible for planning, the purchasing departments sticking to orientations by architects, and construction workers trained to use the material provided effectively.

The need for *solutions based on cooperation* challenges management with issues of institutionalizing sustainability management, so that ecological and social objectives become an integral part of all management areas in all business functions. Sustainability management is often dealt with in a separate internal group or department within the company. Without a sustainability manager to lobby and cooperate with other managers, there is a danger that environmental and social interests will not be integrated into business thinking. Sustainability management is mostly introduced through project groups or permanent cross-functional teams (e.g., production, marketing, and waste handling).

In practice, various ways of institutionalizing sustainability expertise into management demonstrate that no single answer to the question about *how to organize* sustainability management exists. At worst, sustainability management can become an isolated and poorly funded function. At best, organizational structures exist that define social and environmental key performance indicators which are linked to executive compensation. This only happens if there is support and commitment from top management, who provides resources to ensure that sustainability management practices are integrated throughout the organization and in the core business model and processes. Other good examples are "sustainability committees" consisting of decision-makers from various functions (e.g., production, R&D, marketing, procurement, strategy) coming together on a regular basis in order to drive sustainability management practices in the company, as well as across value chains. But the necessity of collaboration does not end at organizational boundaries – it is also the engagement with external stakeholders which enables holistic solutions for sustainability (Spitzeck and Hansen 2010).

4.2 *Collaboration with Value Chain Partners and Other Societal Actors*

Once the company has a clear sustainability strategy, it can leverage results by engaging with partners along the value chain (suppliers and customers). In order to create sustainable products, it might be necessary to brainstorm with suppliers in the

Table 7.1 Examples of multi-stakeholder initiatives for solving sustainability-related problems

Issue	Resolution	More information
Child labor	Fair Labor Association	www.fairlabor.org
Conflict diamonds	Kimberley Process	www.kimberleyprocess.com See also the film Blood Diamond
Deforestation	Forest Stewardship Council	www.fsc.org
Overfishing	Marine Stewardship Council	www.msc.org

value chain. British retailer Marks & Spencer, for example, asked suppliers of light bulbs to come up with more energy efficient lighting solutions, which reduced energy costs significantly (Spitzeck 2009b). The same company encourages its clients to wash their clothes at 30°, as it realized that the highest emissions impact regarding washing powder rests with the consumer.

Other companies need to engage with their value chain and other societal actors because of risks. Nike was confronted with the risk of child labor in the value chain and cofounded the Fair Labor Association (FLA) in order to eradicate this social problem in the whole industry. In most cases, different actors come together in order to resolve an issue (see examples in Table 7.1).

No company or organization could resolve these issues alone. Only by cooperating on issues such as overfishing can they be addressed effectively. The research by MIT and BCG (Haanaes et al. 2012) also identified that sustainability leaders collaborate more intensively with customers, suppliers, governments, industry associations, local communities, NGOs, and competitors.

4.3 *Towards Transdisciplinary Collaboration*

In fact, sustainability-related challenges are often transdisciplinary in nature, meaning that not a single sector (e.g., industry, politics, society) nor a single discipline (e.g., business administration, environmental science) or business function (e.g., marketing, production, procurement) can solve them alone. Transdisciplinary collaboration in which various sectors and disciplines work together on solutions is necessary (e.g., researchers from universities engage with industry experts in order to implement more sustainable solutions in practice). Different generic approaches, more or less close to transdisciplinary collaboration, exist (Fig. 7.4).

Question Take the energy company which you selected before and analyze which forms of collaborations they pursue on the three levels: within the organization (e.g., different corporate functions), across the value chain (e.g., with energy producers or consumers), and on a broader societal level (e.g., NGO collaboration). Which forms of collaboration seem to have real impact and which are rather superficial or even greenwashing?

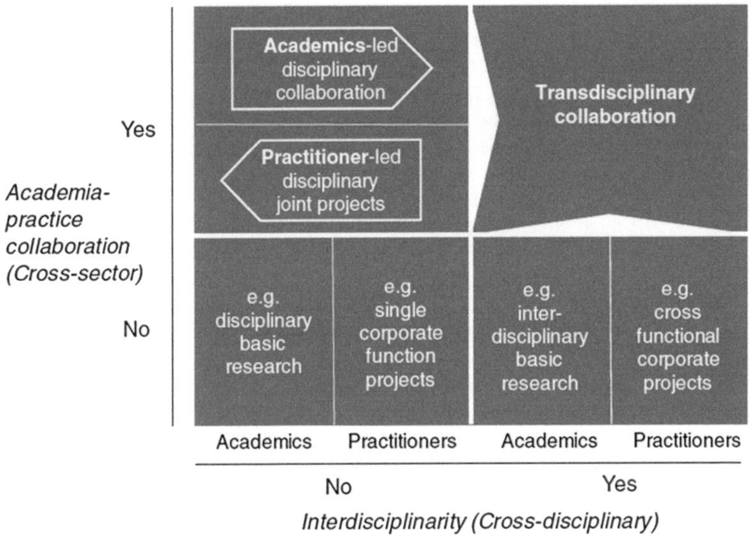

Fig. 7.4 Approaches to transdisciplinary collaboration in the context of corporate sustainability (Schaltegger et al. 2013a, b, c, p. 225)

5 Conclusions

Sustainability is one of the largest, if not the largest, challenge to today's industry and economies. If sustainability challenges are not sufficiently addressed in the near future, existing economic practices, as well as societal well-being, will become increasingly constrained. Sustainability management is a concept which has developed over the last two decades to support companies in overcoming trade-offs between economic, environmental, and social interests. To make sustainability management successful in corporations, collaboration is key, not only within the organization but, in a transdisciplinary sense, also across organizations, sectors, and disciplines.

Check your initial notes. Do you recognize the similarity between your list and the content of this chapter? What do sustainability leaders do differently from other companies? Examine how they implement social and environmental indicators in their management system to evaluate eco-/socio-effectiveness and efficiency. They engage with partners in the value chain such as suppliers and customers, as well as with other societal actors. Why do they do it? Sustainability management lowers risks (e.g., protests, bad publicity) and creates opportunities (e.g., higher prices, lower costs) and thus can create competitive advantages for sustainability leaders. At the same time, their actions create value for communities, the environment, and society as a whole – a notion that has recently been termed "creating shared value" (Porter and Kramer 2011). So in sum, sustainability management attempts to create value for the company and society alike.

References

BMU & BDI (Bundesministerium für Umwelt and Bundesverband der Deutschen Industrie), (German Federal Ministry for the Environment, German Federation of Industry Associations) (eds) Schaltegger S, Kleiber O, Müller J, Herzig C (authors) (2002) Sustainability management in business enterprises. Concepts and tools for a sustainable organisation development. BMU/BDI/CSM, Berlin

Carroll AB (1979) A three-dimensional conceptual model of corporate performance. Acad Manage Rev 4(4):497–505

Carroll AB (1991) The pyramid of corporate social responsibility: towards the moral management of organizational stakeholders. Bus Horiz 34(4):39–48

Dyllick T, Hockerts K (2002) Beyond the business case for corporate sustainability. Bus Strateg Environ 11(2):130–141

Elkington J (2004) Enter the triple bottom line. In: Henriques A, Richardson J (eds) The triple bottom line, does it all add up? Earthscan Publications, London, pp 1–16

Freeman E (1984) Strategic management. A stakeholder approach. Pitman, Marshfield

Haanaes K, Reeves M, Von Strengvelken I, Audretsch M, Kiron D, Kruschwitz N (2012) Sustainability nears a tipping point. In: MIT Sloan Management Review and Boston Consulting Group (ed) Research report. MIT Sloan Management Review, Boston

Hansen EG (2010) Responsible leadership systems: an empirical analysis of integrating corporate responsibility into leadership systems. Gabler, Wiesbaden

Hansen EG, Schaltegger S (2013) Made with 100 % organic. A sustainable entrepreneurship perspective on the diffusion of organic clothing. Corp Gov Int J Bus Soc 13(5):583–597

Hansen EG, Große-Dunker F, Reichwald R (2009) Sustainability Innovation Cube. A framework to evaluate sustainability-oriented innovations. Int J Innov Manag 13(4):683–713. doi:10.1142/S1363919609002479

Hansen EG, Sextl M, Reichwald R (2010) Managing strategic alliances through a community-enabled balanced scorecard: the case of Merck Ltd, Thailand. Bus Strateg Environ 19(6):387–399

Porter M, Kramer MR (2011) The big idea: creating shared value. Harv Bus Rev 89(1–2):62–77

Schaltegger S (2010) Unternehmerische Nachhaltigkeit als Treiber von Unternehmenserfolg und Strukturwandel, Wirtschaftspolitische Blätter, 57. Jg., Nr. 4, 495–503

Schaltegger S, Burritt RL (2005) Corporate sustainability. In: Folmer H, Tietenberg T (eds) The international yearbook of environmental and resource economics 2005/2006. Edward Elgar, Cheltenham, pp 185–222

Schaltegger S, Synnestvedt T (2002) The link between green and economic success: environmental management as the crucial trigger between environmental and economic performance. J Environ Manage 65(4):339–346

Schaltegger S, Petersen H, Burritt R (2003) An introduction to corporate environmental management: striving for sustainability. Greenleaf, Sheffield

Schaltegger S, Lüdeke-Freund F, Hansen EG (2012) Business cases for sustainability: the role of business model innovation for corporate sustainability. Int J Innov Sustain Dev 6(2):95–119. doi:10.1504/IJISD.2012.046944

Schaltegger S, Beckmann M, Hansen EG (2013a) Transdisciplinarity in corporate sustainability: mapping the field. Bus Strateg Environ 22(4):219–229. doi:10.1002/bse.1772

Schaltegger S, Harms D, Hörisch J, Windolph SE, Burritt R, Carter A, Truran S, Crutzen N, Ben Rhouma A, Csutora M, Tabi A, Kokubu K, Hirotsugu K, Haider M, Kim J, Lee K-H, Moneva JM, Ortas E, Alvarez-Etxeberria I, Daub C, Schmidt J, Herzig C, Morelli J (2013b) International corporate sustainability barometer. a comparative study of 11 countries. Centre for Sustainability Management, Lüneburg. ISBN ISBN 978-3-942638-38-8

Schaltegger S, Harms D, Hörisch J, Windolph S (2013c) Corporate sustainability barometer 2012. Praxisstand und Fortschritt des Nachhaltigkeitsmanagements in den größten Unternehmen Deutschlands. Centre for Sustainability Management, Lüneburg

Schmidheiny S, BCSD (Business Council for Sustainable Development) (1992) Changing the course. MIT Press, Cambridge

Spitzeck H (2009a) Organizational moral learning: what, if anything, do corporations learn from NGO critique? J Bus Ethics 88(1):157–173

Spitzeck H (2009b) Marks & spencer – the business case for plan A. Cranfield University, School of Management, UK (ecch case study no. 709-018-1M)

Spitzeck H, Hansen EG (2010) Stakeholder governance – how do stakeholders influence corporate decision-making? Corp Gov Int J Bus Soc 10(4):378–391

Stead J, Stead W (2008) Sustainable strategic management: an evolutionary perspective. Int J Sustain Strateg Manage 1(1):62–81

van Marrewijk M (2003) Concepts and definitions of CSR and corporate sustainability. Between agency and communion. J Bus Ethics 44(2/3):95–105

von Weizsäcker E, Lovins A, Lovins L (1997) Factor four. Doubling wealth, halving resource use. Earthscan, London

von Weizsäcker EU, Hargroves K, Smith M, Desha C, Stasinopoulos P (2009) Factor five. Transforming the global economy through 80 % improvements in resource productivity. Earthscan, London

Zadek S (2004) The path to corporate responsibility. Harv Bus Rev 82(12):125–132

Chapter 8
Sustainable Development in Economics

Michael von Hauff

Abstract The new paradigm of sustainable development has not yet penetrated all of the business sciences. Although the concept has found its way into management studies in business administration programs, as well as into marketing and other subdisciplines, most economists have ignored the topic. Two contrasting positions are identifiable in the relatively few publications to date that deal with this subject: The position advocated in neoclassical economics stands in irreconcilable opposition to that of "ecological" economics. The proponents of both disciplines, however, initially start with the premise of intergenerational equity, which states that the lifestyles of current generations may not jeopardize the quality of life of future generations (WCED, Our common future, Oxford, 1987, p. 43).

Keywords Neoclassical economics • Ecological economics • Substitution rule • Weak sustainability • Strong sustainability • Balanced sustainability • Post-growth economy

1 Introduction

The new paradigm of sustainable development has not yet penetrated all of the business sciences. Although the concept has found its way into management studies in business administration programs, as well as into marketing and other subdisciplines, most economists have ignored the topic. Two contrasting positions are identifiable in the relatively few publications to date that deal with this subject: The position advocated in neoclassical economics stands in irreconcilable opposition to that of "ecological" economics. The proponents of both disciplines, however, initially start with the premise of intergenerational equity, which states that the lifestyles of current generations may not jeopardize the quality of life of future generations (WCED 1987, p. 43).

M. von Hauff (✉)
TU Kaiserslautern, FB Wirtschaftswissenschaften, Gottlieb-Daimler-Str., Geb. 42, D-67663 Kaiserslautern, Germany
e-mail: hauff@wiwi.uni-kl.de

H. Heinrichs et al. (eds.), *Sustainability Science*, DOI 10.1007/978-94-017-7242-6_8

The proponents of neoclassical economics pursue intergenerational equity through the preservation of a stock of capital (natural resource + real capital). Accordingly, this concept allows for the substitution of nonrenewable natural capital by reproducible real capital as long as the total remains the same. This aspect is fundamentally rejected by the proponents of ecological economics. The proponents of ecological economics demand a model of sustainable development that is ecocentric and recognizes the interrelationships between the two dimensions of ecology and economy. The social dimension, as in neoclassical economics, is largely ignored.

The proponents of ecological economics critique neoclassic theory mainly through a focus on two points: the idea of maximizing utility and the inherent constraints on the individual. Within the framework of ecological economics, an eco-centric view of sustainability holds that the preservation of environmental systems must be the starting point for any argumentation (von Hauff and Kleine 2009, p. 30). The advancement to a post-growth society is viewed as a further development in ecological economics, which is briefly introduced in this chapter.

The opposing positions of neoclassical economics and ecological economics can be overcome by a balanced view of sustainability. Balanced sustainability is not in itself a complete concept. Several essays in this regard emphasize different points and are also briefly discussed here. The essays on balanced sustainability are linked to the greater question of whether sustainable growth is even possible.

First, it is necessary to examine the opposing positions of neoclassical economics and ecological economics regarding sustainable development. This looks at the controversy between weak and strong sustainability. Both approaches share the goal of intergenerational justice to ensure that future generations are not worse off than the generations now living. The neoclassical proponents of weak sustainability seek intergenerational sustainability by preserving the total capital stock (human-made capital + natural capital). In this model, natural capital can be substituted with human-made capital. Consequently, there can be a decrease of natural capital together with growth in human-made capital, as long as the total capital stock remains constant. The proponents of strong sustainability (ecological economists) reject this idea and demand the preservation of natural capital for future generations.

The more recent debate about a post-growth society is included within the framework of ecological economics. The focus of the subsequent section is on how to overcome the controversy between neoclassical economics and ecological economics in the context of balanced sustainability. The most important findings are summarized in the conclusion.

2 Sustainability in the Context of Neoclassical Economics

The neoclassic understanding of sustainability is based, in principle, on the disagreement that arose over the first Club of Rome report on the "The Limits to Growth," which was published in 1972. The report critically called into question the contemporary goal of permanent growth through economic activity. The key insight of the report was that essential nonrenewable resources would be exhausted in a foreseeable period of time in a world economy oriented toward exponential growth. It concluded that there are limitations to growth. As a result, efficient resource models arose within the framework of neoclassical environmental economics, which acknowledged the scarcity of natural resources. Borrowing from social welfare economics, these models differentiate between nonrenewable and reproducible natural resources in addressing the need to ensure the satisfaction of basic human needs (von Hauff and Jörg 2013, p. 53 ff).

The debate has been strongly influenced by Robert Solow with the integration of neoclassical and resource economics in development theory since the 1970s. He positioned himself, as did Josef Stiglitz, as a critic of the "The Limits to Growth" report. He determined, "The world can, in effect, get along without natural resources, so exhaustion is just an event, not a catastrophe (Solow 1974, p. 11)."

In the mid-1980s, he developed the so-called constant capital rule, based on the "Hartwick Rule." The Hartwick rule was proposed by John M. Hartwick and says that the sustainable use of resources in an economy is also dependent on exhaustible resources (1977). The important point it makes is that some resources must be used efficiently and the scarcity rents on current extraction of those resources be fully reinvested in human-made (or manufactured) capital. This is the way to insure the level of consumption by future generations is maintained.

In this respect, Solow presented himself as a representative of the so-called "weak" sustainability, which assumes complete substitutability of natural capital with real capital, if necessary. The dominant neoclassical position on sustainability today assumes a level of utility (the satisfaction associated with consumption) that does not decrease over time. At a minimum, this implies that average utility for future generations must equal the average utility of the current generations. It must be critically noted that utility is very subjective and the wide-ranging variation requires specification (Ott and Döring 2008, p. 102; Panayotou 2000, S. 61)

The range of variation extends from the utilitarian position of happiness, to microeconomics as a function of consumption, to the exercise of capabilities. Utility, specifically in the context of weak sustainability, is interpreted only as a function of consumption. This position presupposes a narrow understanding of consumption, in other words, the consumption of material goods. The consumption of intangible assets – assets like a beautiful sunset – is not included. Therefore, for the proponents of weak sustainability, economic growth is the key to sustainability.

If a constant stock of capital is maintained, average utility will be constant and current utility maximized. The logical question to ask of this theory is, how are the future costs of natural destruction (depletion of natural capital) to be valued today? This leads to the issue of intertemporal equity. In other words, what are the future value of environmental pollution and the consumption of exhaustible resources today? Furthermore, this presupposes the intertemporal allocation of the resource used across generations (von Hauff and Jörg 2013, p. 126). In the context of weak sustainability, this leads to the position that the substitution of natural capital by reproducible real capital is, in principle, without limits. The neoclassical paradigm is based on the optimistic view, which has risen to dominance today because of technological advances and higher efficiencies in the use of the factors of production. Correspondingly, the depletion of a nonrenewable resource can be compensated for with real capital.

Robert M. Solow

- Emeritus Professor of the Massachusetts Institute of Technology (MIT).
- Main research area: economic growth theory.
- In 1956, he released the essay "*A Contribution to the Theory of Economic Growth*," which contains the well-known Solow model.
- In 1987, he received the Nobel Prize for his work on economic growth (Fig. 8.1).

Fig. 8.1 Robert M. Solow

Therefore, it is not the preservation of nature that is important, but rather the safeguarding of overall prosperity. With this reasoning, Solow theorized that economic growth is possible even without natural resources (Solow 1997, p. 267). Under a profit maximizing viewpoint, if there is a better option than the preservation of natural capital, it should be taken. To this extent, projects for environmental and climate protection or for the conservation of nature are obligated in the framework of weak sustainability to prove they are superior to or provide a greater benefit than other investments over the long term.

- ***Task:*** *Briefly explain the substitution rule.*

In a neoclassical argumentation, the arguments of weak and strong sustainability are merged. It suggests that a compensation for diminishing or depleted natural capital by means of real capital is justifiable as long as it does not fall below the defined threshold value of vital capital needed to safeguard strong sustainability (von Hauff and Jörg 2013, p. 128). In this context, a two-step sustainability rule must be differentiated from weak sustainability. Absolute restrictions are required as drawn from the argumentation of ecological economics.

The two-step sustainability rule states that safeguarding the vital stock of natural capital in this way can result in the continued increase in the standard of living according to weak sustainability. The substitution of natural capital with real capital is acceptable as long as it does not fall below the threshold stock of natural capital (safe minimum standards). It is self-evident that it is both risky and controversial to attempt to define a minimum requirement for the natural capital essential for human survival when the reality is characterized by risk and uncertainty. Furthermore, it neglects other functions of the environment, such as the function of human relaxation and regeneration in nature. Consequently, the view of nature in the two-step sustainability concept is based on a utility-oriented understanding of sustainability. It therefore corresponds largely with weak sustainability.

In summary, it can be said that in the framework of formal models, neoclassical economics recognizes the ecological challenges and attempts to portray environmental and resource issues as an allocation problem. Neoclassical economics also reacts to the intertemporal dimension of the environmental issues and aligns itself with the position of weak sustainability. This is by no means a new development but, rather, has its origins in the approaches developed back in the 1970s. That is to say, neoclassical economics is based on a definition of sustainability characterized by safeguarding a level of utility that never decreases with the passage of time.

3 Sustainability in the Context of Ecological Economics

The proponents of ecological economics developed strong sustainability in opposition to weak sustainability. Ecological economics was inspired by the work of Nicholas Georgescu-Roegen, Kenneth Boulding, and William Kapp and was introduced into the economic debate in the 1970s (von Hauff and Kleine 2009, p. 29ff). This concept initially became established in the USA during the 1980s. In 1987, the International Society for Ecological Economics (ISEE) was founded (Rogall 2012, p. 119). Ecological economics is defined by its focus on the environmental model of sustainable development, which takes the categories ecology and economy into account.

Consequently, the economic subsystem must be returned to the biophysical environment and, especially in industrialized countries, grow no further, because the economy depends on the availability of natural resources and the carrying capacity of the natural sink function. If the progressive burdening or destruction of nature continues, natural capital could become a limiting factor of production. In this respect, human awareness is required with the will to preserve ecological systems as the basis of life for future generations and to subordinate economic self-interests. However, the third dimension of sustainable development, the social dimension, is neglected, as in the neoclassical economic theory (von Hauff and Kleine 2009, p. 30).

In contrast to the neoclassical economists, the major advocates of ecological economics, like Daly and Costanza, soundly reject the substitution rule. Rather than substitutability, they propose the complementarity of natural and real capital, to the extent that production depends on natural capital. Complementarity is when a natural resource is required for the production of goods. In some cases, there are functions of natural capital in the production of goods that cannot be provided by real capital (Costanza et al. 1997, p. 5ff).

The proponents of strong sustainability promote the idea of a "steady-state economy" (Daly 1991, p. 35ff). A steady-state economy, or balanced economy, as defined by Daly, is an economic system supplied with a constant stock that is sufficient to provide the material goods for a "good life." This is the reason why the economic system is viewed as a subsystem of the environmental system. The economy is dependent – as mentioned above – on resource availability and the capacity of nature sinks (Ott and Döring 2008, p. 145).

- ***Question:*** *Please explain, why does ecological economics support the model of an economic system as a subsystem within the environmental system?*

In light of global problems like the exponential population growth, increasing pollution, and degradation of the environment, human-made climate change and the sharply increasing level of consumption of nonrenewable resources, a reduction in the demands made on the ecological system in economic processes is considered to be essential. This may be the only way to preserve nature as an integrated system and reduce the (not exactly calculable) risk of a negative backlash from the ecosystems to the economy and society. This is a major requirement, according to Costanza, for the equitable distribution of the use of natural resources across the generations (Costanza et al. 1997, p. 83).

Ecological economics is not only about the elimination of negative external effects by means of internalization strategies such as statutes, bans, eco-taxes, bilateral negotiations, or certificates as advocated in neoclassical environmental economics. This is based largely on the risk of irreversible damage to ecosystems, something barely mentioned in neoclassical environmental economics. In this context, one of the major critiques expressed by ecological economists is that neoclassical theory, with its one-sided emphasis on marginal equilibrium analysis, is not prepared to account for the integration of complex phenomena, as required in the ecological real world.

Under the framework of ecological economics and steady-state economic models, there has been a series of publications on the subject of the post-growth economy or post-growth society in recent years. The following discussion focuses on selected fundamental concepts. In general, for proponents of the post-growth economy, it is all about the justification and configuration of a growth-free economy. One of the most important proponents of this is Tim Jackson from Great Britain. The central aim, in his opinion, is to achieve prosperity without growth, something he believes is not only fiscally and ecologically possible, but essential.

The post-growth economy focuses, especially, on stronger local and regional production and consumption. This refers, in other words, to a "small is beautiful" economy. A shortening of the value-adding chain reduces the structural pressure for growth in many ways. This enables a creative subsistence economy that will contribute to the strengthening of the post-growth economy. Another important criterion is sufficiency, which aims at achieving a decrease in consumption and requires a return to the essentials.

4 Overcoming the Controversy Between Neoclassical and Ecological Economics

The two approaches of weak and strong sustainability still stand in opposition to each other. Efforts were made relatively early to bridge the differences in the contrary positions. The contributions of Steurer (2001), for example, deserve mention (von Hauff and Jörg 2013, p. 130ff). This effort also includes the concept of "sustainable economics" as presented, for example, by Rogall (2012, p. 190). However, these approaches do not by any means present a uniform position. For example, while Steurer promotes strong sustainability, Hedinger tends to focus on weak sustainability. The approaches nevertheless combine the strengths of both concepts.

In this respect, it is appropriate to return to the category of vital natural capital. As already established in the two-step sustainability rule, there is no substitute for essential natural capital and it must not be endangered. On the other hand, the substitution of nonessential natural capital is permitted. The problem, as has already been mentioned, is in defining acceptable limits on the use of essential natural capital. The issue involves accounting for the unpredictability and risk, which is not always clear or easy to define.

At a minimum, an intact ecosystem multiplies the prosperity of humans and is therefore indispensable. Sometimes real and natural capital are substitutable and, in other cases, complementary. If this insight is applied to the central point of the controversy between neoclassical economics and ecological economics, the following becomes evident: Neither a general halt to growth (ecological economy) nor infinite quantitative growth (neoclassical economy) seems to be warranted. Pearce claims that growth, in addition to the many positive effects like the strengthening of the social security systems, stabilization of the labor markets, and the increase in government revenues with the associated greater scope for government expenditures on things like education and research, also facilitates efforts to preserve the environment for motivational, structural, and financial reasons (Pearce 1991, p. 11).

Through improved efficiency, frugality, recycling, structural improvements, and increased use of alternative renewable resources, it must be possible to decouple the two factors of growth and environmental quality from one another. Decoupling, for example, may refer to a rising GDP and a lowering of emissions. In this case, a clear target for the lowering of emissions (absolute decoupling) should be defined. The advocates of balanced sustainability promote an economic and ecological optimization. Balanced sustainability envisions a slowing of growth or a stop to growth solely as a potential result of the ecological restructuring of the society. In effect, the aim is for the harmonization of growth and environmental quality.

- ***Task:** Please discuss how the supporters of balanced sustainability integrate the two opposing positions of weak and strong sustainability.*

Considering some successful examples of the substitution of natural capital with real capital, it does not appear necessary to reject the substitution rule fully. There are some situations in which real capital can be substituted for natural capital (e.g., cultivated, managed forests). However, the idea of unlimited substitution, as is permitted in the framework of weak sustainability, does not support the requirement for intergenerational equity. This is predicated on the unpredictable risk associated with the rising consumption of essential natural capital.

It can be assumed that technological progress will continue in the future, and new alternatives for the substitution of natural resources will be found. Nevertheless, there is no reason to assume with any regularity that technological innovation will be in a fundamental position to fully and equivalently replace natural capital. Consequently, every ecological function must have an "artificial substitute." For example, to substitute a forested area, fully equivalent functions must be provided. There can be no downside to the substitute that was not also present in the original (von Hauff and Jörg 2013, p. 131). The substitute must verifiably exist and not just be promised in the "escape avenues of technological daydreams" (Ott and Döring 2008, p. 156). If happiness and moral values are included in the equation, it may be assumed that future generations may not even want the substitution of nature. With that in mind, it can be said that strong sustainability leaves open more options than weak sustainability, and it gives future generations more "freedom to choose."

In summary, balanced sustainability takes on a mediating role in this debate, in which not only a constant total stock of capital, but, because of the limited substitutability of natural capital, many of the essential components of natural capital (climate balance, global life cycles, ozone layer, and balanced ecological system like forests, lakes and rivers, biodiversity, etc.) must also be maintained at a constant level. The maintenance of a minimum reserve is proposed for less vital stocks (Pearce 1991). It must also be acknowledged in this context that not all of the relevant information necessary to facilitate the identification and definition of clear limits on essential natural capital is available.

5 Conclusions

The sustainable development paradigm, which was introduced at the Earth Summit in Rio de Janeiro in 1992 and recognized by the world community as a paradigm for the twenty-first century, has been received with very different perceptions among experts in economic theory. It is possible to discern among the continuum of perceptions at least three different orientations and justifications. These are then further differentiated, but all fundamentally share a common starting point. Intergenerational equity is the common starting point, which means the ability of future generations to satisfy their needs should not be less than that of the current generation. However, departing from this starting point, we quickly come to a divergence of various perceptions.

The first reaction to the Club of Rome report "The Limits to Growth" became the neoclassical argumentation regarding sustainable development. This perception is largely characterized by the paradigm of weak sustainability. Of primary importance in this paradigm is the maintenance of a stable stock of capital (natural capital + real capital) for future generations. Consequently, it is quite possible that the result is a substitution process between natural and real capital. One variation in weak sustainability is the two-step sustainability rule, which acknowledges a need to preserve some "essential stocks" of natural capital.

Ecological economics diverges at this point in that it clearly differentiates itself by fundamentally questioning the substitution rule. Accordingly, it promotes the view that the production of goods and consumption must be placed within the boundaries of the environmental system. If the limits of the ecological systems are exceeded, human existence is placed at risk. The economic sphere is therefore a subsystem of the environment. This became the basis for the paradigm of strong sustainability, which states that the relationship of the environment to the economy cannot be characterized by substitution, but rather is a complementary one. Daly proposed the steady-state economy, which is linked to the more recent concept of the post-growth economy. It promotes a growth-free economy, something which demands a very thorough transformation to a completely different economic system.

The opposing positions can be overcome under the terms of balanced sustainability. This proposes the existence of various ecosystems. In some cases, certain ecosystems are characterized by an essential stock of natural capital. No further claims should be placed on such systems. In other cases, there are ecosystems that hold a large and renewable stock of natural capital. It is possible to continue using these as a source of prosperity. To this extent, the relationship between the economy and the environment is sometimes characterized by complementarity and, at other times, by substitution.

References

Costanza R, Perrings C, Cleveland CJ (1997) The development of ecological economics. Edward Elgar Publishing, Cheltenham

Daly HE (1991) Steady-state economics, 2nd edn. Island Press, Washington, DC

Hartwick JM (1977) Intergenerational equity and the investing of rents from exhaustible resources. Am Econ Rev 67(5):S. 972–S. 974

Maedows D et al (1972) The limits to growth – a report for the Club of Rome's Project on the predicament of mankind. Universe Books, New York

Ott K, Döring R (2008) Theorie und Praxis starker Nachhaltigkeit, 2d edn. Metropolis Verlag, Marburg

Panayotou T (2000) Economic growth and the environment, CID working paper no. 56, environment and development paper no. 4. Harvard. https://ideas.repec.org

Pearce DW (1991) Blueprint 2: greening the world economy. Earthscan, London

Rogall H (2012) Nachhaltige Ökonomie – Ökonomische Theorie und Praxis einer nachhaltigen Entwicklung, 2d edn. Springer, Heidelberg

Solow RM (1974) The economics of resources or the resources of economics. Am Econ Rev 64:1–14

Solow RM (1997) Georgescu-Roegen versus Solow-Stiglitz-Reply. Ecol Econ 22:267–268

Steurer R (2001) Paradigmen der Nachhaltigkeit. Zeitschrift für Umweltpolitik und Umweltrecht 24/4

von Hauff M, Jörg A (2013) Nachhaltiges Wachstum. Oldenbourg Verlag, München

von Hauff M, Kleine A (2009) Nachhaltige Entwicklung. Springer, München

WCED (World Commission on Environment and Development) (1987) Our common future. Oxford University Press, Oxford

Chapter 9
Sustainable Development and Law

Marjan Peeters and Thomas Schomerus

Abstract Since the emergence of the concept of sustainable development, lawyers across the globe are trying to come to grips with its legal status and the potential legal consequences (See Bosselmann, Sustainability law. Ashgate Publishing, 2008; French, Sustainable development. In: Fitzmaurice M, Ong DM, Merkouris P(eds) The research handbook on international environmental law. Edward Elgar, 2010, and Barstow Magraw D. Hawke LD, Sustainable development. In: Bodansky D, Brunnéé J, Hey E (eds) Oxford handbook of international environmental law. Oxford University Press, 2007). Nowadays, the concept of sustainable development is represented in legally binding texts at international, European, and national levels. Taking EU law as an example, both the Treaty on the European Union (TEU) and the Treaty on the Functioning of the European Union (TFEU) refer to sustainable development in several articles. This clearly means that sustainable development is part of EU law. The real question, however, is whether this reference to sustainable development in binding law has any significant consequence for legal practice. Can, for instance, the Court of Justice of the European Union annul a decision of the European Commission should this decision be qualified as conflicting with sustainable development? Such a far-reaching and dramatic annulment is most unlikely under EU law, while the potential legal consequences of sustainable development will probably be more subtle. This chapter provides insight into the appearance of sustainable development in international and EU law and gives observations on its possible legal effects and the importance of national decision-making in view of sustainable development.

Keywords Sustainable development • EU law • International law

M. Peeters (✉) • T. Schomerus
Leuphana University of Lüneburg, Scharnhorststraße 1, 21335, Lüneburg, Germany
e-mail: marjan.peeters@maastrichtuniversity.nl; schomerus@leuphana.de

H. Heinrichs et al. (eds.), *Sustainability Science*, DOI 10.1007/978-94-017-7242-6_9

1 Exploring Legal Consequences of Sustainable Development

Since the emergence of the concept of sustainable development, lawyers across the globe are trying to come to grips with its legal status and the potential legal consequences.[1] Nowadays, the concept of sustainable development is represented in legally binding texts at international, European, and national levels. Taking EU law as an example, both the *Treaty on the European Union (TEU)* and the *Treaty on the Functioning of the European Union (TFEU)* refer to sustainable development in several articles. This clearly means that sustainable development *is* part of EU law. The real question, however, is whether this reference to sustainable development in binding law has any significant consequence for legal practice. Can, for instance, the Court of Justice of the European Union annul a decision of the European Commission should this decision be qualified as conflicting with sustainable development? Such a far-reaching and dramatic annulment is most unlikely under EU law, while the potential legal consequences of sustainable development will probably be more subtle. This chapter provides insight into the appearance of sustainable development in international and EU law and gives observations on its possible legal effects and the importance of national decision-making in view of sustainable development.

2 Sustainable Development in International Law

2.1 *Treaties*

International law concerns the law between states. The most extensive legal reference to sustainable development in international law is an explicit mention in a range of treaties, in particular environmental treaties, thereby making it legally binding within the treaties' context. The binding nature of treaties refers only to their parties, which can be states or international organizations. Industries and citizens (private actors) are not bound by treaties, except under very specific circumstances such as the criminal responsibility of individuals before the International Criminal Court.[2] State action influencing private actors might be necessary in order to reach compliance with treaty obligations. As a concrete example: if a country has ratified a treaty containing the obligation to reduce greenhouse gas emissions, it can impose duties on its private actors in order to comply with that international obligation.

A general characteristic of environmental treaties is their often vague terms. For example, the United Nations Framework Convention on Climate Change from 1992 (UNFCCC) hardly contains any binding substantive obligations for the treaty par-

[1] See Klaus Bosselmann, Sustainability Law, Ashgate Publishing, 2008; Duncan French, Sustainable Development, in: Maglosia Fitzmaurice, David M. Ong, Panos Merkouris (2010) *The Research Handbook on International Environmental Law,* (Chap 3) Edward Elgar, and Daniel Barstow, Magraw, Lisa D. Hawke, Sustainable Development, in: Daniel Bodansky, Jutta Brunnéé and Ellen Hey (2007), *Oxford Handbook of International Environmental Law,* Oxford University Press.

[2] See http://www.icc-cpi.int/en_menus/icc/about%20the%20court/Pages/about%20the%20court.aspx.

ties but introduces an institutional framework for further decision-making, to be conducted by so-called Conferences of the Parties (COP). As a guide to such further decision-making, the UNFCCC provides principles, among which the following mentions the right to sustainable development:

> The Parties have a right to, and should, promote sustainable development. Policies and measures to protect the climate system against human-induced change should be appropriate for the specific conditions of each Party and should be integrated with national development programmes, taking into account that economic development is essential for adopting measures to address climate change. (Art. 3(4) UNFCCC)[3]

The above text leaves ample room for interpretation and, consequently, no clear substantive rule can be derived from the reference to sustainable development. It is hence up to the treaty parties to promote interpretation and further policymaking, with the aim to conclude binding commitments such as in the Kyoto Protocol of 1997.

- ***Task***: *Describe the direction Article 3(4) gives for national policymaking by the treaty parties. Under what circumstances can differentiated commitments among treaty parties be developed?*

2.2 *Soft Law*

Since there are huge political barriers to concluding clear commitments within treaties, many international legal texts appear in the form of so-called soft law. These are nonbinding documents with a very different legal status from treaties. For sustainable development, the most important soft law document is the 1992 UN Rio Declaration. Although not much attention is given to the socioeconomic dimension, this declaration is generally seen as the most important and comprehensive international governmental document for sustainable development. The first principle proclaims that:

> Human beings are at the centre of concerns for sustainable development. They are entitled to a healthy and productive life in harmony with nature.

Sustainable development is also emphasized in Principle 8:

> To achieve sustainable development and a higher quality of life for all people, States should reduce and eliminate unsustainable patterns of production and consumption and promote appropriate demographic policies.

The environmental law literature draws attention to principles possibly deriving from the concept of sustainable development, such as that of intergenerational equity, of the sustainable use of nature, and of integrating environmental considerations into economic and development plans.[4]

---

[3] The text of the UNFCCC can be found at http://unfccc.int/files/essential_background/background_publications_htmlpdf/application/pdf/conveng.pdf.

[4] See French, o.c., Philippe Sands, International Environmental Law, third ed. (2012) p. 206–216, the International Law Association (ILA) New Delhi Declaration of Principles of International Law Relating to Sustainable Development, published in *International Environmental Agreements: Politics, Law and Economics* 2: 211–216, 2002, and Resolution No. 7/2012 from the Committee on International Law on Sustainable Development reaffirming the Delhi Declaration (http://www.ila-hq.org/en/committees/index.cfm/cid/1017).

In his article *Losing the Forest for the Trees: Environmental Reductionism in the Law*,[5] Klaus Bosselmann highlights the close connection between environmental and sustainability laws. Favoring a comprehensive legal concept of sustainability, he criticizes the limited approach of environmental law. Bosselmann points out that "environmental laws and policies have saved some 'trees', but the 'forest' is being lost as critical global issues including climate change, biodiversity loss, and our ecological footprint continue to worsen. Existing laws and policies mitigate the ecological damage inflicted by industrial economies and western lifestyles." He proposes "a sustainability approach to law that aims for transformation rather than environmental mitigation."

It would, however, be unbalanced to put sustainable development only into a green perspective: the core aim of sustainable development is to reach intergenerational and intragenerational equity. In the latter case, the aim is to reach equity between the poor and the rich, thereby recognizing that developing countries are in need of economic growth. The real challenge of sustainable development is to find a proper balance between environmental, social, and economic concerns. This is reflected in Principle 3 of the Rio Declaration:

> The right to development must be fulfilled so as to equitably meet developmental and environmental needs of present and future generations.

- ***Task**: Discuss the interrelation between environmental law and sustainability law. Should the protection of the environment be the primary goal of sustainability law?*

2.3 Courts

The question arises whether the International Court of Justice (or other dispute settlement mechanisms such as tribunals) might increase the legal weight of the concept of sustainable development. The limited case law thus far shows that judges are reluctant to fill the gap left by international politicians. It also shows that they refrain from giving an adequately substantive interpretation of sustainable development for it to become a leading standard in case decisions for parties unable to find a resolution. In a dispute between Hungary and Slovakia concerning a major hydraulic project (the Gabčíkovo-Nagymaros Project), which was relevant for *development purposes* (including the generation of hydro energy) but with *environmentally negative consequences*, the International Court of Justice decided in its judgment of 25 September 1997 that sustainable development is only a concept and hence not a binding principle of international law. The court considered:

> This need to reconcile economic development with protection of the environment is aptly expressed in the concept of sustainable development. (para 139)

[5] Bosselmann, Sustainability 2010, 2, 2424–2448 (free access under http://www.mdpi.com/2071-1050/2/8/2424)

The Court then ruled that it is for:

> (…) the Parties themselves to find an agreed solution that takes account of the objectives of the Treaty, which must be pursued in a joint and integrated way, as well as the norms of international environmental law and the principles of the law of international watercourses. (para 140)

As stated by Cesare Romano, the court essentially threw the dispute back into the parties' lap, leaving them to negotiate a solution to their conflict.[6] The concept of sustainable development was clearly not applied by the court as a tool to solve the dispute. The International Court of Justice Vice-President Weeramanty is, however, of the impressive opinion that sustainable development must be seen as *a principle of modern international law androoted in history*.[7] He feels strongly that *pristine and universal values* must be integrated into the "corpus of living law".[8] In his words:

> Sustainable development is thus not merely a principle of modern international law. It is one of the most ancient of ideas in the human heritage.[9]

As a general principle of international law, sustainable development would be applicable even without having been agreed upon in a treaty. According to Judge Weeramantry, the principle of sustainable development primarily means that the court must hold an even balance between environmental and developmental considerations.[10] Nevertheless, the core question remains as to how, in specific cases, this balance can be struck.

The idea that sustainable development should to some extent play a role in judicial decisions has also been expressed by the Appellate Body to the World Trade Organization (WTO). The preamble to the WTO agreement from 1994 contains a mix of economic, social, and environmental values, referring to sustainable development as follows:

> The Parties to this Agreement, Recognizing that their relations in the field of trade and economic endeavour should be conducted with a view to raising standards of living, ensuring full employment and a large and steadily growing volume of real income and effective demand, and expanding the production of and trade in goods and services, while allowing for the optimal use of the world's resources in accordance with the objective of sustainable development, seeking both to protect and preserve the environment and to enhance the means for doing so in a manner consistent with their respective needs and concerns at different levels of economic development, (…)[11]

In 1995, the WTO's Appellate Body argued in the Shrimp/Turtle case that sustainable development "must add colour, texture and shading to our interpretation."[12]

---

[6] Cesare Romano, *The peaceful settlement of international environmental disputes,* (2000), p. 255–256

[7] The separate opinion can be found at http://www.icj-cij.org/docket/files/92/7383.pdf; see also Bosselmann (2008), p. 12.

[8] See the separate opinion, pp. 105–106 (bottom page number).

[9] See the separate opinion, p. 107 (bottom page number).

[10] See the separate opinion, p. 85 (bottom page number).

[11] Agreement establishing the World Trade Organization, http://www.wto.org/english/docs_e/legal_e/04-wto.pdf

[12] The ruling from 12 October 1998 is published on http://www.wto.org/english/tratop_e/dispu_e/58abr.pdf; see para 153.

While the opinion of Weeramantry and the WTO Appellate Body reasoning favor a more important role for sustainable development in judicial decisions, the real effect of this concept and its precise legal status remain hard to grasp. The concept of sustainable development can be given further texture by more detailed provisions or definitions in treaty texts and also in specific judicial cases which are suitable for a further interpretation and application of that concept. The Committee on International Law formulated its potential development as follows:

> Recourse to the concept of 'sustainable development' in international case law may, over time, reflect a maturing of the concept into a principle of international law, despite a continued and genuine reluctance to formalise a distinctive legal status.[13]

- ***Question***: *Can the concept of sustainable development be considered a principle of international law?*

3 Sustainable Development in EU Law

EU law, particularly the TEU and the TFEU, refers to sustainable development. As long ago as 1991, the Treaty of Maastricht mentioned "sustainable and non-inflationary growth respecting the environment" and "the fostering of sustainable economic and social development of the developing countries, and more particularly the most disadvantaged among them."[14] The 1998 Treaty of Amsterdam underlined sustainable development as one objective of European integration. Today, sustainable development is specified in Article 3 of the TFEU as follows:

> The Union shall establish an internal market. It shall work for the sustainable development of Europe based on balanced economic growth and price stability, a highly competitive social market economy, aiming at full employment and social progress, and a high level of protection and improvement of the quality of the environment. It shall promote scientific and technological advance.

Here, the focus lies on the sustainable development of *Europe*. The international and intragenerational relevance of sustainable development are referred to in Article 3(5) of the TEU as follows:

> In its relations with the wider world, the Union shall uphold and promote its values and interests and contribute to the protection of its citizens. It shall contribute to peace, security, the sustainable development of the Earth, solidarity and mutual respect among peoples, free and fair trade, eradication of poverty and the protection of human rights, in particular the rights of the child, as well as to the strict observance and the development of international law, including respect for the principles of the United Nations Charter.

[13] The Committee on International Law on Sustainable Development reaffirming the Delhi Declaration (http://www.ila-hq.org/en/committees/index.cfm/cid/1017), guiding statement 1

[14] The Maastricht Treaty introduced these references into Article 2 and Article 130u of the European Community Treaty.

This text deals with the "sustainable development of the Earth," apparently focusing on the concept's environmental dimension. The eradication of poverty is mentioned separately. Moreover, in Article 21(d), which provides general principles for EU external relations, eradicating poverty is given as the primary aim.[15] Within the TEU, the three dimensions of sustainable development are incorporated, while, in places where external action by the EU is concerned, the needs of the poor become a priority.

None of the abovementioned TEU statements have great legal relevance. The TEU does not provide a definition for "sustainable development," leaving the precise meaning difficult to determine. It is left to EU institutions to create the necessary clarity in defining sustainable development with legislative and administrative acts.[16] Since the terminology is vague, and given the traditional discretion the courts give to the legislature when decision-making involves political, economic, and social choices or complex assessments and evaluations, it would be exceptional if the courts attached any direct consequences to the mention of sustainable development in the TEU articles.

One legal rule deserves specific attention: the external integration obligation in Article 11 of the TFEU, requiring the integration of environmental protection into all Union policies and activities. Such integration must be done with a view to promoting sustainable development.[17] On the one hand, it is unlikely that EU courts will interfere in governmental decision-making on the grounds that *the promotion of sustainable development* has been insufficient.[18] The courts may, on the other hand, intervene if it is clear that the integration of environmental protection requirements, for instance, into transport measures, has been disregarded or if such environmental integration has completely neglected the promotion of sustainable development. One interpretation of the article is the requirement for the acting institutions to justify their compliance with Article 11 of the TFEU. The effect of Article 11 will be

[15] The article reads: "The Union shall define and pursue common policies and actions, and shall work for a high degree of cooperation in all fields of international relations, in order to: (…) foster the sustainable economic, social and environmental development of developing countries, with the primary aim of eradicating poverty (…)" (Art. 21(2)(d) TEU).

[16] For a discussion of EU policymaking in view of sustainable development, see Ludwig Krämer, EU Environmental Law, 7th edition (2012), Sweet & Maxwell, 9–11.

[17] The full text of Article 11 of TFEU is: "Environmental protection requirements must be integrated into the definition and implementation of the Union's policies and activities, in particular with a view to promoting sustainable development." An extensive examination of the external integration principle has been given by Nele Dhondt, Integration of Environmental Protection into other EC policies, 2003, Europa Law Publishing. Regarding the incorporation of "sustainable development" into the external integration rule, she argues that it aims at the reconciliation of ecological objectives with socioeconomic ones, which is, in fact, the same as the meaning of sustainable development in international law (p. 72).

[18] Jans and Vedder discuss Article 11 of the TFEU mainly in view of the legitimacy of acts in light of the environmental objectives. They state in line with Nele Dhondt (o.c. p. 183) that only in very exceptional cases will a measure be susceptible to annulment. They do not elaborate specifically whether such an annulment can be foreseen for short falling action in view of promoting sustainable development. Jans, Vedder, European Environmental Law, 4th edition, (2012) Europa Law Publishing, pp. 25–27

that institutions must clarify whether and how they have completed this assessment, thereby offering an opportunity for sustainability science and, particularly, integrated assessments to fulfill a role in supporting EU policy and law.

- ***Task**: Discuss the potential significance of Article 11 of TFEU in view of developing product standards for transport fuels, including biofuels and the role played by sustainability science.*

4 An Illustrative Court Case

This section will highlight a court decision that illustrates how "sustainable development" can become part of case law. The case at hand concerns a so-called preliminary ruling by the European Court of Justice (ECJ).[19] This ruling occurs when a national court of an EU member state is in doubt about the interpretation of the EU law. In order to clarify the issue, the national court submits questions to the ECJ, according to which the court will provide guidance. In the Acheloos River case, a Greek court submitted many questions related to inter alia the Water Framework Directive (WFD) and the Habitats Directive (nature conservation). Conflict emerged between a Greek regional authority and the Greek Minister for the Environment. It concerned measures relating to the partial diversion of upper waters of the river Acheloos to the river Pineios in Thessaly which would have negative consequences for the water status. This diversion was intended to serve other interests, in this case, the drinking water supply, irrigation, and renewable energy (hydropower).

It is important to recognize that sustainable development is explicitly mentioned in the WFD. The directive allows a body of surface water to move from high to good status when this is the result of "new sustainable human development activities" and when, inter alia,

> - the reasons for those modifications or alterations are of overriding public interest and/or the benefits to the environment and to society of achieving the objectives set out in paragraph 1 are outweighed by the benefits of the new modifications or alterations to human health, to the maintenance of human safety or to sustainable development, and
>
> - the beneficial objectives served by those modifications or alterations of the water body cannot for reasons of technical feasibility or disproportionate cost be achieved by other means which are a significantly better environmental option.[20]

The court ruled, inter alia, that the fact that it is impossible for the receiving river basin or river basin district to meet its needs in terms of drinking water, electricity production, or irrigation from its own water resources is not a sine qua non for a diversion to be legal.[21] The EU court made clear that such action would nonetheless only be lawful if it meets the criteria as put out in Article 4 of the directive.[22] The national court must determine whether those criteria have been met. A specific

[19] C-43/10, decision from 11 September 2012. See also Article 267 TFEU.

[20] See Article 4 of the Water Framework Directive, 2000/60.

[21] Para 69

[22] Para 69

aspect in this case was that consent for the river diversion had been given prior to 22 December 2009, the deadline for the adoption of management plans for river basin districts. It is important to understand that obligatory management plans will give further substance to subsequent decision-making regarding the diversion of water ways. In other words: governmental authorities must put forward sufficient evidence for any deterioration of water quality permitted under Article 4 of the WFD. Water management plans will most probably play an important role for courts' assessments of the compatibility of activities causing deterioration of the water quality status with the conditions of the directive. Both governmental planning decisions regarding specific river basins and public participation, obligatory in the development of such plans, are important steps toward determining how sustainable development can support specific activities contributive to developmental purposes but less conducive for environmental protection. This particular case illustrates how environmental protection, specifically water quality protection, may be less important than other interests contributing to sustainable development. For an in-depth understanding of the balance struck by EU governmental authorities between development goals and water protection goals, adopted water management plans and, if available, case law must be examined.

- ***Task**: Discuss the potential role of courts in view of sustainable development promotion. Should courts limit themselves to procedural concerns, such as public participation provisions for water management plans, or should they go further by intervening in substantive decision-making?*

5 Conclusion

Since sustainable development is an integrative concept requiring a balance of environmental, social, and economic interests, its legal consequences are hard to determine. First and foremost, the interpretation of the term "sustainable development" must be made in a political process, such as the ordinary EU legislative procedure as prescribed by the EU treaties. It is hence primarily through *legally designed* governmental processes that concrete policies, legislation, and administrative decisions will be made, allowing the overall idea of sustainable development to materialize in specific situations.

The Water Framework Directive is one example for sustainable development being considered a criterion for allowing deviation from the highest environmental protection goal with regard to the promotion of other interests. Since member states are obliged to adopt water management plans, thereby respecting public participation requirements, it is exactly in this process where a balance can be struck between environmental protection and other concerns. Sustainability science and integrated assessments will be needed in order to support this governmental decision-making. Courts will probably refrain from intervening in governmental authorities' substantive decision-making related to sustainable development issues. The International Court of Justice qualifies sustainable development as a concept, although one judge, the vice president, has provided an alternative view, arguing that sustainable development should be seen as a principle of modern international environmental law rooted in history. For the near future, the most likely role for courts is to intervene in (national) governmental decision-making for those either not in compliance with public participation obligations or not having taken interests or goals, as prescribed by directives, into account. At the EU level, courts might even find noncompliance with the obligation of integrating environmental considerations in EU policies with regard to the promotion of sustainable development. In this sense, some "sustainable development case law" might emerge.

The examples of court cases show that the law is challenged by sustainability, whether it is seen as a simple concept or a binding principle of international and European law. Legal science is challenged to examine how to integrate sustainability into the law and, subsequently, how to apply this in specific legal procedures. In line with this, it is necessary to further develop sustainability law as part of the wider field of sustainability sciences.

References

Barstow Magraw D, Hawke LD (2007) Sustainable development. In: Bodansky D, Brunnéé J, Hey E (eds) Oxford handbook of international environmental law, (chapter 26). Oxford University Press, Oxford/New York

Bosselmann K (2008) The principle of sustainability. Ashgate, Farnham

Bosselmann K (2010) Losing the forest for the trees: environmental reductionism in the law, sustainability 2:2424–2448 (free access under http://www.mdpi.com/2071-1050/2/8/2424)

Dhondt N (2003) Integration of environmental protection into other EC policies. Europa Law Publishing, Groningen

French D (2010) Sustainable development. In: Fitzmaurice M, Ong DM, Merkouris P (eds) The research handbook on international environmental law, (chapter 3). Edward Elgar, Cheltenham

Jans JH, Vedder HHB (2012) European environmental law, 4th edn. Europa Law Publishing, Groningen/NL

Krämer L (2012) EU environmental law, 7th edn. Sweet & Maxwell, London

Romano C (2000) The peaceful settlement of international environmental disputes. Kluwer Law International, New York

Chapter 10
Finance and Sustainability

Olaf Weber

Abstract The connection between the financial industry and sustainable development is indirect. The industry channels financial capital into different industries and therefore has an *indirect* effect on sustainable development through these industries. Depending on which client is financed the impact can be positive or negative. Therefore, the financial industry has developed strategies, products, and services to manage sustainability issues. Most of the products and services focus on risk management connected with risks that are material for financial institutions instead of managing risks for sustainable development. However, sustainability issues are dealt with in internal operations, credit risk management, socially responsible investing, and impact finance. Though products and services connected with sustainability are still marginal, their ratio is increasing and social banks that focus exclusively on sustainable products and services are growing significantly. Key challenges in the field of sustainable finance are to scale up the respective products and services, to focus on the creation of positive impacts on sustainable development through finance, to involve executive management representatives in sustainability issues, and to increase research about the connection between finance and sustainable development.

Keywords Sustainable finance • Sustainable banking • Indirect impact • Social bank • Socially responsible investing • Sustainable credit risk management

1 Sustainability Problems and Finance

Financial institutions are intermediaries. They channel capital into different industries and therefore have an *indirect* effect on sustainable development through these industries. This indirect impact is much more significant than the same industries' direct impact (see Fig. 10.1); for example, studies suggest that the indirect

O. Weber (✉)
School for Environment, Enterprise and Development (SEED), University of Waterloo, 200 University Avenue West, Waterlooo N2L 3G1, Canada
e-mail: oweber@uwaterloo.ca

H. Heinrichs et al. (eds.), *Sustainability Science*, DOI 10.1007/978-94-017-7242-6_10

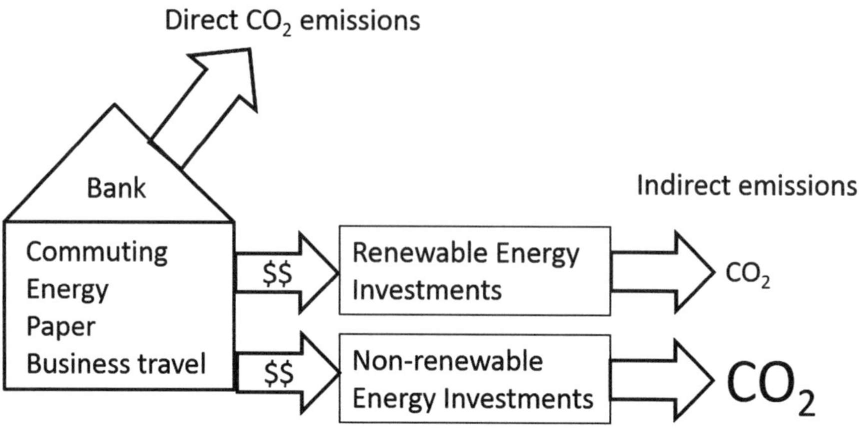

Fig. 10.1 Direct and indirect impacts of the financial sector

emissions created through financing industries are 50 to 200 times higher than direct emissions caused by financial institutions (van Gelder et al. 2008; Weber 2011).

Finance decisions can have negative and/or positive impacts on sustainable development (Wiek and Weber 2014). In the past (and current) financial crisis, it became obvious that the financial sector can have significant negative impacts on the economy (Greider 2011; Herzig and Moon 2013). The same is true for the impact on sustainable development. Currently, sustainability issues do not play a significant role for conventional financial decision-making, with the exception of issues that pose credit or investment risks. Thus, financial institutions invest in a great number of industries and projects with negative effects on society, environment, and the economy, under a long-term perspective. On the other hand, there are positive impacts that are achieved through niche products that proactively seek out loans and capital for industries that support sustainable development, such as renewable energy, health care, or education (see next section below or Jeucken 2004; Vandekerckhove et al. 2011).

2 Solution Options: Sustainable Finance

Classifying financial returns as sustainable conventionally means that they provide long-term, high financial returns. Financial institutions are classified as sustainable if they are able to sustain their business. However, relating sustainable banking to basic concepts of sustainability and sustainable development in a richer sense reveals two complementary perspectives.

Box 10.1: Sustainable Finance Definition
Following the Brundtland Report (1987), *sustainable* finance is finance that meets the social, environmental, and livelihood needs of the present generation without compromising the ability of future generations to meet their own needs and that creates a fair balance between societies in the north and the south.

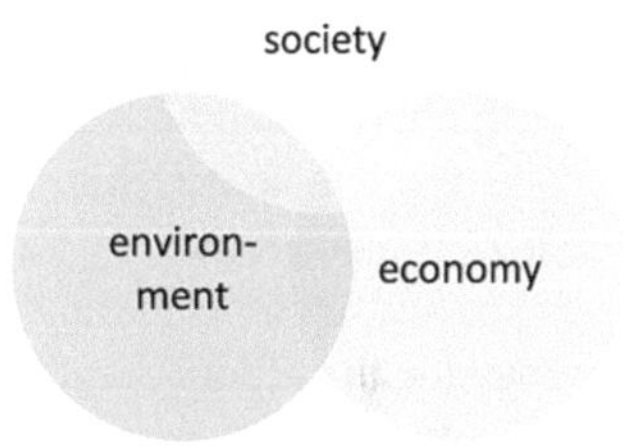

Fig. 10.2 The *triple-bottom-line* (TBL) concept of sustainable business. The triple-bottom-line concept of sustainable business takes environmental, social, and economic issues equally into account in business (see Elkington 1998)

The definition of sustainable finance in Box 10.1 points to a balanced and fair development across generations and nations. It establishes an active role of finance with regard to sustainable development. It emphasizes the need for the contribution to development to be just and sustainable, instead of a one-dimensional (monetary) benefit for the financial sector. This focus on a positive contribution to sustainable development was already discussed in the book *Financing Change* (Schmidheiny and Zorraquin 1996), published in 1996. Since then, sustainable products and services like impact investing (Geobey et al. 2012) or socially responsible investing (Hamilton et al. 1993) have been developed. However, a general strategy as to how the financial sector might contribute to sustainable development is missing, and there is only a small body of literature available for the "sustainability case" of finance.

The second useful perspective is already indicated in the definition above (in the term "needs"), but has more explicitly been developed as the triple-bottom-line (TBL) concept of sustainability (Fig. 10.2).

If we transfer the TBL concept from business to finance, environmental and social criteria should be used as criteria in lending and investment decisions and in other banking operations. To date, these criteria are mostly used to mitigate risks for banks. But they should be used to create a positive impact on sustainable development as well.

The members of the Global Alliance for Banking on Values (GABV), for instance, follow a sustainable finance approach that conducts banking in a way that fosters sustainable development (Korslund 2012). Though the total balance sheet of these banks is still very small, they could increase the number of lenders by more than 50 % between 2007 and 2011 (Weber and Remer 2011).

A closer look at the industry demonstrates that, in addition to managing direct social and environmental impacts caused by the business operations of financial institutions, the main products and services of sustainable banking are sustainable credit risk management, sustainable project finance, socially responsible investment, responsible investments, impact finance, and social banking. We describe these products and services in the following section.

Financial institutions strive to reduce the environmental impacts of their operations by reducing the use of water, paper, travel impacts, and energy. On the social side, they manage their relationships with employees, communities, and other stakeholders. A standard framework for measuring and reporting direct environmental impacts is integrated into the Global Reporting Initiative's financial sector supplement (the Global Reporting Initiative 2011).

One of the main activities of banks is the loan business, and thus credit risk management is a major activity for guaranteeing the business success of a bank. In order to be successful, lenders must rate those factors that influence the borrower's ability to repay the loan (Saunders 1999a, b; Caouette et al. 1998; Fitch 1997). Recently, in addition to standard criteria that are used to analyze borrowers, environmental and social risks have been analyzed in comprehensive studies (Goss and Roberts 2011). The results suggest that there is a correlation between credit risks and sustainability risks of borrowers and that the integration of indicators that measure sustainability risks improves the predictive validity of credit rating systems (Weber et al. 2010). Therefore, systems that assess sustainability credit risks have become more popular in the financial sector (Weber 2012). The US Security Exchange Commission already demands the disclosure of climate risks being material for the value of securities. Consequently, these risks will become material for the lending and investment portfolios of banks and financial institutions, and sustainability aspects, at least those that are related to climate change, will be taken into account by the financial sector.

Project finance involves large, legally independent projects, often in fields such as natural resources and infrastructure (Esty 2004). This type of financing grew significantly over the last couple of decades, and projects are critically observed by environmental and other civic organizations (Missbach 2004). Key aspects are sustainability impacts (Hadfield-Hill 2007), stakeholder relations (Stern 2004), and international environmental regulations (Ong 2011). As a sustainability guideline for project finance, the Equator Principles, a voluntary code of conduct, were proposed in 2003 for assessing and managing sustainability standards in project finance transactions (Lawrence and Thomas 2004).

Linked to the former two solution options is the internalization of externalities in different industries. Driven by regulations or stakeholder pressure, different industries will internalize sustainability issues, previously treated as externalities. The European Union Emissions Trading Scheme (ETS) (Rogge et al. 2011) is a first step into this direction. Firms involved in the ETS have to integrate the value of CO_2

emissions or offsets into their balance sheet. Hence, these positions have to be taken into account in any lending or investment decision of financial institutions.

Socially responsible investing (SRI) and responsible investing (RI) are business fields in sustainable banking that increased significantly over the last decade. In the USA, the assets of socially responsible investment products and services have increased by about 9 % annually since 2007. Overall, $33.3 trillion in assets were under management in the USA in 2012 (Social Investment Forum Foundation 2013). SRI integrates nonfinancial indicators, such as environmental, social, or sustainability indicators, into investment decisions and management for managing sustainability risks of investing. SRI tries to perform similarly or to outperform conventional benchmarks rather than creating a sustainability impact. Though the impact of SRI on sustainable development is rarely analyzed, it is argued that SRI could be able to push firms in a more sustainable direction to be attractive to investors. However, as long as SRI is relatively small, it might not be able to have a strong impact on the financial market (Weber 2006). Because institutional investors such as pension funds are powerful players, it will be important to enable them to conduct sustainable finance as well. There is already a lively discussion about the relationship between the fiduciary duty of institutional investors and responsible investment, as well as a discussion about the materiality of sustainability risks for institutional portfolios (Bauer et al. 2005). A movement of institutional investors into a sustainable way of finance would influence the majority of corporations significantly because of the significant market power of institutional investors.

A newer development in sustainable finance is impact finance. It uses the concept of blended returns (Emerson 2003; Nicholls 2009) that declares positive social, environmental, and sustainability impacts compatible with financial returns. In contrast to SRI, it uses investments for creating a positive impact on sustainable development instead of applying sustainability criteria for risk management. Impact investing (Bugg-Levine and Emerson 2011), microfinance (Morduch 1999), and social banking (Weber and Remer 2011) can be subsumed under the umbrella of impact finance. In contrast to the financial products described above, impact finance gives societal impacts a higher priority than financial returns.

- ***Task**: Review sustainability reports of banks and financial institutions (e.g.,www.globalreporting.org) and analyze them with respect to sustainability issues. Focus on whether and how the reports present the impact of products and services on sustainability impacts.*

3 Open Issues: Challenges of Sustainable Finance

It is still open as to whether the financial sector is willing to take responsibility for sustainable development. On the one hand, we find involvement in SRI or impact investing, but on the other hand, main representatives of the sector neglect their indirect impacts; and sustainable products and services are implemented reactively rather

Table 10.1 Key figures of SRI and social banking compared with conventional financial institutions

Type of product	Amount in $ billion	Comparison group	Amount in $ billion	Percentage
SRI assets under management in the USA	3,140	Total assets under management in the USA in 2012	33,300	11.3
Social Banking Loans	35	Total loans of members of World Council of Credit Unions in 2011	1,016	3.44

Social Investment Forum Foundation (2013), World Council of Credit Unions, (2012)

than proactively. As could be seen during the financial crisis, the financial industry mainly concentrates on itself and does not take impacts on other industries or the society into account. Furthermore, regulations regarding compliance and responsibility of financial sector representatives were weakened rather than enforced. With the exception of credit unions and cooperative banks, the financial industry lost the role of being an intermediary between financial capital, the economy, and the society and became a ruler of the economy instead of being its servant. Therefore, without accepting responsibility about where financial capital is invested, the conventional financial sector will not integrate sustainable finance into its core business.

A key challenge is scaling up sustainable finance. The percentage of socially responsible investment products in asset management portfolios of conventional banks is usually below 2 %. The total amount of loans of the members of the Global Alliance for Banking on Values has been $35 billion in 2012. Compared to the global multitrillion dollar financial industry, these amounts are small, as Table 10.1 demonstrates. Though conventional banks such as the Royal Bank of Canada (Royal Bank of Canada 2012) started to conduct impact investing, the concept is not perceived as a core financial product by the conventional financial industry.

Another challenge is to assess the impact of sustainable finance. Approaches that measure the indirect impact of the financial sector on sustainable development will be needed to analyze both positive and negative impacts (Wiek and Weber 2014). So far, sustainability reporting mostly concentrates on the internal direct impacts of the financial institutes' operations or on philanthropic engagement and community relations. Even the Global Reporting Initiative's financial sector supplement, which provides a standard for the sector's sustainability reporting, uses only 3 out of 82 indicators to demonstrate the impact of products and services on sustainable development (Weber 2013). Neither the Equator Principles for project finance nor the principles for responsible investment propose how to measure the impact of following the guidelines on the sustainability impact of project finance or institutional investments.

In order to analyze the impact of finance on sustainable development, both academia and industry have to shift their focus away from purely analyzing the business case for sustainability in finance. To date, sustainable finance has mainly been researched as a business opportunity (Galema et al. 2008), as a way to manage risks (Weber, et al. 2010), and with respect to its connection to corporate social responsibility (Carroll 1999; Matten and Moon 2005; Porter and Kramer 2006). Only a few studies have focused on strategic changes in the financial sector for becoming sustainable (Geobey and Weber 2013; Ingham et al. 2013; Wiek and Weber 2014).

- ***Task***: *Compare the missions and visions of social banks that are members of the Global Association for Banking on Values (gabv.org) with those of conventional banks.*

Another challenge is the involvement of executive management representatives in sustainability issues. Though more or less each financial institution of a certain size has some person or department in charge of environmental or sustainability issues, there are only a few cases in which sustainable finance is implemented in the general strategy or the management board of financial institutions, or executive compensation is connected to sustainability performance. However, studies suggest that corporate governance plays an important role for the sustainability of the sector (de Graaf and Stoelhorst 2009).

The success of sustainable finance is strongly related to the success of sustainability industries. In times of high earnings of the renewable energy industry, financial institutions were involved in the success by financing a sector that provided attractive returns. This changed in recent years because of market and regulative issues. It will be important to see whether the financial sector will be able to support the sustainability industry actively in the future through looking for investment opportunities in earnest.

Though they mainly follow a business niche approach, social finance, impact investing, and microfinance are drivers of innovation in sustainable finance (Weber 2005). Successful concepts such as SRI and impact investing are often adopted by conventional banks and support the sustainable business of the big players in the sector. Therefore, the important question is whether social finance and impact investing as the next important sustainability drivers in the sector will be successful in the future.

4 Conclusions

The sustainable finance approach connects sustainability with financial issues. It is beyond dispute that capital is needed to enable sustainable development. The world has become painfully aware of how strong the influence of the financial industry is on all aspects of the economy and society during the recent financial crisis. However, the industry is able to channel capital into activities that benefit society as well. Therefore, the integration of the financial industry and the financial market into the sustainable development discussion will be crucial for the success of sustainable development.

More broadly, the presented perspective throws financial and economic aspects into sustainability science. These aspects, though essential for sustainable development, have not been taken into consideration in sustainability science so far.

- ***Task**: Analyze the sustainability science literature, for instance, papers published in Sustainability Science, with respect to the integration of financial issues.*

References[1]

Bauer R, Koedijk K, Otten R (2005) International evidence on ethical mutual fund performance and investment style. J Bank Financ 29:1751–1767
*Bugg-Levine A, Emerson J (2011) Impact Investing – transforming how we make money while making a difference. Jossey-Bass, San Francisco
Caouette JB et al (1998) Managing credit risk: the next great financial challenge. Wiley, New York
Carroll AB (1999) Corporate social responsibility – evolution of a definitional construct. Bus Soc 38(3):268–295
de Graaf FJ, Stoelhorst JW (2009) The role of governance in corporate social responsibility: lessons from Dutch finance. Bus Soc. doi:10.1177/0007650309336451
Elkington J (1998) Cannibals with forks: the triple bottom line of 21st century business. New Society Publishers, Gabriola Island
*Emerson J (2003) The blended value proposition: integrating social and financial returns. Calif Manag Rev 45(4):35–51
Esty BC (2004) Why study large projects? An introduction to research on project finance. Eur Financ Manage 10(2):213–224. doi:10.1111/j.1354-7798.2004.00247.x
Fitch TP (1997) Dictionary of banking terms. Barron's, Hauppauge
Galema R, Plantinga A, Scholtens B (2008) The stocks at stake: return and risk in socially responsible investment. J Bank Financ 32:2646–2654
Geobey S, Weber O (2013) Lessons in operationalizing social finance: the case of Vancouver City Savings Credit Union. J Sustain Financ Invest 3:124–137
Geobey S, Westley FR, Weber O (2012) Enabling social innovation through developmental social finance. J Soc Entrep 3(2):151–165. doi:10.1080/19420676.2012.726006
Goss A, Roberts GS (2011) The impact of corporate social responsibility on the cost of bank loans. J Bank Financ 35(7):1794–1810
Greider W (2011) Needles in a Haystack, Comment, The Nation
Hadfield-Hill S (2007) The greening of project finance. Geogr Compass 1(5):1058–1075. doi:10.1111/j.1749-8198.2007.00050.x
Hamilton S, Jo H, Statman M (1993) Doing well by doing good? The investment performance of socially responsible mutual funds. Finan Anal J 49:62–66
Herzig C, Moon J (2013) Discourses on corporate social ir/responsibility in the financial sector. J Bus Res (0). doi:http://dx.doi.org/10.1016/j.jbusres.2013.02.008
Ingham M, Grafé-Buckens A, Tihon A (2013) Bank-based microfinance: from peripheral to integrated responsibility toward sustainability. Strateg Chang 22(1–2):107–119. doi:10.1002/jsc.1925
*Jeucken MHA (2004) Sustainability in finance. Eburon, Delft
Korslund D (2012) Strong and straightforward: the business case for sustainable banking. Global Alliance for Banking on Values, Zeist, p 16
Lawrence RF, Thomas WL (2004) The equator principles and project finance: sustainability in practice? Nat Resour Environ 19(2):20

[1] *The sources highlighted with an asterisk (*) are recommended as further readings.*

Matten D, Moon J (2005) Corporate social responsibility. J Bus Ethics 54(4):323–337. doi:10.1007/s10551-004-1822-0
Missbach A (2004) The equator principles: drawing the line for socially responsible banks? An interim review from an NGO perspective. Development 47(3):78–84
Morduch J (1999) The microfinance promise. J Econ Lit 37(4):1569–1614. doi:10.2307/2565486
Nicholls A (2009) 'We do good things, don't we?': 'Blended value accounting' in social entrepreneurship. Acc Organ Soc 34(6–7):755–769. doi:10.1016/j.aos.2009.04.008
Ong D (2011) International environmental principles and project finance. In: Leader S, Ong D (eds) Global project finance, human rights and sustainable development. Cambridge University Press, New York, pp 67–104
Porter ME, Kramer MR (2006) Strategy & society: the link between competitive advantage and corporate social responsibility. [Article]. Harv Bus Rev 84(12):78–92
Rogge KS, Schneider M, Hoffmann VH (2011) The innovation impact of the EU emission trading system – findings of company case studies in the German power sector. Ecol Econ 70(3):513–523. doi:10.1016/j.ecolecon.2010.09.032
Royal Bank of Canada (2012) RBC announces $20 million in commitments to new social and environmental initiative. Retrieved 13 Feb 2012, from http://www.rbc.com/newsroom/2012/0124-social-finance.html
Saunders A (1999a) Credit risk measurement: new approaches to value at risk and other paradigms. Wiley, New York
Saunders A (1999b) Credit risk measurement: new approaches to value at risk and other paradigms. Wiley, New York
Schmidheiny S, Zorraquin F (1996) Financing change: the financial community, eco-efficiency, and sustainable development. MIT Press, Cambridge, MA
Social Investment Forum Foundation (2013) Report on socially responsible investing trends in the United States – executive summary. Social Investment Forum Foundation, Washington, DC, p 19
Stern S (2004) International project finance: the Ilisu dam project in 2004 and the development of common guidelines and standards for export credit agencies. [Article]. J Struct Proj Financ 10(1):46–54
The Global Reporting Initiative (2011) Sustainability reporting guidelines & financial services sector supplement. The Global Reporting Initiative, Amsterdam
van Gelder JW, Scheire C, Kroes H, Denie S (2008) Financing of fossil fuels and renewable energy by Canadian banks. Profundo, Amsterdam, p 294
*Vandekerckhove W, Leys J, Alm K, Scholtens B, Signori S, Schäfer H (2011) Responsible investment in times of turmoil. Springer, New York
Weber O (2005) Sustainability benchmarking of European banks and financial service organizations. Corp Soc Responsib Environ Manag 12:73–87
Weber O (2006) Investment and environmental management: the interaction between environmentally responsible investment and environmental management practices. Int J Sustain Dev 9(4):336–354
Weber O (2011) Social banking: products and services. In: Weber O, Remer S (eds) Social banks and the future of sustainable finance. Routledge, London, pp 96–121
Weber O (2012) Environmental credit risk management in banks and financial service institutions. Bus Strateg Environ 21(4):248–263. doi:10.1002/bse.737
Weber O (2013) Measuring the impact of socially responsible investing. Retrieved from http://ssrn.com/abstract=2217784; doi:10.2139/ssrn.2217784
*Weber O, Remer S (eds) (2011) Social banks and the future of sustainable finance. Routledge, London
Weber O, Scholz RW, Michalik G (2010) Incorporating sustainability criteria into credit risk management. Bus Strateg Environ 19(1):39–50. doi:10.1002/bse.636
Wiek A, Weber O (2014) Sustainability challenges and the ambivalent impacts of the financial sector. J Sustain Financ Invest 4(1):9–20
World Council of Credit Unions (2012) 2011 statistical report. World Council of Credit Unions, Madison, p 4

Chapter 11
Sustainability: Politics and Governance

Harald Heinrichs and Frank Biermann

Abstract The article gives an overview of global sustainability policy and politics. It is shown how international policy making on sustainable development has progressed from environmental policy toward recent approaches of Earth system governance. Key challenges of international sustainability politics are discussed, and institutional and instrumental options to improve sustainability policy are presented. The article ends with an outlook of the need for cosmopolitan policy making on sustainable development.

Keywords Sustainability policy • Politics • Earth system governance • Cosmopolitanism

1 Sustainable Development as Political Challenge

Development toward a sustainable (world) society remains an ongoing challenge. Numerous global assessments on ecological, economic, and social dynamics published around the 2012 UN Conference on Sustainable Development in Rio de Janeiro ("Rio + 20")[1] indicate, among other things, that global greenhouse gas emissions are increasing, biodiversity loss is accelerating, social inequality is growing, and economic instability threatens societal cohesion and political stability (e.g., United Nations 2013; UNEP 2012; WWF 2012). Looking at long-term ecological, economic, and social developments through key indicators such as population growth, gross domestic product, declining fish stocks, nitrogen input, individual motorized mobility, or even the proliferation of McDonald's restaurants as a proxy for mass consumption, one can observe exponential growth rates from the

[1] Twenty years after the important conference on environment and development in Rio de Janeiro in 1992, the global sustainability community met again in Rio de Janeiro to take stock and look ahead: http://www.uncsd2012.org/

H. Heinrichs (✉)
Institute of Sustainability Governance (INSUGO), Leuphana University, Lüneburg, Germany
e-mail: harald.heinrichs@uni.leuphana.de

F. Biermann
Institute for Environmental Studies, VU University Amsterdam, Amsterdam, Netherlands
e-mail: Frank.biermann@vu.nl

H. Heinrichs et al. (eds.), *Sustainability Science*, DOI 10.1007/978-94-017-7242-6_11

beginning of the industrial revolution in the eighteenth century (Steffen et al. 2004). Material wealth has globally increased, but it remains highly unequally distributed between and within countries. One key reason for the present unsustainable development is the globalization of resource-intensive economic growth and a consumerist lifestyle. As indicated by numerous data, progress in eco-technological innovations has been far outstripped by economic growth. The positive effects of economic development opened up new opportunities for hundreds of millions of people in emerging countries such as China, Indonesia, Brazil, and Mexico and contributed to economic growth in many developed countries, yet were also accompanied by adverse socioeconomic and ecological effects. Overall, the global community has not succeeded in fulfilling the goal of the 1992 Rio Conference on environment and development: to achieve sustainable development with equal optimization of economic growth, social well-being, and ecological stability.

This balance sheet indicates that the manifold actions taken by business, civil society, and policy making around the world have not managed to reverse fundamentally unsustainable dynamics. In order to better understand why only limited progress on improving sustainability has been made, we describe in the following: first, the emergence of sustainable development as a global political issue and the development from environmental policy to sustainability governance. We then introduce key conceptual perspectives for understanding, analyzing, and framing policy making for sustainable development in a globalized world. We conclude this overview by demonstrating that sustainability needs to be recognized as an essentially *political* issue, which requires policy making in a cosmopolitan perspective.

2 From Environmental Policy to Sustainable Development

The foundations of today's *sustainability policy* are rooted in environmental policy. During the 1960s, environmental problems such as air or water pollution were diagnosed as adverse side effects of industrialization (McNeill 2001). At the same time, the global character of environmental problems became visible: similar environmental problems appeared in all industrialized countries, and it also became clear that environmental pollution did not stay within borders but evolved into a transnational and increasingly global issue. In the 1970s, the first global scenario studies pointed out the limits of nonrenewable resources, as well as natural sinks, with regard to continuous (material) economic growth (Meadows et al. 1972).

The new scientific findings were accompanied by changing values – at first in smaller parts of society – and changing political appraisal, all leading to the development of environmental policies as a distinct policy domain at both national and international levels (Jänicke et al. 2003). New policy principles were formulated and concretized, new institutions were designed, and complex new instruments were developed. Since its beginnings in pioneering countries such as Sweden and the United States, modern environmental policy has spread and developed through policy learning and diffusion of concepts and approaches around the world.

Over time, limits of first-generation environmental policy instruments and institutions became visible and led to further conceptual developments. It became apparent:

- That sectoral environmental policies were not sufficient to grasp interconnected environmental problems
- That environmental protection in the form of simply cleaning up environmental pollution has its limits
- That questions of social development, such as poverty and demography, are central

Political answers to these questions mark the renewal of environmental policy and the development of sustainable development in the 1980s. Innovations in environmental policies such as policy integration, preventive, production-integrated environmental protection, and strategic environmental policy in the context of ecological modernization (Jänicke 2008) led to more proactive environmental policy making. At the same time, social and economic development challenges gained in relevance. The so-called Brundtland Commission (1984–1987) synthesized these discourses and propagated the idea of "sustainable development," which finally led to the Agenda 21 adopted by 179 states at the 1992 UN Conference on Environment and Development.

These processes in the 1990s constitute the beginning of sustainability policy. Sustainability policy goes beyond environmental policy. Since then, sustainability policy has developed – conceptually, institutionally, and instrumentally – from the local to the national to the international level with varying degrees of ambition and success around the world. Within nation states, as well as internationally, there are significant differences and variations in values, interests, power potentials, and solution orientation regarding (un)sustainability (Meadowcroft 2008). The political debate on sustainable development is coined by heterogeneous interpretations, definitions, and controversies around its concretization (Grunwald and Kopfmüller 2006). Like any other political issues, sustainable development became an object of political struggle. The depth and breadth of societal transformation needed for sustainable development pose significant challenges to sociopolitical decision-making. Despite its complexities, fueled by uncertainty and ambivalent evaluations, a basic conceptual understanding, typical instruments, and institutional approaches have emerged.

In many countries, especially in Europe, a multidimensional understanding of sustainability has become accepted (Swanson et al. 2013). Even though policy content and arenas vary between different policy levels, there is a tendency toward working on interconnected policy issues. This perspective is accompanied by instrumental developments, including new instruments such as sustainability strategies, sustainability assessment, and communicative and cooperative approaches (Box 11.1).

Box 11.1: Analyzing Sustainability Policy: Case Study Germany

What significance does sustainability have in policy making and administration in Germany?

That was the key question asked in a cooperative study between WWF Germany and the Institute for Sustainability Governance at Leuphana University, Lüneburg. Surveys in Germany show that sustainability in policy making – in contrast to official government rhetoric – is prioritized only to a limited degree. Based on interviews within all ministries and document analysis, the study reveals that German sustainability policy is not as good as it seems. Even though there are efforts to cooperate with civil society concerning sustainability challenges, the coordination between ministries is not very developed in daily practice – despite institutions such as the state secretary commission for sustainable development, which has been established to improve the coordination between ministries. The cooperation between the national and state level within the federal German system is even less established. The main reason hereby is the fear of the states that the central government would interfere within state decision-making. The same logic holds for the parliamentary advisory board on sustainable development, where members of other parliamentary commissions fear that the sustainability advisory board could intervene in their domains. However, the cross-cutting and long-term character of sustainability requires exactly this – a higher degree of integration and coordination. Regarding public procurement, the study indicates that aspects of sustainability are partly considered, but in practice too often included only to a limited degree. In sum, all too often is sustainability superseded by more short-term, single-issue priorities in daily policy making in the German government.

(Heinrichs and Laws (2012): Mehr Macht für eine nachhaltige Zukunft. Politikbarometer zur Nachhaltigkeit in Deutschland. WWF, Berlin; Heinrichs et al. 2013) (Fig. 11.1).

Fig. 11.1 WWF study

In order to cope with integrative and long-term challenges of sustainability policy, Agenda 21 already requests all nations to develop sustainability strategies, including goals and indicator systems. Also, sustainability strategies have become a key instrument of sustainability policy around the world (Meadowcroft 2007), and sustainability assessments have gained importance (Grunwald and Kopfmüller 2007) in order to help evaluate policy decisions in advance.

Finally, participative approaches play a significant role (Heinrichs 2011). Sustainable development has been understood from the beginning as a collective search, learning, and collaborative design process. For this, a cooperative, initiating, and moderating state is needed that is willing and able to include non-state actors from business and civil society. On the other hand, more regulative policy instruments – such as mandatory sustainability reporting for the private and public – are much less used. Even though multiple policy instruments, from emission trading up to biosphere reserves, are directly relevant to sustainability, instruments addressing sustainability policy as a cross-cutting and long-term challenge are less developed and implemented and have a tendency to turn into "soft" policies.

Sustainability must be adequately institutionalized in order to become fully effective. Due to its historical links to environmental policy, sustainability policy is often integrated into existing environmental institutions. In some cases, environmental ministries simply became ministries of sustainable development. However, due to the cross-cutting and long-term character of sustainable development, it seems advisable to develop institutional mechanisms that fulfill integrative and coordinative tasks (Lafferty 2004).

In the past years, innovative approaches have been developed and implemented in this regard. On different political-administrative levels, there are coordinating entities on sustainable development, for example, state secretary commissions, parliamentary commissions, or municipal units. At the United Nations level, a "Commission on Sustainable Development" was established in 1992 to monitor and guide the implementation of Agenda 21. In 2012, governments decided to replace this commission, which was widely felt as not having achieved its goals, with a new institution within the UN system that would function at a higher level and have new competences (Biermann 2013).

Overall, institutionalization so far has been too weak and is overall insufficient to make sustainable development a top priority of policy making. The institutional architecture, as well as the existing instruments, seems to be inadequate to drive the sustainability transition as it would be necessary given the ongoing unsustainable trends.

Even though an extension of sectoral environmental policy, beyond environmental policy integration toward integrative sustainability policy, can partly be diagnosed, sustainability is still "in statu nascendi." Short-term pressures for political action pose serious challenges for long-term thinking and action. Importantly, beyond the normative requirements of a strengthened sustainability policy, a more detailed, theory-based understanding of drivers, blockages, and potential solution pathways is needed.

3 Sustainable Development and Earth System Governance

In recent years, the discourse on environmental policy and governance has been further developed into a new perspective that takes the entire Earth system as an object of political efforts: "Earth system" governance. This paradigmatic shift from environmental pollution to an Earth system perspective has been pioneered by the natural sciences. Increased scientific efforts in global research programs, generally supported by vastly increased computing power available to researchers, led to an improved understanding of both the complex interdependencies in the Earth system and the rapidly growing planetary role of the human species. Scientific research brought quickly increasing evidence concerning past developments in planetary history, including the nonlinearity of processes, potentials for rapid system turns, and complex interrelationships between components of the system. The relative stability of the global climate during the Holocene era – the last 10,000 years during which the development of human civilization was brought about – seemed almost a fortunate exception. The Earth system appeared more and more as being marked by interconnectedness and fragility (see, in more detail, Biermann 2014).

Equally visible became the vast and global impact of the human species. The first mass extinctions of larger mammals might be related to early hunter societies. Human influence has grown since the Neolithic revolution with the development of agriculture and husbandry. Today, at the height of industrialization, humanity has fully evolved as a geological force, able to influence global geobiophysical systems (Steffen et al. 2011). This development has been aptly symbolized by Paul Crutzen and Eugene Stoermer's call to declare the end of Holocene and the beginning of a new epoch in planetary history – the Anthropocene.

Earth system governance as a social science paradigm is a response and a reaction to these developments. The notion of Earth system governance accepts the core tenet of these new approaches in science, that is, the understanding of the Earth as an integrated, interdependent system transformed by the interplay of human and nonhuman agency. Yet the focus of Earth system governance is not "governing the Earth," or the management of the entire process of planetary evolution. Earth system governance is different from technocratic visions of what is sometimes referred to as "Earth system management" or even Earth system "engineering."

Instead, Earth system governance is about the human impact on planetary systems. It is about the societal steering of human activities with regard to the long-term stability of geobiophysical systems. As such, Earth system governance is essentially a social science research program within the larger strand of governance theory in the social sciences. Cooperation and, at times, integration with natural science programs are useful and important. Yet the foundation of Earth system governance is firmly within the social sciences.

The notion of Earth system governance now underpins a 10-year global research initiative under the auspices of the International Human Dimensions Programme on Global Environmental Change (IHDP). This initiative – the Earth System Governance Project – was launched in 2009 and is scheduled to last until 2018. The

Project has evolved into a broad, vibrant, and global community of researchers who share an interest in the analysis of Earth system governance and in the exploration of how to reform the ways in which human societies (fail to) steer their coevolution with nature at the planetary scale. More than 2500 colleagues are subscribed to the Earth System Governance newsletter, and about 200 researchers belong to the group of lead faculty and research fellows closely affiliated with the Project. The term "Earth system governance" generates about 400,000 Google hits today.

The Earth system governance research alliance has put forward a science plan that shall help guide research in this domain based on a joint analytical framework (Biermann et al. 2009). This analytical framework revolves around five dimensions of effective governance, which are interrelated yet can be studied apart as well: the analytical problem of *agency* in Earth system governance, including agency that reaches beyond traditional state actors; the overall *architecture* of Earth system governance, from local to global levels; the *accountability* and *legitimacy* of Earth system governance; the problem of (fair) *allocation* in Earth system governance; and, finally, the overall *adaptiveness* of individual governance mechanisms and processes and of the overall governance system (see Biermann 2007; Biermann et al. 2009; Biermann 2014, in more detail).

- ***Questions:*** *What are key differences between environmental, sustainability, and Earth system governance?*
- ***Task:*** *Please investigate current developments in sustainability policy on international, regional, national, state, and municipal levels. Discuss with your fellow students to what extent the current institutions and instruments are effective or not effective in advancing sustainable development.*

4 Sustainability Policy as Cosmopolitan Challenge

Climate change, resource scarcity, volatile economic dynamics, social inequality, and demographic change are interconnected problems of (un)sustainable development. Ambitious sustainability policy is needed in order to approach the so-called great transformation (WBGU 2011).

To some extent, this transformation is happening, as evidenced by the development of environmental policy and its extension toward sustainable development over the past four decades. However, the velocity of transformation is insufficient with regard to the unsustainable trends diagnosed by (natural) sciences. Sustainability policy needs to be further developed. Conceptual approaches and empirical analysis, like the work done within the Earth system governance research alliance, and practical innovations realized in research and development projects open up new perspectives and opportunities. Sustainability policy requires as precondition the systematic institutionalization of sustainability in politics and administration. Sustainability strategies are required with monitoring and reporting systems, structural and procedural elements which allow for horizontal integration of policy fields, the vertical coordination within the multilevel political-administrative systems, the cooperation with non-state actors as well as the management of diverging forms of knowledge and knowledge claims, as well as the consideration of short-, medium-, and long-term perspectives in decision-making. These basic institutional and instrumental elements challenge some key characteristics of democratic and bureaucratic policy making, e.g., the short-termism in election cycles or the specialization of units in administrations, yet are key to developing and implementing policies for interconnected problems of sustainable development. Even though nation states will need to continue to stand at the center of sustainability policy, it is also becoming clear that sustainability is inherently a cosmopolitan topic. With regard to our globalized and interconnected world, the social sciences also need to go further beyond methodological nationalism and open up international, transnational, and cosmopolitan perspectives.

Further Reading

McNeill JR (2001) Something new under the sun. An environmental history of the twentieth century. Norton & Company, New York

Meadowcroft J, Bregha F (2009) Governance for sustainable development: meeting the challenge ahead. PRI Initiative, Ottawa

References

Biermann F (2007) 'Earth system governance' as a crosscutting theme of global change research. Glob Environ Chang Hum Policy Dimen 17:326–337

Biermann F (2013) Curtain down and nothing settled: global sustainability governance after the Rio+20 Earth summit. Environ Plann Gov Policy 31(6):1099–1114

Biermann F (2014) Earth system governance. World politics in the anthropocene. MIT Press, Cambridge

Biermann F, Betsill MM, Gupta J, Kanie N, Lebel L, Liverman D, Schroeder H, Siebenhüner B, with contributions from Conca K, da Costa Ferreira L, Desai B, Tay S, Zondervan R (2009) Earth system governance: people, places and the planet. Science and implementation plan of the earth system governance project. ESG report 1. The Earth System Governance Project, Bonn, IHDP. Available at www.earthsystemgovernance.org

Grunwald A, Kopfmüller J (2006) Nachhaltigkeit. Campus, Frankfurt am Main

Grunwald A, Kopfmüller J (2007) Die Nachhaltigkeitsprüfung: Kernelemente einer angemessenen Umsetzung des Nachhaltigkeitsleitbilds in Politik und Recht. Wissenschaftliche Berichte, FZKA 7349. Forschungszentrum Karlsruhe, Karlsruhe

Heinrichs H (2011) Participation: empowerment for sustainable development. Springer, Dordrecht/Heidelberg/London/New York

Heinrichs H, Laws N (2012) Mehr Macht für eine nachhaltige Zukunft. Politikbarometer zur Nachhaltigkeit in Deutschland. WWF, Berlin

Heinrichs H, Laws N, Riedel H (2013) Challenges and opportunities for sustainable development in Germany. Verlag Bertelsmann Stiftung, Gütersloh

Jänicke M (2008) Megatrend Umweltinnovation. Oekom, München

Jänicke M, Kunig P, Stitzel M (2003) Umweltpolitik: Lern- und Arbeitsbuch. Dietz, Bonn

Lafferty WM (ed) (2004) Governance for sustainable development: the challenge of adapting form to function. Edward Elgar, London

McNeill JR (2001) Something new under the sun. An environmental history of the twentieth century. Norton & Company, New York

Meadowcroft J (2007) National sustainable development strategies: features, challenges and reflexivity. Eur Environ 17:152–163

Meadowcroft J (2008) Who is in charge here? Governance for sustainable development in a complex world. In: Newig J, Voß J-P, Monstadt J (eds) Governance for sustainable development. Routledge, New York

Meadows DL, Meadows DH, Randers J (1972) Limits to growth. A report for the club of Rome's project on the predicament of mankind. Universe Books, New York

Steffen W et al (2004) Global change and the earth system. A planet under pressure. Springer, New York

Steffen W, Grinevald J, Crutzen P, McNeill J (2011) The anthropocene: conceptual and historical perspectives. Phil Trans R Soc A 369:842–867

Swanson D, Halle M, Garcia Schmidt A, Esche A (2013) Global trends in sustainable development – a view from the RMP 2013 global search process. Bertelsmann-Verlag, Gütersloh

UNEP (United Nations Environment Program) (2012) Global environment outlook 5 – environment for the future we want. UNEP, Nairobi

United Nations (2013) The millennium development goals report. United Nations, New York

WBGU (2011) World in transition. A social contract for sustainability. WBGU, Berlin

WWF (2012) Living planet report 2012. WWF, Gland

Chapter 12
Sustainability Communication

Daniel Fischer, Gesa Lüdecke, Jasmin Godemann, Gerd Michelsen, Jens Newig, Marco Rieckmann, and Daniel Schulz

Abstract Communication as pivotal part of the human condition plays an essential role in bringing sustainability-related issues onto society's agenda. Sustainability communication does not represent a somewhat discrete and self-contained theoretical approach, but rather draws on a wide range of disciplines, their bodies of knowledge, and their methodological approaches to illuminate the drivers and barriers of a broader and deeper societal engagement with the idea of sustainability. This chapter introduces students to the study of communication processes in the context of sustainable development. It suggests analyzing sustainability communication using a typology of three different communication modes: communication *of*, *about*, and *for* sustainability. The typology is applied in an illustrative way from the perspectives of two particular subsystems familiar to students, the *educational system* as well as the *media system*. The chapter concludes with an overview of relevant literature in the field of sustainability communication. The recommended readings cover three different types of literature highly relevant to students' future studies in this field: introductory readings, practice-oriented readings, and current research.

D. Fischer (✉)
UNESCO Chair Higher Education for Sustainable Development,
Leuphana University Lüneburg, Scharnhorststr. 1, 21335 Lüneburg, Germany
e-mail: daniel.fischer@leuphana.de

G. Lüdecke • J. Newig • D. Schulz
Institute for Sustainability Communication, Leuphana University Lüneburg,
Scharnhorststr. 1, 21335 Lüneburg, Germany
e-mail: gesa.luedecke@leuphana.de; jens.newig@leuphana.de; daniel.schulz@leuphana.de

J. Godemann
Agricultural, Nutritional and Environmental Science, Justus-Liebig University Gießen,
Senckenbergstr. 3, 35390 Gießen, Germany
e-mail: jasmin.godemann@fb09.uni-giessen.de

G. Michelsen
Faculty of Sustainability, Leuphana University, Scharnhorststraße 1,
21335 Lüneburg, Germany
e-mail: michelsen@uni.leuphana.de

M. Rieckmann
Department I / Education and Pedagogical Sciences, University of Vechta,
Driverstr. 22, 49377 Vechta, Germany
e-mail: marco.rieckmann@uni-vechta.de

H. Heinrichs et al. (eds.), *Sustainability Science*, DOI 10.1007/978-94-017-7242-6_12

Keywords Communication of sustainability • Communication about sustainability • Communication for sustainability • Media communication • Education for sustainable development

1 A Communication Perspective on Sustainable Development: Origins and Approaches

> *Fish or humans may die because swimming in the seas and rivers has become unhealthy. The oil-pumps may run dry and the average climatic temperature may rise or fall. As long as this is not the subject of communication it has no social effect. (Luhmann* 1989*, p. 28 f.)*

This quote by German sociologist Niklas Luhmann illustrates the constitutive role of communication for the societal engagement with environmental and sustainability issues in particular, as well as for the organization of our society in general: what issue becomes relevant within a society depends on peoples' mutually shared representations of the social and natural world. In this context, communication is understood as a social process in which common orientations are interchanged. "The necessity of communication can be found in the human condition: each consciousness is isolated, our neurophysiological, cognitive, emotional processes are mutually unobservable and there is no direct access to the thoughts, attitudes and intentions of the other. It is through communication that 'the interior is exteriorised', that we can inform each other, that we become social creatures. Communication is thus the principle of societal organisation itself" (Ziemann 2011, p. 90).

Sustainable development, understood as a societal process of *exploration*, *learning*, and *transformation* (Godemann and Michelsen 2011), poses particular challenges for communication processes. Global sustainability issues are characterized by high complexity, uncertainty, and ambivalence. Furthermore, sustainable development is a task that requires the combined efforts of many actors in order to be successful. Communication is thus essential for developing a mutual understanding of which actions to take and to ensure an effective implementation of those measures (Newig et al. 2008).

> From the perspective of sustainability science, the **task of sustainability communication** lies in introducing an understanding of the world, that is of the relationship between humans and their environment, into social discourse, developing a critical awareness of the problems about this relationship and then relating them to social values and norms. Scientific knowledge and scientific discourse play a central role in this undertaking to the extent that they contribute to strengthen or relativize the various positions and perspectives. (Godemann and Michelsen 2011, p. 6)

Sustainability communication has its **origin** in environmental communication. With the emergence of the debate on sustainable development in the context of the

Rio Summit of 1992, however, communication about environmental issues has contextualized itself more and more in the broader framework of sustainability. Today, sustainability communication also draws on existing scholarship on risk and science communication (Adomßent and Godemann 2011).

From a **policy perspective**, sustainability communication is merely one of the many different instruments available to facilitate the transformation toward sustainable development. Research and policymaking in the field of sustainable development have brought ample experience with both "hard" and "soft" policy instruments. Hard instruments include marketization and regulation approaches, comprising of legislative, regulatory, and juridical, as well as financial and market instruments (Kaufmann-Hayoz et al. 2012). While "hard" instruments often have the benefit of legal control and entail formalized coordination processes, "soft" and persuasive instruments are considered to be more flexible and versatile. Approaches from the field of sustainability communication are commonly categorized as examples of "soft" or persuasive policy instruments. Today, most scholars advocate for a combination of "hard" instrumental approaches and "soft" persuasive measures in order to change both structural and institutional arrangements *as well as* social norms and people's willingness to adopt new attitudes and behavioral patterns (Jackson and Michaelis 2003).

With regard to a **theoretical framing** of sustainability communication, it must be stated that there is still no genuine framework available that would allow for such a theory. Instead, sustainability communication draws from a broad field of different scientific disciplines, each with its own theoretical principles and knowledge. These comprise, among others, systems theory and the epistemology of constructivism, approaches in media theory and in communication theory, as well as psychology and sociology (for further reading, see Chap. 5).

- *Task: Form study groups with your peer students. Think of a case from the field of sustainability communication, for example, a recent wildlife protection campaign. Each group member is then assigned to a different theory and studies this theory on his/her own. Finally, get back together again, present to each other the essentials of your theories, and apply them to the case example. What would your theory contribute to your case example? What implications, benefits, and limitations does your theoretical perspective bring to the case?*

Another perspective on sustainability communication is to consider the issues it deals with and the venues it takes place. Prominent issues and contents in sustainability communication comprise themes such as biodiversity, consumption, mobility, climate, energy, corporate social responsibility, and conservation (Godemann and Michelsen 2011). Venues of sustainability communication can be found in all societal subsystems (Luhmann 1977) such as civil society, education, mass media, science, politics, and economy (Newig et al. 2013). Two of these subsystems (education and the media) will be discussed more closely in the third and fourth sections of this chapter to show how sustainability communication can be applied in real life settings.

2 Communication *of*, *about*, and *for* Sustainability

In the previous section, sustainability communication has been introduced as a term for communication approaches explicitly designed to facilitate sustainable development. This communication perspective can be considered as communication *for* sustainability, since it is clearly directed toward *advocating* sustainability. Apart from a transformative directedness toward achieving sustainable development, communication on sustainability can take/occurs in two more refined modes that can be analytically distinguished: While communication *for* sustainability (CfS) has the main objective of facilitating societal transformation toward the normative goals of sustainable development, other perspectives of communication focus on sharing concepts and frames in the context of sustainable development (communication *about* sustainability, CaS) or transferring information from a sender to a receiver in order to bring a certain motivation across (communication *of* sustainability, CoS) (Godemann and Michelsen 2011; Newig 2011; Newig et al. 2013).

CaS can mainly be described as a many-to-many communication mode with nonhierarchical, horizontal structures. Its purpose is discourse oriented by sharing concepts or frames about sustainability. CoS, in contrast, is a sender-receiver-oriented and one-to-many communication mode. Information toward an objective is getting passed to an (indistinct) audience. The intention behind this communication mode is mainly sender oriented and thereby objective oriented (see Table 12.1). The notion of communication *for* sustainability (CfS) adds the dimension of normative directedness to the distinction between CaS and CoS. It comprises approaches that are openly underpinned by a transformative agenda that seeks to enhance capacity for change. The nature of these **modes** of sustainability communication will be presented in the following section in greater detail. In what follows, we will give examples from two societal subsystems, wherein these modes of sustainability communication are applied and analyzed.

Table 12.1 Communication *about* sustainability in comparison to communication *of* sustainability

	Direction/mode of communication	Function	Measures of effectiveness
Communication *about* sustainability (CaS)	Deliberative; horizontal, many to many	Deliberation; production of intersubjective/shared concepts/frames	Discourse oriented: quality of discourse; compatibility of concepts to sustainability
Communication *of* sustainability (CoS)	Transmissive; sender-receiver, one to many	Transmission; transfer of information toward an objective	Sender oriented: achievement of sender's communication objective

Newig et al. (2013)

2.1 *Communication* about *Sustainability*

CaS refers to processes in which information, interpretations, and opinions regarding sustainability issues are *exchanged* and *debated*. Issues are going to be transformed and framed in horizontal communication that can take place on many different levels, ranging from interpersonal face-to-face interaction up to the mediated level of mass communication (Neidhardt 1993). CaS constitutes our perception of sustainability issues, as it serves important functions of framing issues and structuring facts, arguments, and claims by creating a common understanding of the issue at stake, of the goals that should be pursued, and of who needs to take action. Such processes are not necessarily harmonious and inclusive, but instead can be seen as "controversially structured fields of symbolic interaction in which a variety of actors struggle to establish their respective interpretation of problems, their causes and remedies" (Brand 2011, p. 57).

How might the effectiveness or quality of CaS be assessed? One indicator is the amount of attention that an issue receives from the mass media (Newig 2011; Bonfadelli 2010). A second, procedural indicator refers to who has access to the discourse and influences the framing processes. Effectiveness then refers to structural conditions and the design of communication processes (Rowe and Frewer 2005). A third aspect concerns the (potential for) communication exchange between spheres, or subsystems, of communication (Weingart et al. 2000). An indicator of communication effectiveness would measure the extent to which the discourse in one subsystem (e.g., science) is compatible with discourses in other subsystems (e.g., the political system) and how likely it is to transfer important aspects from one subsystem to another so that, eventually, action toward sustainable development can be taken (Egner 2007).

2.2 *Communication* of *Sustainability*

CoS, by contrast, is intentional, instrumental, or managerial. It focuses on the primarily mono-directional, sender-receiver flow of communication, in which the sender pursues a certain objective of communication (Newig 2011). Scientists, NGOs, educators, companies, and journalists seek to gain the attention of decision-makers or the broader public in order to provide information about sustainability-related phenomena. As the demands of society for sustainable action grow, actors may see the need for CoS as a measure to defend or legitimize their behavior. Corporate sustainability reporting is one example of this type of communication.

Specific functions of CoS are to inform and educate individuals and to achieve some type and level of social engagement and action (Moser 2010). In this respect, it takes an elitist stance, making a central distinction between experts and laypersons in respect to their sustainability-related knowledge and capacities (Nerlich et al. 2010).

Since CoS has clear intentions as to its desired effects, it lends itself to assessment in terms of its effectiveness. Have the recipients been reached? Have they understood the message? Have they, perhaps, changed their values and behavior? Again, referring to science, this mode of communication, in which "experts" educate "lay" people, is increasingly being criticized (Nerlich et al. 2010). Recently, serious failures in communication regarding climate change have stunned public debate. Take, for example, the IPCC's erroneous scenario of Himalayan glacier melting by 2035, which IPCC officials continued to uphold under doubtful circumstances. This contributed to declining public confidence in climate scientists (Leake and Hastings 2010). Not only is the privileged position of science eroding; increasingly, the dominant quest for behavioral change at the individual level (which has only had very limited success) is questioned in favor of dialogue and discourse (Barth 2012). Thus, CoS approaches the sphere of communication *about* sustainability.

- *Task: Before you continue reading, think of an* ***example*** *in which sustainability communication represents either CoS or CaS. How can your example be described and in which context does it appear?*

2.3 *Communication* for *Sustainability*

While the distinction between CaS and CoS refers to the direction and the initiators of information flows, the concept of CfS shifts emphasis to the normative aspect of sustainable development. In this sense, communication is not just about providing sustainability-related information and raising awareness for sustainability issues. Its objective is to facilitate societal *transformation* toward the normative goals of sustainable development. In terms of direction and senders, CfS may share elements of CoS and CaS, including the knowledge generation, (social) learning (Barth 2012), and collaboratively developing solutions for sustainability problems. The effectiveness of CfS relates to its impact in terms of measurable action toward sustainable development.

CfS has counterparts in which sustainability-related communication may, in fact, (be intended to) neglect or even obstruct sustainable development. Since hardly anyone is openly "against" sustainability, this is obviously done by symbolically subscribing to sustainability while pursuing hidden non-sustainable agendas (e.g., "greenwashing" in sustainability reporting).

Although the boundaries between these different types of communication modes are barely selective, they could provide a useful analytical lens when looking at the numerous communication processes to be found in the context of sustainable development.

To give the reader an idea of how sustainability communication might exemplarily be described from the perspective of a particular subsystem, the *educational system* and the *media system* may serve as illustrative examples and will be elaborated in the following.

3 Sustainability Communication in the Media System

During the past 20 years, mass media have played an increasingly significant role in bringing forward and establishing the concept of sustainability in societal discourse. The media function as an observer of society, picking up dynamics within one subsystem (e.g., economy) and delivering them into other societal spheres (e.g., political agenda). Ideally, the media constitute a public sphere that not only represents all relevant voices within society but also offer a venue for discourse about issues. The mass media mainly focus on CoS in a sense that journalists report on topics such as scientific findings or political summits. Communication typically follows a monodirectional "one-to-many" mode, with little access to feedback loops that could possibly initiate discourse in the sense of CaS. However, certain interactive TV formats or the publication of letters to the editor do present opportunities for CaS. As the Internet (especially new social media formats) continues to gain popularity, mass media outlets are increasingly experimenting with more interactive forms of communication with even further potential for CaS. When media outlets openly and actively work toward stimulating reflection and behavioral change, it becomes possible to consider this type of communication mode as CfS.

- *Task: Skim over different newspapers for sustainability topics. Also, look at TV coverage and websites and analyze their approach to sustainability. Try to attribute each coverage to a form of sustainability communication (of, about, for), and identify differences between the media forms. Then, discuss with your fellow students the potential aims and intentions, and think what media communication modes you can imagine for CoS, CaS, and CfS: How would you develop a format for sustainability communication with either CoS, CfS, or CaS intention?*

4 Sustainability Communication in the Education System

Key sustainability concerns, such as peace, environmental protection, or development cooperation, have been advocated by different educational camps. Three different traditions can be distinguished, reflecting the overall distinction between different perspectives on sustainability communication outlined above: facts-based (CoS), pluralistic and deliberative (CaS), and more transformative (CfS) traditions in approaching seminal societal concerns in educational contexts. Since the Rio Summit of 1992, the notion of education *for* sustainable development (ESD) has received remarkable political support on an international level, cumulating in the launch of a United Nation's world decade on ESD (2005–2014) and a Global Action Programme on ESD starting in 2015 in continuation of the decade. ESD is commonly viewed as an integrative framework that has the potential to forge alliances between different adjectival educations. In the scholarly and policy debate, ESD is considered to prefer competency-based emancipatory approaches over behavioral-based instrumental approaches and to thus reflect principles of communication *about* sustainability (Vare and Scott 2007). However, empirical findings show that the competence approach proposed by and favored in the scholarly discussion has not yet translated into practice on a full scale and that ESD practice often reflects principles of communication *of* rather than *about* sustainability (see Newig et al. 2013).

- *Task*: *Think of an illustrative field in the context of sustainable development (e.g., waste disposal, climate change). Now select an educational setting of your choice, for example, a lesson in high school, a guided tour through a national park, or an exhibition in the community center of a rural small town. Develop three different approaches that reflect the different perspectives of CoS, CaS, and CfS. What implications do the perspectives have for the choice of methods, contents, learning objectives, and pedagogy?*

5 Summary

This chapter outlined that communication, as a pivotal part of the human condition, also plays an essential role in bringing sustainability-related issues onto society's agenda. Sustainability communication does not represent a somewhat discrete and self-contained theoretical approach, but rather draws on a wide range of disciplines, their bodies of knowledge, and their methodological approaches to illuminate the drivers and barriers of a broader and deeper societal engagement with the idea of sustainability. It thereby pursues a transformative agenda and reflects the normative principles inherent to sustainable development (CfS). The two given examples from media and the educational system also show that a critical issue for sustainability communication is the understanding of different communication modes that are linked to different motivations and communicative objectives. The multiple communication venues suggest an overall shift from CoS toward a more horizontal, participatory communication mode of CaS within most subsystems. Considering your own background as an undergraduate or graduate student, in what way does this matter to the focus of your study? How can sustainability communication affect, impede, or promote efforts in your field of work, and how can you approach, analyze, and employ sustainability communication in your future working contexts?

Further Reading

The following reading tips give you an overview of relevant literature in the field of sustainability communication. The recommended readings cover three different types of literature: introductory readings, practice-oriented readings, and current research:

Introductory Readings

Godemann J, Michelsen G (eds) (2011) Sustainability communication: interdisciplinary perspectives and theoretical foundation. Springer, Berlin

→ This seminal edited book develops a theoretical and empirical framework for sustainability communication. It integrates interdisciplinary perspectives from communications theory, psychology, sociology, educational sciences, systems theory and constructivism. Furthermore, it provides methods and concepts in a range of fields, such as corporate practice, education and media.

Moser SC, Dilling S (eds) (2007) Creating a climate for change. Communicating climate change and facilitating social change. Cambridge University Press, Cambridge

→ This book takes a comprehensive look at communication and social change specifically targeted to climate change. The contributors of this book have diverse backgrounds from government and academia to non-governmental and civic sectors of society.

Weingart P, Engels A, Pansegrau P (2000) Risks of communication: discourses on climate change in science, politics, and the mass media. Public Underst Sci 9(3):261–283

→ This article discusses the linkage between science, politics and the media with a focus on Germany. The article shows that there are specific discourse dynamics common to each of the three spheres, as well as some important disparities among them, which leads to different communication characteristics within each subsystem.

Practice-Oriented Reading

→ This practical 'how-to' guide provides a feasible starting point for creating and running effective campaigns. It is authored by Chris Rose, an environmental campaigner with vast experience in the non-governmental field. The book features several case studies and addresses key steps, strategies and tools for overcoming obstacles in communicating sustainability-related issues.

Rose C (2010) How to win campaigns: communications for change, 2nd edn. Earthscan, London/Washington, DC

Current Research

Newig J, Schulz D, Fischer D, Hetze K, Laws N, Lüdecke G, Rieckmann M (2013) Communication regarding sustainability: conceptual perspectives and exploration of societal subsystems. Sustainability 5(7):2976–2990. Available at: http://www.mdpi.com/2071-1050/5/7/2976

→ This recent contribution to the scholarly debate on sustainability communication provides an exploration of sustainability communication in six societal subsystems, based on the distinction between communication *of* and *about* sustainability. It shows how most subsystems undergo a shift from communication *of* towards communication *about* sustainability and discusses implications for future efforts in the field of sustainability communication.

Environmental Communication: a Journal of Nature and Culture; ISSN: 1752–4032 (Print), 1752–4040 (Online)

→ This peer-reviewed scholarly journal relates to the latest developments in the fields of environmental and sustainability related education, communication, social marketing, journalism, behavioral science, risk communication, public relations, health communication, governmental and corporate public awareness, as well as campaigns around the world.

References

Adomßent M, Godemann J (2011) Sustainability communication: an integrative approach. In: Godemann J, Michelsen G (eds) Sustainability communication: interdisciplinary perspectives and theoretical foundation. Springer, Berlin, pp 27–38

Barth M (2012) Social learning instead of educating each other. GAIA 21(2):91–94

Bonfadelli H (2010) Environmental sustainability as challenge for media and journalism. In: Gross M, Heinrichs H (eds) Environmental sociology: European perspectives and interdisciplinary challenges. Springer, Dordrecht/New York, pp 255–278

Brand K-W (2011) Sociological perspectives on sustainability communication. In: Godemann J, Michelsen G (eds) Sustainability communication: interdisciplinary perspectives and theoretical foundations. Springer, Dordrecht/New York, pp 55–68

Egner H (2007) Surprising coincidence or successful scientific communication: how did climate change enter into the current public debate? GAIA 16(4):250–254

Godemann J, Michelsen G (2011) Sustainability communication – an introduction. In: Godemann J, Michelsen G (eds) Sustainability communication: interdisciplinary perspectives and theoretical foundation. Springer, Berlin, pp 3–11

Jackson T, Michaelis L (2003) Policies for sustainable consumption. Sustainable Development Commission, London

Kaufmann-Hayoz R, Brohmann B, Defila R, Di Giulio A, Dunkelberg E, Erdmann L, Weiß J (2012) Societal steering of consumption towards sustainability. In: Defila R, Di Giulio A, Kaufmann-Hayoz R (eds) The nature of sustainable consumption and how to achieve it. Results from the focal topic "From knowledge to action – new paths towards sustainable consumption". Oekom, München, pp 113–142

Leake J, Hastings C (2010) World Misled over Himalayan Glacier Meltdown. The Times, 17 January 2010

Luhmann N (1977) Differentiation of society. Can J Sociol 2(1):29–53

Luhmann N (1989) Ecological communication. University of Chicago Press, Chicago

Moser SC (2010) Communicating climate change: history, challenges, process and future directions. Wiley Interdiscipl Rev 1(1):31–53

Neidhardt F (1993) The public as a communication system. Public Understanding of Science 2(4):339–350

Nerlich B, Koteyko N, Brown B (2010) Theory and language of climate change communication. Wiley Interdiscipl Rev 1(1):97–110

Newig J (2011) Climate change as an element of sustainability communication. In: Godemann J, Michelsen G (eds) Sustainability communication: interdisciplinary perspectives and theoretical foundation. Springer, Berlin, pp 119–128

Newig J, Voss J-P, Monstadt J (2008) Governance for sustainable development: coping with ambivalence, uncertainty and distributed power. Routledge, London

Newig J, Schulz D, Fischer D, Hetze K, Laws N, Lüdecke G, Rieckmann M (2013) Communication regarding sustainability: conceptual perspectives and exploration of societal subsystems. Sustainability 5(7):2976–2990

Rowe G, Frewer LJ (2005) A typology of public engagement mechanisms. Sci Technol Hum Values 30(2):251–290

Vare P, Scott WR (2007) Learning for a change: exploring the relationship between education and sustainable development. J Educ Sustain Dev 1(2):191–198

Ziemann A (2011) Communication theory and sustainability discourse. In: Godemann J, Michelsen G (eds) Sustainability communication: interdisciplinary perspectives and theoretical foundation. Springer, Berlin, pp 89–96

Chapter 13
Sustainability and Science Policy

Uwe Schneidewind, Mandy Singer-Brodowksi, and Karoline Augenstein

Abstract What is the role and importance of science policy for a transformation toward a more sustainable society? In what ways can science policy influence science and innovation systems? More specifically, how can science policy create the institutional conditions needed for developing a sustainability science? Where do we see the strongest impetus for a reorientation of science policy toward sustainable development? These are the guiding questions of the following chapter, which provides an insight into science policy – a policy field that is quite often underestimated yet decisive for sustainable development.

Drivers and incentives for a stronger society orientation in the science system are delineated for the case of the German science system, which serves as an example for many other European science systems.

Keywords Sustainability-oriented science policy • "Mode-2 science" • Innovation policy • Institutional embeddedness of sustainability science • Science system transformation

1 The Importance of Science for a Transformation to Sustainability

Why is science policy important for transformation toward a more sustainable society?

Given the growing environmental burden on a global scale and the overstepping of planetary boundaries (cf. Rockström et al. 2009), humanity in the twenty-first century is facing radical change: it is imperative to guarantee a good life for nine billion people within ecological limits. This goal cannot be reached by continuing today's economic and societal development patterns. Rather, a "great transformation"

U. Schneidewind • M. Singer-Brodowksi (✉) • K. Augenstein
Wuppertal Institute for Climate, Environment and Energy,
Döppersberg 19, 42103 Wuppertal, Germany
e-mail: uwe.schneidewind@wupperinst.org; mandy.singer-brodowski@wupperinst.org; karoline.augenstein@wupperinst.org

H. Heinrichs et al. (eds.), *Sustainability Science*, DOI 10.1007/978-94-017-7242-6_13

(WBGU 2011) of global technological, economic, societal, and cultural developments is needed. This amounts to a highly complex system innovation.

New forms of knowledge will be needed for ecological monitoring and ecological problem analysis, as well as for the development of technological, economic, and social innovations. Therefore, science and the production of knowledge play an important role and will be critical to whether or not the goal of a great transformation can be achieved.

Thus, with the growing importance of knowledge production in the twenty-first century, the field of science policy appears in a new light. Historically, working toward sustainable development has mainly been viewed as a task for environmental policy and, recently, for development of social and economic policy as well – while science policy has not played a vital role. At the moment, this is beginning slowly to change. It is more and more recognized that a sustainability-oriented science policy is at least equally important for the needed transformation processes.

- ***Question***: *Why is science and research becoming so important for sustainability transitions?*

2 Three Perspectives on Science and Sustainability: Being Aware of the Institutional Embeddedness of Science

What kind of science is needed to achieve sustainable development? This question needs to be answered on three different levels (see Fig. 13.1).

1. **The concrete research fields of sustainability science**. What are the topics and dimensions that a sustainability science needs to address? Early on, the so-called Earth sciences played a central role regarding their knowledge of geological, ecological, and meteorological processes, in order to understand the current dynamics of human-induced global environmental change. It soon became clear

Fields of Sustainability Science (=) e.g., Earth Science, Engineering, Economics, Social and Cultural Sciences,...)
Methodology of Sustainability Science (=) Transdisciplinarity)
Institutional Setting of Sustainability Science (=) Science Policy)

Fig. 13.1 Three perspectives on science and sustainability

that, if the objective of science is not only to analyze the ecological state of the global system but also to contribute to the development of sustainability-oriented transformation processes, new fields of science would have to be included: there is a need for technological knowledge and innovations, knowledge on economic processes, as well as social and cultural dynamics. Therefore, sustainability science today is a highly interdisciplinary field (Kates et al. 2001; Clark and Dickson 2003; Jerneck et al. 2011; Wiek et al. 2012, 2015; Miller et al. 2014).

2. **The methodology of sustainability science**. The scientific discourse on sustainability and the role of science has revealed that a sustainability science requires not only a combination of different academic disciplines but also a new mode of knowledge production (Nowotny et al. 2001). Modern societies are shaped by traditional scientific knowledge production and continue to exist based on this type of knowledge. In sociological literatures, this phenomenon and the problems that arise from it are, for instance, discussed in terms of "reflexive modernity" (Beck et al. 1994). New demands and requirements related to the production of knowledge are emerging in this context. A new type of science, which integrates knowledge from different academic disciplines, as well as practical and contextual knowledge of concrete actors, is referred to as "Mode-2 science" by Nowotny and Gibbons (Box 13.1).

Especially with regard to sustainability-oriented transformation processes, this new mode of science plays a decisive role (Wiek et al. 2012). Apart from traditional system knowledge (e.g., about the functioning of ecosystems, technological processes, or societal dynamics), there is a need for target knowledge about desirable futures and transformation knowledge that provides orientation for actors in the respective practical contexts of their activities (Fig. 13.2) (see Chap. 3 in this book).

Usually, actors outside academia are more likely to possess target and transformation knowledge – even if not formalized or generalized – which makes it necessary

Box 13.1: Helga Nowotny

Helga Nowotny (* 9 August 1937 in Vienna) is a sociologist with a focus on the interface of science and society, science and technology studies, and science policy. She published a number of books and journal articles on the topics of scientific controversies and technological risks, social time, coping with uncertainty, self-organization in science, and gender relations in science. Nowotny was Professor at the ETH Zurich and a founding member and president of the European Research Council. She has been and continues to be a member of many international advisory boards and selection committees in the field of science and research policy.

Fig. 13.2 Transition research including different forms of knowledge (Source: http://wupperinst.org/en/our-research/transition-research/)

to cooperate with these actors on an equal footing. The participation of nonacademic actors creates challenges for the design of research processes, but the field has developed a set of robust coping strategies (Bergmann et al. 2012; Lang et al. 2012). Forms of science that manage to integrate the different types of knowledge are referred to as transdisciplinary science.

3. **The institutional setting of science**. Whether the relevant fields of a sustainability science will be dealt with sufficiently and whether new forms and modes of knowledge production can be established depend to a large extent on the institutional framework conditions of the science system (Talwar et al. 2011; Lyall and Fletcher 2013). They determine what kind of research and which research fields are generally eligible for funding, and they also shape the incentive and reputation mechanisms, which provide orientation to scientists with regard to their selection of research questions and methods. There is considerable evidence that the existing institutional framework conditions of most national science systems hamper the development of sustainability science and researchers in this field tend to be marginalized (Jahn et al. 2012: 1). Science policy, therefore, plays a central role, because it has a significant impact on institutional conditions in the science system. In fact, it is the responsibility of science policy to guarantee that knowledge is produced, which helps societies to develop in more sustainable ways (Sarewitz 2009).

- ***Questions***:
  1. *Which kind of interplays can one differentiate between science and sustainability?*
  2. *What are the characteristics of these interplays?*

3 Institutional Reforms and Their Relevance for Supporting Sustainability Science

What research questions are dealt with by scientists? What methods do they apply? What type of research is being funded? All of these issues are decided in the context of established institutional settings: the incentive and reputation mechanisms in the science system, the way that financial resources are allocated, and the system structures by which politics influence the science system.

Currently, many of these institutional framework conditions hamper the development of transdisciplinary sustainability science:

- Incentive structures in the science system are organized within academic disciplines. Career pathways are determined by excellence in theory and methods of a scientist's respective discipline. Over the past decades, an integration of neighboring disciplines can be observed, especially between the natural and engineering sciences (see Simon et al. 2010, p. 9). However, building bridges between the natural and engineering sciences on the one hand and economic, social, and cultural sciences on the other hand was superimposed by the trend toward disciplinary specialization (Weingart 2014: 155 ff.), while interdisciplinary approaches across these fields are important for sustainability science. This is due to a lack of incentives, and scientists working at this interface usually do not have access to an academic career and established funding structures. Many countries have only begun to build long-term interdisciplinary research capacities (for the case of the UK, see Lyall et al. 2013). Particularly for transdisciplinary researchers in the field of sustainability science, this lack of incentives is more challenging and the institutional answers to it are quite at the beginning (see Yarime et al. 2012).
- There is a strong technological bias in private as well as public research funding. Technological solutions are an important element on the way toward sustainable development – if they are embedded in economic, social, and cultural developments in suitable ways. The focus on technologies of many research funding programs can be explained by the fact that direct economic opportunities can be expected from technological R&D projects. Over the last decade, research funding structures have largely served the development of technological innovations, which is in line with scientific findings that national economies can gain a competitive advantage by investing in innovations (see Fealing et al. 2011; Martin 2012; Knie and Simon 2010).

- ***Question***: *What are the main reasons for the importance of institutional reforms in the science system?*

4 The Nature and Impact of Science Policy

How can science policy influence the science and innovation system? The preceding section has shown that there are different institutional elements with an impact on scientific knowledge production, which may foster or hamper the development of transdisciplinary sustainability science. Not all of these institutional elements can be controlled and directed by the political process, e.g., reputation systems and relevance criteria in individual scientific communities. The field of science studies shows that epistemic communities in academia usually remain within disciplinary boundaries and evade nonacademic steering processes and thus also societal expectations (e.g., with regard to the development of transdisciplinary sustainability science) (see Gläser and Lange 2007, p. 441). Principles and routines of academic autonomy are a central element of the science system, which partly subverts or counteracts political steering efforts and which therefore needs to be balanced within new forms of science system governance (see Knie and Simon 2010, p. 36).

Nonetheless, science policy can exert influence on the science system in various ways, especially in national science systems that are mainly publicly funded – which is the case for most European countries and overall EU research funding.

It is thus worthwhile to take a closer look at science policy and the concrete policy instruments in this field. First, it can be observed that, over the past decades, science policy has increasingly been discussed together with innovation policy and that today, science and innovation policy have emerged as a common policy field: science policy and innovation studies (see Martin 2012, p. 1220).

Over the past 20 years, science policy has focused on the introduction of new steering instruments for scientific institutions and on the role of new actors at the interface of science, economy, politics, and society, the so-called intermediaries (e.g., policy consultancies).

Overall, an increase can be observed in third-party funding and, at the same time, a strengthening of academic autonomy through new steering instruments, especially at the level of governing boards of universities. This has led to universities becoming more "responsive" (cf. Jansen 2010, p. 47), i.e., they are better able to react quickly to external demands (e.g., developing in more market- and application-oriented ways).

Science policy could be utilized in such a situation by setting external incentives. This can be done in a number of ways (see also the following section for a more detailed discussion).

- Policy shapes *fundamental political paradigms* that provide orientation to the science system (e.g., "science as a driver for strengthening competitiveness," "academic autonomy," etc.). These paradigms have an impact on the activities and the topical focus of scientists and research institutions.

- *Funding policy* is a central starting point for political steering efforts. Through the allocation of financial means to specific research programs and institutions, the overall topical and methodological focus can be influenced.
- Established scientific institutions can be influenced by new *steering mechanisms*: e.g., indicator-based steering, target agreements, appointing advisory boards, or steering committees.

These science policy instruments range from the European level, across national policies, to the level of entities below the nation state level. The role and importance of the different levels vary according to the respective national science system structure. In Europe, at the turn of the millennium, the "Lisbon Strategy" has been of key importance. The EU has committed itself to the goal of becoming the most competitive knowledge-based economy and, to that end, to invest 3 % of annual GDP in R&D funding. As a result, steadily increasing budgets are available at the EU level for the so-called Research Framework Programmes. They are the central science political steering instrument at the European level. The 8th Research Framework Programme (2014–2020, "Horizon 2020") is explicitly addressing the grand societal challenges and, to some degree at least, seems to move away from a purely economic focus on increasing competitiveness.

At the national level and below, apart from program funding, instruments of institutional funding are available as well. These can be used by political actors to exert direct influence on the capacity of specific research areas. Furthermore, there are indicator-based incentive instruments, e.g., performance-based indicators, which can be used to measure a research institution's output (e.g., in terms of number of graduates or publications) and allocate funds accordingly.

With regard to all of these science policy instruments, an orientation along key societal challenges and issues of a more sustainability-oriented science plays only a minor role. A fundamental reorientation is needed.

- ***Question***: *What are instruments of science policy and how can they influence the science system?*

5 Science Policy for Sustainability Transitions

What would be the type of new science policy that could create better institutional framework conditions for a sustainability science?

As a first step, the *guiding visions of science* policy would have to change.

At the global level, efforts in this direction are made in the context of the large international science organizations' joint research initiative "Future Earth" (http://www.icsu.org/future-earth/). This research program places significant value on interdisciplinary approaches for dealing with global sustainability challenges and cooperation with nonacademic actors. The "coproduction" of knowledge and "codesign" of research projects are called for and supported by the Future Earth program. It will be a major challenge to implement these new guiding principles in national research funding programs.

Similarly, an orientation along grand societal challenges also characterizes the EU's 8th Research Framework Programme (Horizon 2020[1]). It thus goes beyond the strategy of the 7th Research Framework Programme, which centered on increasing the EU's competitiveness.

The new paradigm also has an impact on *research program politics*. The Horizon 2020 program will, for instance, cover a 7-year period and includes an expenditure of almost 80 billion euros. In order to preclude that the "grand societal challenges" are primarily defined from an economic perspective, it will be vital that civil society organizations have an opportunity to participate in the development of the program's details and concrete structure. Over the past years, there have been various initiatives at the European level that have aimed at a more profound involvement of civil society organizations in specific fields of research. An actual science policy instrument in this regard can be civil society research funds,[2] i.e., research funds that can be shaped to a significant extent by civil society stakeholders. Due to a relatively small volume and high barriers posed by the application process, these new approaches to research funding remained ineffective in the 7th Research Framework Programme.

Funding for transdisciplinary research – at global, European, or national levels and below – proves to be useful for fostering sustainability science when it includes *structural incentives*, e.g., aiming at the establishment of interdisciplinary structures at universities, new research institutions, and career opportunities in the field of inter- and transdisciplinary sustainability science. Internationally, a lot of initiatives can be identified that strengthen transdisciplinary approaches for sustainability science, i.e. the EPSCoR (Experimental Program to Stimulate Competitive Research) funding program of the National Science Foundation in the USA, which aims among other funding strategies especially at building interdisciplinary Sustainability Research Networks (SRN).

Another good example for devising incentives is the Asia-Pacific ProSPER.Net, an alliance of leading universities that encourage each other to integrate sustainability into courses and curricula. Within the network there is also a focus on developing indicators of a sustainable university to enable the measurement of concrete progress at the institutional level (see Fadeeva and Mochizuki 2010).

Eventually, *traditional steering instruments* should also be used to foster a reorientation along sustainability goals. Examples could be the integration of sustainability-related aspects into target agreements within universities or the definition of sustainability-oriented performance indicators. Some of the German "Länder" are currently experimenting with these kinds of instruments.

- ***Task***: *Please mention a few developments on European, national, and regional levels of the science system that support the orientation toward more sustainability.*

---

[1] http://ec.europa.eu/research/horizon2020/index_en.cfm

[2] http://ec.europa.eu/research/science-society/index.cfm?fuseaction=public.topic&id=1298

6 Drivers of Progressive Science Policies

Where do we see the most powerful impetus for sustainability-oriented change in science policy coming from?

Science policy is only slowly beginning to adapt to the demands of sustainability science. Therefore, it is interesting to see where the needed impulses for change could most likely originate. In Germany, the issue of sustainability is high on political and societal agendas (i.e., the so-called Energiewende), and this also influences the science policies. The German government has proclaimed that it will invest around 500mio€ per year in research programs for sustainability during the legislative period 2013–2017. This amount has increased continuously over the last few years and is an effort to translate the European demands for a science oriented toward the "grand challenges." Furthermore, the German Ministry of Education and Research has launched an overall initiative, "Sustainability in Science," to strengthen the research communities' own capacity for reorientation toward sustainability research, education, and management. The German case, therefore, can be an interesting case for identifying the key drivers of reorientation toward sustainability in science policy and the science system in general (cf. Schneidewind and Augenstein 2012; Schneidewind and Singer-Brodowski 2013).

Key drivers are civil society organizations, students, scientific foundations, and pioneer initiatives by individual "Länder" or research institutions, which utilize their autonomy in order to improve conditions for sustainability science:

One of the most important pressure groups calling for a change in science policy is made up of **civil society organizations**. For instance, a large number of German environmental and development organizations, churches, and labor unions founded a platform called "Forschungswende" in 2012. In May 2013, they published ten core requirements for a future science and research policy. The first requirement is more participation by civil society in science, for instance, by active involvement in the formulation of research questions and programs and participation in committees or boards of publicly financed institutions. These claims were also integrated in the German coalition agreement of 2013.

Civil society initiatives for sustainable development are also carried out by **students**. They can be important catalysts for change in universities, because they are not bound to institutional structures and routines. In contrast, students' creativity and openness can create a culture of change within universities. An outstanding example in this respect can be found in the UK. A study on attitudes of freshmen students and the impact of sustainability criteria on their choice of university (cf. Drayson et al. 2012), as well as a university ranking initiated by students, the "People & Planet Green League," has attracted substantial media attention and created considerable pressure on UK universities to improve their sustainability performance.

Foundations can be important actors supporting innovative orientations of universities and other research institutions. They can fund risky pilot projects and, by this, contribute to new forms of knowledge production and diversity in the science system. This can also create momentum for change in science policy. Examples are the Stockholm Resilience Centre, which was founded by the Swedish Mistra Foundation, or the Mercator Research Institute on Global Commons and Climate Change (MCC), founded by the Mercator Foundation together with the Potsdam Institute for Climate Impact Research and the Technical University of Berlin.

In many countries, **entities below the national level**, e.g., the German "Länder," are responsible for universities and science policy. These entities can thus become important pioneers for a more sustainability-oriented science policy by utilizing available steering instruments, in order to achieve a paradigmatic and programmatic reorientation of their science policy. From 2011 to 2013, some of the larger German "Länder," such as North Rhine-Westphalia, Baden-Württemberg, and Lower Saxony, have made such efforts.

Finally, innovative **sustainability research institutes and pioneer universities** are important actors for a reorientation of science policy. They demonstrate potential and opportunities, which can be strengthened by politics, and thus provide important starting points for change. Due to its traditional heritage of environmental and sustainability policies going back to the 1970s and 1980s, Germany has a strong network of such pioneering institutions – ranging from independent sustainability research institutes (e.g., the Öko-Institut (Institute for Applied Ecology), the Institute for Ecological Economy Research (IÖW), etc.) to universities with a focus on sustainability issues (e.g., the universities of Lüneburg, Kassel, and Oldenburg). Over the past years, these institutions have increasingly cooperated in networks, in order to strengthen their pioneering role and impact on political agenda-setting processes.

Although good initiatives and drivers of a transformation in existing science policy strategies toward fostering sustainability science were illustrated mainly by the example of the German science system, at the international level many interesting endeavors can be observed as well. At the institutional level, the process of redesigning Arizona State University is worth mentioning (see Crow and Dabars 2014). Last but not least, the successful networks of civil society (i.e., the AASHE – Association for the Advancement of Sustainability in Higher Education) are important examples for driving the sustainability agenda forward in the science system as a whole.

- ***Question***: *Who are the key drivers of sustainability-oriented science policies in Germany and how do they take effect?*

7 Conclusion

Science policy can be a decisive driver for sustainable development in modern knowledge societies. However, in its current form, today's science policy is barely oriented toward this end. Thus far, only early attempts and experiments can be observed in this respect, some of which have been discussed in this contribution. It is therefore important that science policy in general is discovered and further developed as an important policy field with regard to the goals of sustainable development.

Further Reading

Fealing KH, Lane JI, Marburger JH, Shipp SS (eds) (2011) The science of science policy: a handbook. Stanford Business Books, Stanford

Jahn T, Bergmann M, Keil F (2012) Transdisciplinarity: between mainstreaming and marginalization. Ecol Econ 79:1–10. http://dx.doi.org/10.1016/j.ecolecon.2012.04.017

Lang DJ, Wiek A,Bermann M,Stauffacher M, Martens P, Moll P, Swilling M, Thomas Christopher J (2012) Transdisciplinary research in sustainability science: practice, principles, and challenges. Sustain Sci 7(1):25–43. http://dx.doi.org/10.1007/s11625-011-0149-x

Martin BR (2012) The evolution of science policy and innovation studies. Res Policy 41(7):1219–1239. http://dx.doi.org/10.1016/j.respol.2012.03.012

Schneidewind U, Augenstein K (2012) Analyzing a transition to a sustainability-oriented science system in Germany. Environ Innov Soc Trans 3(2012):16–28

References

Beck U, Giddens A, Lash S (1994) Reflexive modernization: politics, tradition and aesthetics in the modern social order. Polity Press, Cambridge

Bergmann M, Jahn T, Knobloch T, Krohn W, Pohl C, Schramm E (2012) Methods for transdisciplinary research. A primer for practice. Campus, Frankfurt/New York

Clark WC, Dickson NM (2003) Sustainability science: the emerging research programm. Proc Nat Acad Sci U S A 100(14):8059–8061. http://dx.doi.org/10.1073/pnas.1231333100

Crow MM, Dabars WB (2014) Towards interdisciplinarity by design in the American Research University. In: Weingart P, Padberg B (eds) University experiments in interdisciplinarity. Obstacles and opportunities. Transcript, Bielefeld, pp 13–36

Drayson R, Bone E, Agombar J (2012) Student attitudes towards and skills for sustainable development. A report for the Higher Education Academy. Download: http://www.heacademy.ac.uk/resources/detail/sustainability/esd_First_Year_Students_report

Fadeeva Z, Mochizuki Y (2010) Higher education for today and tomorrow: university appraisal for diversity, innovation and change towards sustainable development. Sustain Sci 5(2):249–256

Gläser J, Lange S (2007) Wissenschaft. In: Benz A, Lütz S, Schimank U, Simonis G (eds) Handbuch governance: theoretische Grundlagen und empirische Anwendungsfelder. VS, Wiesbaden, pp 237–251

Jansen D (2010) Von der Steuerung zur Governance: Wandel der Staatlichkeit? In: Simon D, Knie A, Hornbostel S (eds) Handbuch Wissenschaftspolitik. VS, Wiesbaden, pp 39–50

Jerneck A, Olsson L, Ness B et al (2011) Structuring sustainability science. Sustain Sci 6:69–82

Kates RW et al (2001) Environment and development – sustainability science. Science 292(5517):641–642. http://dx.doi.org/10.1126/science.1059386

Knie A, Simon D (2010) Stabilität und Wandel des deutschen Wissenschaftssystems. In: Simon D, Knie A, Hornbostel S (eds) Handbuch Wissenschaftspolitik. VS, Wiesbaden, pp 26–38

Lang DJ, Wiek A, Bergmann M, Stauffacher M, Martens P, Moll P, Swilling M, Thomas C (2012) Transdisciplinary research in sustainability science – practice, principles and challenges. Sustain Sci 7(Supplement 1):25–43

Lyall C, Fletcher I (2013) Experiments in interdisciplinary capacity-building: the successes and challenges of large-scale interdisciplinary investments. Sci Public Policy 40:1–7. http://dx.doi.org/10.1093/scipol/scs113

Lyall C, Bruce A, Marsden W, Meagher L (2013) The role of funding agencies in creating interdisciplinary knowledge. Sci Public Policy 40:62–71. http://dx.doi.org/10.1093/scipol/scs121

Martin BR (2012) The evolution of science policy and innovation studies. Res Policy 41(7):1219–1239. http://dx.doi.org/10.1016/j.respol.2012.03.012

Miller TR, Wiek A, Sarewitz D, Robinson J, Olsson L, Kriebel D, Loorbach D (2014) The future of sustainability science: a solutions-oriented research agenda. Sustain Sci 9(2):239–246

Nowotny H, Scott P, Gibbons M (2001) Re-thinking science. Knowledge in the public in an age of uncertainty. Polity Press, Cambridge

Rockström J et al (2009) A safe operating space for humanity. Nature (461):461–472. http://dx.doi.org/10.1038/461472a

Sarewitz D (2009) A tale of two sciences. Nature 462:566

Schneidewind U, Augenstein K (2012) Analyzing a transition to a sustainability-oriented science system in Germany. Environ Innov Soc Transit 3:16–28. http://dx.doi.org/10.1016/j.eist.2012.04.004

Schneidewind U, Singer-Brodowski M (2013) Transformative Wissenschaft. Klimawandel im deutschen Wissenschafts- und Hochschulsystem. Metropolis, Marburg

Simon D, Knie A, Hornbostel S (2010) Einleitung. In: Simon D, Knie A, Hornbostel S (eds) Handbuch Wissenschaftspolitik. VS, Wiesbaden, pp 9–10

Talwar S, Wiek A, Robinson J (2011) User engagement in sustainability research. Sci Public Policy 38(5):379–390

WBGU (German Advisory Council on Global Change) (2011) (eds) World in transition. Social contract for sustainability. Flagship report: Berlin, Download: http://www.wbgu.de/en/flagship-reports/fr-2011-a-social-contract/

Weingart P (2014) Interdisciplinarity and the New Governance of Universities. In: Weingart P, Padberg B (eds) University experiments in interdisciplinarity. Obstacles and opportunities. Transcript, Bielefeld, pp 151–174

Wiek A, Ness B, Brand FS, Schweizer-Ries P, Farioli F (2012) From complex systems analysis to transformational change: a comparative appraisal of sustainability science projects. Sustain Sci 7(Supplement 1):5–24

Wiek A, Harlow J, Melnick R, van der Leeuw S, Fukushi K, TakeuchiK, Farioli F, Yamba F, Blake A, Geiger C, Kutter R (2015) Sustainability science in action – a review of the state of the field through case studies on disaster re-covery, bioenergy, and precautionary purchasing. Sustain Sci 10(1):17–31

Yarime M, Trencher G, Mino T, Scholz RW, Olsson L, Ness B, Frantzeskaki N, Rotmans J (2012) Establishing sustainability science in higher education institutions: towards an integration of academic development, institutionalization, and stakeholder collaborations. Sustain Sci 7(Supplement 1):101–113. doi:10.1007/s11625-012-0157-5

Chapter 14
Justice and Sustainability

Sonja Klinsky and Aaron Golub

Abstract This chapter explores the relationships between justice and sustainability theory and argues that despite being entwined, practical and theoretical challenges prevent an easy or complete integration of these two concepts. Specifically, we examine how the multiscalar, multigenerational, and multidimensional characteristics of sustainability interact with ideas of just processes or just outcomes. Using insights from justice theory, sustainability science, and the social psychology of justice, we suggest we ask questions like: what would a just multiscalar and multigenerational sustainability process look like? If social context changes how people use or understand ideas of justice, what should justice look like in complex sustainability challenges that extend across traditional boundaries? We suggest that although these issues present persistent theoretical challenges, past and ongoing efforts – such as environmental justice work or international climate negotiations – provide some lessons and guidance about strategies for assisting this integration in practice. Overall this chapter suggests that although fully integrating justice and sustainability may not be entirely possible, sustainability without a consideration of justice would be nonsensical from a normative perspective and difficult to achieve strategically. This leaves us with a strong rationale to insist on the centrality of justice in any sustainability effort, but with an awareness of the complexities of doing so.

Keywords Justice • Sustainability • Procedural justice • Just outcomes • Social psychology of justice

---

S. Klinsky (✉)
School of Sustainability, Arizona State University,
PO Box 875402, Tempe, AZ 85287-5302, USA
e-mail: Sonja.klinsky@asu.edu

A. Golub
Toulan School of Urban Studies and Planning, Portland State University,
Portland, OR, USA
e-mail: agolub@pdx.edu

H. Heinrichs et al. (eds.), *Sustainability Science*, DOI 10.1007/978-94-017-7242-6_14

This chapter focuses on the integration of justice within sustainability theory and practice. It will be argued that the multiscalar, multigenerational, and multidimensional characteristics of sustainability problems pose profound theoretical and practical challenges for the integration of justice into sustainability thought and practice which have yet to be fully addressed. However, past and ongoing efforts to work through these challenges in sustainability and related approaches suggest that there are ways to assist this integration in practice.

1 State of the Art

Justice and sustainability have a long history of integration. Considerations of justice appear in many aspects of the sustainability problem and solution formulation. Recognition of the importance of considering future generations – often framed as an expression of intergenerational justice – is core to many definitions of sustainability. Concerns for justice are easily seen in formational works, such as in the Brundtland Report, in which sustainable development is predicated upon the ability for future generations to meet their own needs (Brundtland 1987). Intragenerational forms of justice, including fair processes and questions of social equity, are also widely recognized. For instance, in his widely cited principles for sustainability assessment, Gibson (2006) includes standards for intragenerational equity, intergenerational equity, and democratic governance, all of which touch on central questions of justice. Many others have explored the overlaps between justice and sustainability in their practice in specific communities (e.g., Aygemen et al. 2003). There is also evidence that striving for justice improves outcomes along other significant sustainability dimensions, such as environmental conservation and economic performance (Aygeman et al. 2003).

Sustainability has emerged as a normative frame used to define an improved decision-making process yielding improved long-term outcomes (Gibson 2006; Wiek et al. 2012; Miller et al. 2014). From this normative perspective, justice is inherently integrated into sustainability: by definition, an action or approach deemed "sustainable" must address justice. As some have pointed out, this integration may not be as smooth as it first appears (Beckerman 1999). In this short chapter, we explore the theoretical integration of justice and sustainability and argue that if we are genuinely attempting to create sustainable and just societies, then we need to think carefully about how these concepts intersect. We start with a brief discussion of the challenges of defining justice before we turn to the integration of sustainability and justice.

1.1 Defining Justice

Western efforts to answer the core justice question "what is owed to whom" go back thousands of years – to the very origins of society itself. Efforts notwithstanding, debates remain unresolved on many issues of justice in both theory and practice. Justice is "slippery" because it is profoundly social. As Walzer argued, "there cannot be a just society until there is a society" (Walzer 1983): it is impossible to define justice independently from its social context. As understandings of social life are neither static nor universal, isolating single and comprehensive understandings of justice is difficult, if not impossible (Miller 2002; Walzer 1983). There are some general principles, however, to guide sustainability practice, which do reflect core ideas of justice. In this brief review, we focus on three core understandings of justice particularly relevant to sustainability: just processes, just outcomes, and the social psychology of justice.

1.1.1 Just Processes

Processes refer to the way decisions are made in society: who participates and who is excluded and how robust those processes are for creating meaningful participation and representation. Processes are diverse, from the simple right to vote, to more complex issues like having a voice in policy decisions or cultural representation.

In some schools of justice thought, especially Libertarianism, a just process is the primary prerequisite for justice. That is, a just process will always yield a just outcome. In the words of justice scholar Nozick, "whatever arises from a just situation by just steps is itself just" (1974, 151). This means that inequalities and undesirable outcomes are just, so long as a fair process produced them. Though the process requirement seems weak, it is arguably the foundational driving philosophy for most modern western democracies. The Declaration of Independence and the Constitution of the United States emphasize the roles of rights, freedoms, and responsibilities and do not enter into outcome requirements at all.

For others, process is more complicated, although no less central. For instance, Fraser places representation at the heart of justice. She argues that justice demands that all people have the opportunity to participate as "equals in social life," the penultimate form of which is representation in decision-making, but also acknowledges that threats to this form of justice can come from other injustices, including a lack of sufficient material wealth and inappropriate cultural recognition (Fraser 2009). Despite the differences in these perspectives, they share two common ideas. First, if it is accepted that all people have equal moral worth, then all have a right to be included in decisions that affect their well-being. This is tied to a second proposition that people are best equipped to identify and represent their own best interests.

These two common ideas about just process resonate strongly with both normative and strategic elements of sustainability. From a normative perspective, strong arguments have been made that, by definition, sustainability must include processes of involvement for those directly involved, or, when this is not possible (such as for future generations), clear representation by designated people (Gibson 2006; Dobson 1999). Strategically, it has been argued that ordinary people have values and knowledge that are essential, highlighting the idea that many of the core requirements for sustainability decision-making and implementation are beyond the purview of elites, experts, or elected representatives (Fischer 2000). This places the onus on sustainability initiatives to feature processes that maximize wide-scale involvement and which therefore improve process justice.

1.1.2 Just Outcomes

Outcomes refer to the specific differences and distributions over a population in the attainment of some material measure, such as wealth, income, education, tax burden, etc. In this view, the fairness of a process is only as important as the fairness of the outcomes it creates. Just outcomes can be defined in various ways, ranging from "equality" to more broad understandings of "equity." Equality refers to a strict requirement for equal distributions of some goods, while equity adapts to specific circumstances, conditions, histories, needs, etc. Generally, an appeal to equity refers to focusing resources toward those not already better off. Equity, in the words of urban planning scholar Susan Fainstein, "refers to a distribution of both material and nonmaterial benefits derived from public policy that does not favor those who are already better off at the beginning" (2010, pp. 35–36).

Even among scholars and activists who focus on outcomes, there are a wide range of views of what are just distributions of outcomes which deviate from pure equality. For instance, "sufficientists" feel that just distributions of goods are those that provide the minimum necessary for a productive life for everyone – having more than necessary is not a problem, so long as others have enough. Another approach to distributive justice is that of "communitarians" which starts from the conceptualization of society as a community in which people produce a variety of goods that differ in terms of the social meaning members of society attach to them. Given these differing meanings, there can be no single and just criterion by which all goods are to be distributed. Rather, each good should be distributed in a way corresponding to the social meaning of that good. According to Walzer, regular (relatively unimportant) goods can be distributed through the free market, where distribution is determined by individuals' ability and willingness to pay. In contrast, goods to which a particular society ascribes a distinct social meaning "deserve" their own distributive sphere. Distributing such "special" goods would require a distributive principle different from market exchange, ranging, for example, from equality to distribution based on need (Trappenburg 2000). For example, many societies consider access to medical care too important to leave to the free market and provide care for everyone free of charge.

Ideas about just outcomes resonate clearly with various normative elements of sustainability. Strong arguments have been made that sustainability practice must strive for just distributions of outcomes and social equity (Gibson 2006; Dobson 1999).

1.1.3 Social Psychology of Justice

Philosophical debates about justice provide arguments about what we should do or how society might best accommodate the claims of all of its members, but it provides little insight into how people actually think about justice. In contrast, the social psychology of justice focuses explicitly on what people think about justice and how this contributes to their actions in 'real-world' justice dilemmas.

Two fundamental messages emerge from this research tradition. First, perceptions about justice (process and outcome) are shaped by moral boundaries. In order to consider another's interests in a justice dilemma, the "other" must be seen to have moral standing (Mikula and Wenzel 2000). People use a variety of arguments to draw moral boundaries between those whose claims count and those whose interests are deemed invisible. Second, since justice depends on our relationships and the types of obligations we recognize within these relationships, it is quite common for people to consider different notions of justice as appropriate in different social contexts (Fiske and Tetlock 1997; Wenzel 2004). For instance, the same person might simultaneously use multiple, different arguments about justice when negotiating relationships with her children, her co-workers, and her fellow corporate shareholders (Deutsch 1975).

As will be seen below, this fluidity of justice perceptions raises a range of complications for sustainability and is essential to bear in mind as a feature of both the normative and strategic aspects of sustainability.

1.2 Challenges of Integrating Sustainability and Justice

Justice may be deeply embedded in our notions of sustainability, but there are also theoretical and practical challenges to this integration. In this section, we highlight the multiscalar, multidimensional, and multigenerational nature of sustainability problems and the challenges they pose for justice.

1.2.1 Multiscalar

A classic characteristic of sustainability problems is their tendency to be multiscalar (Sassen and Dotan 2011; Wilbanks 2002). Accordingly, much of the sustainability literature stresses the importance of integrative, multilevel approaches for understanding and addressing sustainability problems (Geels 2010; Smith et al. 2010). Unfortunately, the multiscalar nature of sustainability problems causes profound challenges for the integration of justice and sustainability theory.

First, multiscalar problems raise questions about just process. If a problem necessarily involves people (and nonhumans) at many different scales, how is representation handled fairly? To date, systems addressing multiscalar issues feature jurisdictional divisions of authority between regional, national, or global level authorities and are not designed to integrate representation from multiple scales. What would a just multiscalar decision process look like and how would it be conducted?

Second, the multiscalar nature of sustainability problems can cause justice dilemmas between scales. It is entirely possible, as has been seen with debates about renewable energy siting (Wolsink 2007) or natural gas development (Lindseth 2006), that local and global evaluations of justice and sustainability can conflict.

What may appear just within a global scale may not appear just to those involved most locally and vice versa. At what scale is an outcome to be evaluated from a justice perspective? How should trade-offs between notions or evaluations of justice at different scales be evaluated?

Third, and underlying both previous complications, as a problem exists at multiple scales, it may invoke different understandings of social relationships and thus justice obligations. For instance, in one empirical investigation, people held multiple, occasionally internally inconsistent, ideas about the contours of just allocation of climate change mitigation efforts based on the different relationships they shared with people across social and physical distances (Klinsky et al. 2012). Which ideas of justice should be used to evaluate a multiscalar sustainability issue if the rules of justice considered appropriate shift by social context and understandings of the social context itself vary as the issue is framed by location or scale?

1.2.2 Multidimensional

Recognition of the multidimensional nature of sustainability, and the need to assess multiple aspects of a given problem simultaneously, is another central concept within sustainability thought. At a minimum, sustainability forces us to consider the economic, social, and environmental impacts for any given action or practice. From a social psychological perspective, this multidimensionality deeply complicates justice. If people identify with different dimensions of the same overarching issue, they are likely to hold different notions about what the "just" process or outcome would look like.

The conflicts that emerge from the multidimensionality of sustainability problems are commonly observed in the literature. For instance, questions over resource use often include: is a particular forest, mountain, desert, river, or valley a wellspring of cultural value or spirituality; a place with an inherent value or right to exist; a natural resource to be managed for future consumption; or a crucial link in the economic vitality of a region in the present? While debates about value differences in multidimensional problems are widely acknowledged (Fischhoff 1991; Gregory et al. 1993), the theoretical challenge multidimensionality poses to the integration of justice and sustainability is less commonly addressed. Each of these dimensions may stimulate the use of different justice frameworks. If justice is supposed to arbitrate the relationships of those across a system, how should it deal with sub-systems or dimensions that pull on multiple human relationships?

1.2.3 Cross-Generational

Building on a central concern of long-term health for future human generations and other species and ecological systems, intergenerational equity is a central concept in sustainability thought. Intergenerational equity is commonly associated with *Our Common Future's* definition of sustainable development as, "development that meets the needs of the present without compromising the ability of future generations to meet their own needs" (Brundtland 1987, p. 19). In the literature, various forms of intergenerational equity appear in nearly all conceptions of sustainability and sustainability science (Bebbington 2000; Kates et al. 2005; Gibson 2006; Jurneck et al. 2011). With Brundtland-inspired and future-focused language, Robert Gibson's oft-cited criteria for sustainability assessments include this passage on intergenerational equity: "Favour present options and actions that are most likely to preserve or enhance the opportunities and capabilities of future generations to live sustainably" (2006, p. 174). Golub et al. (2013) argue that intergenerational concerns would also implicate a concern for injustice enduring from past practices which have not been addressed properly. In particular communities, these past injustices, they argue, are significant and would hamper efforts to move forward toward a more sustainable future.

There are significant challenges in addressing intergenerational justice in practice. Intergenerational equity presents significant methodological challenges, for example, for economic discount rate modeling (Asheim and Mitra 2010; Endress et al. 2009). "Equity" between generations might imply a reduction of the welfare of current or future generations in order to balance opportunity with prior generations, a proposition Beckerman (1999) questions. Further, it's impossible to know in advance the preferences of future generations, meaning the exact configurations of what is to be sustained remain contested between different conceptions of the future and the good (Yabareen 2008).

1.3 Solution Options: Lessons on Integration from Environmental Justice and Climate Change Action

While there are intuitive connections between justice and sustainability – and good reasons for placing justice at the heart of sustainability efforts – fully integrating the concepts faces a range of theoretical and practical challenges. Despite these difficulties, several areas of practice have invested efforts toward integration. Two such areas, environmental justice and climate change action, are presented here.

1.3.1 Environmental Justice

The environmental justice (EJ) movement was born in response to environmental and spatial injustices resulting from both unjust processes and unjust outcomes (see inset). It was increasingly recognized that race and class are strongly linked to not only environmental quality but also the strength of environmental regulations, permitting, and site selection (Aygeman et al. 2003). Even where rules were in place to prevent unequal burdens, there were often failures of reporting and enforcement in low-income and minority communities.

Successful struggles by the Civil Rights Movement of the 1960s and later the EJ movement eventually led to the creation of legal protections for civil rights and environmental justice and the extension of the obligations of public agencies to both the inclusiveness of the process itself and the fairness of its substantive outcomes, including environmental issues (Bryner 2002; Lee 1997). These protections arise under the National Environmental Policy Act of 1969 (NEPA), Title VI of the Civil Rights Act of 1964, President Clinton's Executive Order 12898, and subsequent implementing orders from federal agencies. Title VI of the Civil Rights Act provides that: "No person in the United States shall, on the ground of race, color, or national origin, be excluded from participation in, be denied the benefits of, or be subjected to discrimination under any program or activity receiving Federal financial assistance" (42 U.S.C. § 2000d, emphasis added). The prohibition on exclusion extends not only to the substantive benefits that federally funded state and local agencies provide through their programs but also to the inclusiveness of participation in the decision-making process itself. These legal protections have been used to address environmental injustices in a variety of contexts (Bryner 2002).

1.3.2 Climate Change Action

The justice dimensions of climate change negotiations are well recognized (Gardiner et al. 2010) and epitomize many challenges of integrating justice and sustainability. A vast literature has explored the implications of different justice arguments for dividing global mitigation requirements. Some scholars and international negotiators have stressed the importance of per capita allowances of emissions (Baer et al. 2009); others have debated the relevance of historical emissions for determining future commitments (Winkler et al. 2011); still others have focused on the questions of need and access to energy for development (Ott et al. 2004; Winkler et al. 2011). Simultaneously, other scholars have discussed the most appropriate way to evaluate policies across generations (Cole 2008; Howarth 2003) and the definition of "dangerous" climate change. Each of these debates about justice can be tied to different ideas about the relationships – and the obligations that stem from these relationships – among people across both space and time.

Several important lessons about the integration of justice and sustainability have emerged from this complexity. First, justice perceptions are pragmatically central for resolving the collective action dilemma embedded in the challenge of climate change mitigation: no nation is likely to agree voluntarily to actions that it thinks are fundamentally unfair to its interests. Second, no single definition of justice is likely to meet every nation's perception of fair treatment. Instead, current negotiations are focused on assembling packages of actions that are "fair enough" to accommodate a broad range of perceptions.

Defined from this single perspective, justice is likely impossible due to the multiple dimensions in play, and yet perceptions of fairness are essential for resolving the collective action dispute at the heart of negotiations. These observations suggest that this "fuzzier' framework, which acknowledges both the diversity of contexts and the social psychological aspects of justice, may offer scope for the integration of justice and sustainability theory in other contexts as well.

2 Contributions of Justice to Sustainability

This chapter has suggested that, although deeply entwined, fully integrating the concepts of justice and sustainability is difficult for both theoretical and practical reasons. This difficulty does not, however, mean that concepts of justice should be ignored or excluded from sustainability efforts. We suggest that a focus on justice is essential for sustainability even if it is impossible to achieve in multidimensional, multigenerational, and multiscalar contexts. This argument comes from an appreciation of both the normative and strategic elements of sustainability.

From a normative perspective, justice has long been central to sustainability, as has been reflected in common statements about the principles and goals of sustainability (Brundtland 1987; Kates et al. 2005; Gibson 2006; Jerneck et al. 2011). The centrality of justice to the normative claims of sustainability is highlighted by the difficulties one would have in imagining and describing an ideal form of sustainability that actively ignored fairness. For instance, intergenerational justice in the form of efforts to maintain social and ecological integrity for future generations is a central component of the underlying rationale for sustainability. Similarly, if sustainability is, to some extent, an expression of our visions of a "good life" now and in the future, excluding justice from this definition raises pointed questions. Can a "good life" be envisioned, created, and sustained without attention to intragenerational justice? Even if justice is not absolutely possible, recognizing its centrality to sustainability encourages a deeper reflection on the status quo and provides a reasoned set of concepts for suggesting alternative ways of changing the world.

The centrality of justice to sustainability thinking is also apparent from a strategic perspective. If sustainability challenges are at least partially composed of collective action, problems then addressing justice can be essential from a strategic perspective. No one is going to contribute voluntarily to an effort that is seen as a burden if the premise does not have some elements of fairness: in other words, some level of fairness is a requirement for action. Similarly, as seen in the example of environmental justice, taking justice seriously draws attention to stakeholders who may have been systematically excluded from prior consideration and may lead to forms of sustainability that are more genuinely able to integrate social, economic, and environmental components.

The challenge identified in this chapter is that justice is both impossible and essential for sustainability. Due to the theoretical and practical complexities of sustainability and justice across time, space, and dimensions, it is extremely difficult – likely impossible – to fully integrate the two concepts. However, sustainability without a consideration of justice would be nonsensical from a normative perspective and difficult to achieve strategically. We suggest that this leaves us with a strong rationale for including justice as a central component of any sustainability effort, but for doing so in full recognition of the need for careful definition, thought, and attention to the complexities of integrating sustainability and justice.

Questions

1. What does it mean for something to be normative? Why are norms central to the concept of sustainability? What are some core sustainability norms?
2. Why is it argued that justice cannot be defined independently from its social context? Do you agree?
3. Do you think that just processes necessarily make the outcomes just? Why or why not?
4. What is meant by the idea of a "moral boundary"? Can you identify some examples that may occur in a sustainability context?
5. Why does the multiscalar nature of sustainability pose challenges to justice? Can you think of ways of addressing these challenges?
6. What do you think is meant by the suggestion that "fuzzier" ideas of fairness could help resolve complex sustainability dilemmas? Can you think of any other examples in which this may (or may not) be the case?

Box 14.1: The Birth of the Environmental Justice Movement

Significant patterns of disproportionate exposure to environmental hazards in low-income communities were first recognized in a study by the Federal Council on Environmental Quality in 1971. These patterns define the environmental justice frame and led the environmental justice movement to recognize and correct these injustices. Many trace the movement to a particular struggle over environmental harm and remediation in rural North Carolina. In 1973, dumping of polychlorinated biphenyl (PCB) by Ward Transformer Company contaminated the soil along hundreds of miles of roadways in North Carolina. When the state proposed to move the contaminated soil to a waste facility in Warren County, it caused an uproar. Warren County happened to be 75 % African-American, among the highest rates in the state, and had the 97th lowest, out of 100 counties, gross economic product in the state. The uproar turned to protest and civil disobedience, attracting support and attention from across the country and the world. The protests raised awareness of both the spatial and institutional nature of environmental injustices and inspired a US General Accounting Office analysis (1983) which showed that race was the most significant predictor of where toxic waste facilities were located. These results were further confirmed by the landmark work by sociologist Robert Bullard (1983) and the study "Toxic Waste and Race" by the United Church of Christ Commission for Racial Justice in 1987.

Further Reading

Agyeman J, Bullard R, Evans B (2003) Just sustainabilities: development in an unequal world. MIT Press, Cambridge, MA
Dobson A (1999) Fairness and futurity: essays on environmental sustainability and social justice. Oxford University Press, New York
Fraser N (2009) Scales of justice: reimagining political space in a globalizing world. Columbia University Press, New York
Mutz KM, Bryner GC, Kenney DS (eds) (2002) Justice and natural resources: concepts, strategies, and applications. Island Press, Washington, DC

References

Asheim GB, Mitra T (2010) Sustainability and utilitarianism in discounted models of economic growth. Math Soc Sci 59(2):148–169
Baer P, Athanasiou T, Kartha S, Kemp-Benedict E (2009) Greenhouse development rights: a proposal for a fair Global Climate Treaty. Ethics Place Environ 12(3):267–281. doi:10.1080/13668790903195495
Bebbington J (2000) Sustainable development: a review of the international development, business and accounting literature. Aberdeen papers in accountancy, finance & management working paper 00–17
Beckerman W (1999) Sustainable development and our obligations to future generations. In: Dobson A (ed) Fairness and futurity: essays on environmental sustainability and social justice. Oxford University Press, New York, pp 71–92
Brundtland GH (1987) Our common future. Oxford University Press, New York
Bryner G (2002) Assessing claims of environmental justice. In: Mutz KM, Bryner GC, Kenney DS (eds) Justice and natural resources: concepts, strategies, and applications. Island Press, Washington, DC
Bullard RD (1983) Solid waste sites and the black Houston community. Sociol Inq 53:273–273
Clinton WJ (1994) Federal actions to address environmental justice in minority populations and low-income populations, Executive Order 12898 of February 11, 1994. Fed Regist 59:7629–7633
Cole DH (2008) The Stern review and its critics: implications for the theory and practice of benefit-cost analysis. Nat Resour J 48:53
Deutsch M (1975) Equity, equality, and need: what determines which value will be used as the basis of distributive justice? J Soc Issues 31(3):137–149. doi:10.1111/j.1540-4560.1975.tb01000.x
Endress L, Pongkijvorasin S, Roumasset J, Pitafi B (2009) Impatience and intergenerational equity in a model of sustainable growth. Working paper no. 09–6
Fainstein S (2010) The just city. Cornell University Press, Ithaca
Fischer F (2000) Citizens, experts, and the environment: the politics of local knowledge. Duke University Press, Durham
Fischhoff B (1991) Value elicitation: is there anything in there? Am Psychol 46(8):835–847. doi:10.1037/0003-066X.46.8.835
Fiske AP, Tetlock PE (1997) Taboo trade-offs: reactions to transactions that transgress the spheres of justice. Pol Psychol 18(2):255–297. doi:10.1111/0162-895X.00058
Gardiner S, Caney S, Jamieson D, Shue H (eds) (2010) Climate ethics: essential readings. Oxford University Press, Oxford/New York
Geels FW (2010) Ontologies, socio-technical transitions (to sustainability), and the multi-level perspective. Res Policy 39(4):495–510. doi:10.1016/j.respol.2010.01.022

Gibson RB (2006) Beyond the pillars: sustainability assessment as a framework for effective integration of social, economic and ecological considerations in significant decision-making. J Environ Assess Policy Manag 08(03):259–280. doi:10.1142/S1464333206002517
Golub A, Mahoney M, Harlow J (2013) Sustainability and intergenerational equity: do past injustices matter? Sustain Sci 8(2):269–277
Gregory R, Lichtenstein S, Slovic P (1993) Valuing environmental resources: a constructive approach. J Risk Uncertain 7(2):177–197. doi:10.1007/BF01065813
Harlow J, Golub A, Allenby B (2011) An exploration of Utopian themes in sustainable development discourse. Sustain Dev 21(4):270–280
Howarth RB (2003) Discounting and uncertainty in climate change policy analysis. Land Econ 79(3):369–381. doi:10.3368/le.79.3.369
Jerneck A, Olsson L, Ness B, Anderberg S, Baier M, Clark E, Hickler T, Hornborg A, Kronsell A, Lovbrand E, Persson J (2011) Structuring sustainability science. Sustain Sci 6(1):69–82
Kates RW, Parris TM, Leiserowitz AA (2005) Sustainability science. Environ Sci Policy Sustain Dev 47(3):8–21
Klinsky S, Dowlatabadi H, McDaniels T (2012) Comparing public rationales for justice trade-offs in mitigation and adaptation climate policy dilemmas. Glob Environ Chang 22(4):862–876. doi:10.1016/j.gloenvcha.2012.05.008
Lee BL (1997) Civil rights and legal remedies: a plan of action. In: Bullard R, Johnson G (eds) Just transportation: dismantling race and class barrier to mobility. New Society Publishers, Connecticut, pp 156–172
Lindseth G (2006) Scalar strategies in climate-change politics: debating the environmental consequences of a natural gas project. Environ Plan C Gov Policy 24(5):739–754. doi:10.1068/c0513j
Mikula G, Wenzel M (2000) Justice and social conflict. Int J Psychol 35(2):126–135. doi:10.1080/002075900399420
Miller D (2002) Debate – Caney's 'International distributive justice': a response. Polit Stud 50(5):974–977
Miller D (2005) Against global egalitarianism. In: Brock G, Moellendorf D (eds) Current debates in global justice. Springer, Netherlands, pp 55–79
Miller TR, Wiek A, Sarewitz D, Robinson J, Olsson L, Kriebel D, Loorbach D (2014) The future of sustainability science: a solutions-oriented research agenda. Sustain Sci 9(2):239–246
Nozick R (1974) Anarchy, state, and utopia. Basic Books, New York
Ott HE, Winkler H, Brouns B, Kartha S, Mace M, Huq S, Rahman A (2004) South-North dialogue on equity in the greenhouse: a proposal for an adequate and equitable global climate agreement. Deutsche Gesellschaft für Technische Zusammenarbeit (GTZ). Retrieved from http://www.erc.uct.ac.za/Research/publications/04Ott-etal-SouthNorthDiaLogue.pdf
Sassen S, Dotan N (2011) Delegating, not returning, to the biosphere: how to use the multi-scalar and ecological properties of cities. Glob Environ Chang 21(3):823–834. doi:10.1016/j.gloenvcha.2011.03.018
Smith A, Voß J-P, Grin J (2010) Innovation studies and sustainability transitions: the allure of the multi-level perspective and its challenges. Res Policy 39(4):435–448. doi:10.1016/j.respol.2010.01.023
Trappenburg M (2000) In defense of pure pluralism: two readings of Walzer's spheres of justice. J Polit Philos 8:343–362
U.S. General Accounting Office (USGAO) (1983) Siting of hazardous waste landfills and their correlation with the racial and socio-economic status of surrounding communities. General Accounting Office, Washington, DC
United Church of Christ (1987) Toxic wastes and race in the United States. United Church of Christ, New York. http://www.ucc.org/about-us/archives/pdfs/toxwrace87.pdf
Walzer M (1983) Spheres of justice: a defense of pluralism and equality. Basic Books, New York
Wenzel M (2004) Social identification as a determinant of concerns about individual-, group-, and inclusive-level justice. Soc Psychol Q 67(1):70–87. doi:10.1177/019027250406700107

Wiek A, Ness B, Brand FS, Schweizer-Ries P, Farioli F (2012) From complex systems analysis to transformational change: a comparative appraisal of sustainability science projects. Sustain Sci 7(Supplement 1):5–24

Wilbanks TJ (2002) Geographic scaling issues in integrated assessments of climate change. Integr Assess 3(2–3):100–114. doi:10.1076/iaij.3.2.100.13575

Winkler H, Jayaraman T, Pan J, Santihago de Oliveira A, Zhang Y, Sant G, … Raubenheimer S (2011) Equitable access to sustainable development: a paper by experts from BASIC countries. BASIC Expert Group, Beijing/Brasilia/Cape Town/Mumbai

Wolsink M (2007) Wind power implementation: the nature of public attitudes: equity and fairness instead of "backyard motives". Renew Sustain Energy Rev 11(6):1188–1207. doi:10.1016/j.rser.2005.10.005

Yabareen J (2008) A new conceptual framework for sustainable development. Environ Dev Sustain 10:179–192

Chapter 15
Sustainability Ethics

Nils Ole Oermann and Annika Weinert

Abstract The article examines the relationship between ethics and the concept of sustainability. Exemplified by case studies, different sustainability concepts will be applied to various philosophical as well as political discourses related to fundamental and applied ethics. In particular the question will be discussed if there are ethical duties towards future generations. Ecological issues as well as demographics will be ethically examined and related to the discourse of sustainability.

Keywords Ethics of sustainability • Ethical duties towards future generations • Applied ethics • Rawls • Kant

1 Introduction

In recent years *sustainability* has become a key term in discussions[1] about the relationship of human beings with each other and with their environment. Efforts to promote sustainable development, such as government policies or self-imposed *corporate social responsibility* programmes, often have – implicitly or explicitly – an ethical foundation. Since the idea of sustainability is at its core that of a protective relationship towards nature and humankind extending beyond the present to future generations, sustainability always implies ethical standards. If sustainability is understood as a "collective goal modern societies have committed themselves to" (Christen 2011, p. 34), then these societies can be seen to have a duty to act

[1] The perspective on sustainability ethics taken in this chapter is clearly positioned in discussion found in the German-language literature. It was felt that such an approach would complement the better known discussions taking place in what might be called a more Anglo-Saxon tradition. To this extent I am presupposing familiarity with such works as, to mention a few of those that are perhaps more notable: Lisa Newton's *Ethics and Sustainability: Sustainable Development and the Moral Life* (2003), Bryan Norton's *Sustainability: A Philosophy of Adaptive Ecosystem Management* (2005), Christian Becker's *Sustainability Ethics and Sustainability Research* (2011), and Jenneth Parker's *Critiquing Sustainability, Changing Philosophy* (2014).

N.O. Oermann, D.Phil. (Oxon) (✉) • A. Weinert, B.A.
Faculty of Sustainability, Leuphana University Lüneburg, Lüneburg, Germany
e-mail: Oermann@leuphana.de

H. Heinrichs et al. (eds.), *Sustainability Science*, DOI 10.1007/978-94-017-7242-6_15

sustainably. This question of human duties ultimately leads to Immanuel Kant's (1724–1804) second central question of philosophy "What ought I to do?" In such a duty-based ethics, the principle of sustainability seems to be an ethical principle that focuses on responsibility for and justice towards succeeding generations.

An essential contribution that philosophy can make to the sustainability debate consists of structuring the terminological difficulties of this concept. What makes the term "sustainability" problematic – not in spite of but because of its widespread use – is that it leads a "double life" (Grober 2010, p. 17). In everyday use it means something is "lasting", while in academia or politics it is a technical term. All too often and in a variety of contexts, there are references to the societal or economic relevance of sustainability, but what is often missing is a sufficiently clear or consistent understanding of what "sustainable" means. The goal from a philosophical perspective should be to structure these fundamental ambiguities.

After this brief look at the classification of sustainability terminology in practice, we can see how ethics comes into play, since ethics is normally understood as a discipline of practical philosophy that provides evaluation criteria, methodological procedures or principles for the "grounding and critique of action rules or normative statements about how one should act" (Fenner 2008, p. 35). This is where ethics in particular can show "that the idea of sustainability is not only understandable by means of natural science terminology and methodologies but is an action guideline based on a genuinely normative foundation" (Christen 2011, p. 35). Sustainability is not a purely descriptive concept but instead aims at "regulating the relationship between society and its natural surroundings" (Christen 2011, p. 35), that is, not only at describing how contemporary societies actually develop but also at formulating how societies ought to develop and can develop. "The natural limits to human action are not values that can be discovered. There 'are' no limits in a strict sense of the word and they cannot be identified as a separate entity. On the contrary, they are normative guidelines that are agreed upon for the sake of a good life for future generations" (Christen 2011, p. 35).

In addition to furthering theoretical and conceptual clarification, ethics has a practical integration and orientation function. It can contribute to "rationalising practical statements" (Nida-Rümelin 2005, p. 8) by introducing well-grounded actions and claims in decision-making situations and placing statements of opinion on a meaningful justificatory foundation. Such complex decisions are mostly found in so-called dilemma situations. A "dilemma" differs semantically from a "problem" in that a dilemma does not involve a decision between two or more alternatives that might be able to completely solve what was initially a complex problem. It involves the weighing of more or less desirable options. A problem, on the other hand, might have an optimal solution. Ethics is often about dilemma situations in which individuals, groups or whole societies are in need of orientation and a structured decision-making process when weighing alternatives or options in order to identify a feasible course of action. The main task of ethics is then not the solving of monocausal problems but the structuring and classifying of complex dilemmata.

2 What Is Ethics? From Principle to Application

The task of ethics is the systematic and structured development of criteria for evaluating moral action. Aristotle treated ethics as a separate philosophical discipline when he categorised the disciplines of practical philosophy – economics, politics and ethics – from those of theoretical philosophy, namely, logic, mathematics, physics and metaphysics (Pieper 2007, p. 24).

Ethos, Ethics; Morality, Mos

The term "ethics" is derived from the Greek *ethos*, which appears in two variations. A person acts ethically in the broader sense of θος (habits, customs or practices) if they "as a result of their upbringing are used to orientating their actions to moral customs" (Pieper 2007, p. 25f.). In a narrower sense of the word, ethical action is when "out of insight and reflection to do what is good in a given situation" (Pieper 2007, p. 25f) habit becomes θος (character) (Pieper 2007, p. 25f.; Fenner 2008). The word "morality" comes from the Latin "mos" (habits, customs or practices), which encompasses both semantic dimensions of the term *ethos* in the sense of practised behaviours that are then reflected on from an ethical perspective (Fig. 15.1).

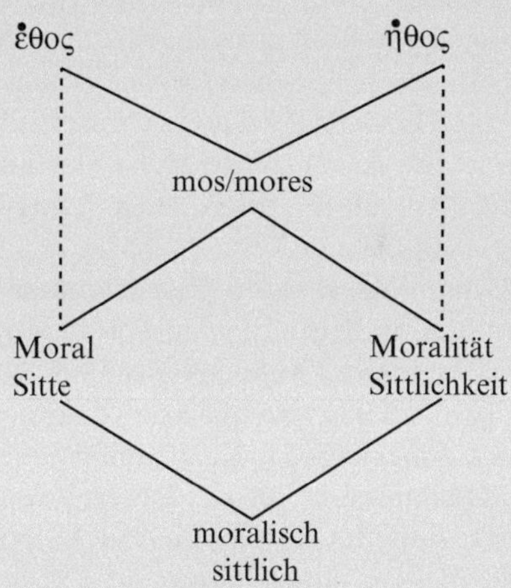

Fig. 15.1 The terminological roots of ethics and morality (Pieper 2007, p. 27)

While the term "morality" is commonly understood as "the *essence* of moral norms, value judgments, institutions", "ethics" describes the "*philosophical investigation* of the area of morality" (Patzig 1971, p. 3, emphasis in the original). In contrast to morality, ethics does not have to do with action itself, but instead it critically reflects on actions and behaviour. An ethics thus understood as critically reflecting on morality can be subdivided into general and applied ethics. The main focus of *general ethics* is "the provision of a set of terminological and methodological instruments with the help of which fundamental problems of morality can be investigated in depth" (Pieper and Thurnherr 1998, p. 10). It can be further divided into three subdisciplines: normative ethics, descriptive ethics and metaethics.

Normative ethics formulates justifiable normative judgements. When, for example, Aristotle asks what makes a life a good life, then there will be a variety of answers depending on one's perspective. That is why normative ethics is in turn subdivided into teleological and deontological approaches. Teleological conceptions of ethics (from the Greek *telos* meaning completion, end or goal) evaluate actions by focusing on ends or goals that are, in a broad sense of the term, "good" (Hübenthal 2006, p. 61). They make a division between moral rightness and nonmoral goodness and determine what is morally right by whether it promotes the best possible nonmoral good (ibid.). The moral judgement of an action is performed then by evaluating its consequences. A prominent example is found in classic utilitarianism, which has its roots in the eighteenth century in England. It is one of the so-called teleological-consequentialist approaches, that is, moral judgement of human action takes as its starting point an evaluation of the consequences of an action. Utilitarianism takes its name from its core value of utility, which is understood as "the extent of happiness, well-being or satisfaction of desires (preferences) effected by an action" (Birnbacher 2006, p. 96). One of the first systematic treatments of utilitarianism is Jeremy Bentham's *An Introduction to the Principles of Morals and Legislation* (1780). Bentham evaluates an action's consequences by means of its so-called gratification value, which is determined by calculating the degree of pain and pleasure of an action for each person affected by its consequences and then adding these individual values to a total collective gratification value, which is the total utility of an action (Höffe 2008). Other major proponents of utilitarianism are John Stuart Mill (1806–1873), Henry Sidgwick (1838–1900) and Richard M. Hare (1919–2002).

Deontological approaches (from the Greek *deon* meaning duty) deny that the right and the morally good are directly or indirectly dependent on an abstract good. In contrast to teleological approaches to ethics, an action is judged not on the basis of its consequences but instead on its characteristics, typically moral duties. An example is Kant's imperative-based ethics, in which duty is seen as an action required by reason. Descriptive ethics, the second subcategory of general ethics, provides an empirical description of norm and value systems without itself making moral judgements. This is its similarity to metaethics, the third subdiscipline, which in contrast to descriptive ethics does not describe which specific moral judgements are made but instead focuses on a meta-level of ethical reflection structuring dilemmata.

Applied ethics forms the second large category of approaches to ethics. It provides a "systematic application of normative-ethical principles to fields of human action, occupational fields and specific subject areas" (Thurnherr 2000, p. 13). It makes use of "justified universal statements about the good life of the individual or about just coexistence in a community" as formulated in normative ethics and then applies them to specific social areas (Fenner 2010, p. 11). Due to the variety of different problems and fields of action in its corollary disciplines, applied ethics has developed several specific types of ethics, such as medical, science, technology, legal and media ethics. Whether it is constructed on the basis of Kant's imperative-based ethics, on a utilitarian ethics or on a different normative approach, applied ethics needs a corollary discipline that provides it with a foundation of empirical knowledge of its respective field.

Sustainability ethics can be understood as an area of applied ethics that as part of a larger discourse about sustainability examines ethical problems and attempts to structure them with the goal of offering guidelines in specific situations. However, in contrast to, for example, business, technology or medical ethics, it is not a "hybrid ethics", since such approaches to ethics have recourse to their corollary disciplines, while sustainability ethics is based on a principle, with "principle" meaning "insights, norms and goals that are methodologically the starting point of a theoretical structure or a system of action guidelines" (Kambartel 1995, p. 341).

In this sense ethics is looking for "an overriding principle of morality as a final unifying principle …, from which one can derive all specific norms or be able to criticise them with this standard" (Fenner 2010, p. 171). Examples of this overriding principle include Kant's categorical imperative or the utilitarian principle of the greatest good for the greatest number. While these examples are all ethical imperatives, ethical dilemmata can be structured by means of substantive principles such as freedom, justice or, following Hans Jonas, responsibility. Sustainability can be made a guiding ethical principle in the second sense and systematically anchored in a similar fashion as the ethics of conviction and responsibility is in Weber's concept (see box below). The latter as a higher principle also do not ground a hybrid ethics but, similar to sustainability, are themselves principles that structure ethics.

Weber's Ethic of Responsibility and Conviction

A sociologist, political economist and legal scholar, Max Weber (1864–1920) drew a distinction in his 1919 lecture *Politics as a Vocation* between an ethic of responsibility and an ethic of conviction. "We must be clear about the fact that all ethically oriented conduct may be guided by one of two fundamentally differing and irreconcilably opposed maxims: conduct can be oriented to an 'ethic of ultimate ends' or to an 'ethic of responsibility'" (Weber 1992, p. 120). The central difference between the two is found in the principles used to evaluate action. A proponent of the ethic of conviction, for Weber, determines the moral value of an action by the conviction, that is, by the good intentions of the agent, while ignoring the foreseeable or specific consequences of the action. If an action undertaken out of conviction has negative consequences, then they are not attributed to the agent but to "the world" or "God's will" (Weber 1992, p. 120). By contrast the advocate of an ethic of responsibility takes the position that a person is liable for the consequences of his actions and so he attributes them to the agent (Weber 1992). The Weberian comparison was taken up again by, among others, Hans Jonas (1903–1993), who reformulated the ethic of responsibility as an "ethics of the future" under the "principle of responsibility" (Fig. 15.2).

Fig. 15.2 Max Weber

Just as Weber subordinates "all ethically oriented action" (Weber 1992, p. 120) to either the principle of conviction or of responsibility, sustainability ethics can be subordinated to the "principle of sustainability". Sustainability ethics understood as an area of ethics under the principle of sustainability comprises then not only abstract normative principles but can also become a guiding principle through the application of ethical principles to the lives of human beings. Sustainability ethics understood in this way will not remain abstract but will always refer to concrete, practical dilemmata. The task of such an ethics in general and of sustainability ethics in particular cannot be to solve ethical dilemmata much less to give paternalistic answers about the "good" life or "right" action. It can, however, structure the search for such answers by "emphasising its specific philosophical competence in the public process of searching for a solution to a problem" (Bayertz 1994, p. 26). Its contribution is thus primarily a hermeneutic (from the Greek *hermeneus* meaning the interpreter) one, an act of translation between principle and practice, found in the precise definition of terminology and in structuring ethical dilemmata in order to identify real options to take action. The next section will show how this structuring and guiding act of translation can contribute to dealing with dilemma situations.

Question: What perspectives can be used to define the term "ethics"?
Task: Discuss the relationship between ethics and the concept of sustainability. Exchange your views with other students.

3 Sustainability Ethics: Justice and Responsibility Through Time

Sustainability ethics, some believe, can be defined as ethical reflection from the perspective of a clearly defined and practical inter- and intragenerational principle of justice (Rogall 2008). Although this definition seems to delimit the scope and applications of sustainability ethics, it is a relatively recent area of ethics and its contours in sustainability discourse are still largely blurred. There is little consensus about sustainability ethics, but in the relatively small number of publications dealing explicitly with this topic, there is agreement about which sources and values should be at its core. While sustainability in the widely cited definition of the Brundtland Report is anthropological in the sense that it places the needs and rights of future generations in the foreground (Unnerstall 1999), there are proponents of a pathocentric standpoint that advance the thesis that human beings have an obligation to protect other creatures, as they are also bearers of rights. There are also some who take a biocentric position and extend the concept of moral rights even to plants and other natural objects that are incapable of suffering (Schüßler 2008).

From the question what should be at the core of an ethics – only human beings or also other creatures and their natural environment – we can derive the main difference between sustainability ethics in the sense outlined here and the varied approaches of environmental ethics, with which sustainability ethics is too often mistakenly confused. The philosopher Konrad Ott defines environmental ethics in the following way. "Environmental ethics (synonymous with ethics of nature) enquires on one hand into the reasons and the standards (values and norms) that are derived from them that should determine our individual and collective behaviour towards the non-human natural world. On the other it asks how these standards can be implemented" (Ott 2010, p. 8). Its subject is, as Ott writes in another passage, "the relationship of human to non-human" (Ott 1997, p. 58). It thus relativizes the anthropocentric perspective, as found in most classic approaches to ethics, and contrasts it with an eco- or biocentric orientation" (Ott 1997, p. 59–63). This distinction forms a defining characteristic between environmental and sustainability ethics, since the latter applies an anthropocentric perspective to the ethical dilemmata it examines.

A further defining characteristic can be found in how such an ethics is justified. Ott classifies environmental ethics as part of applied ethics and places it in proximity to business ethics and an ethics of technology (Ott 1997), that is, with the classic hybrid ethics that rely on a corollary science. By contrast, sustainability ethics, as already mentioned, is a principle-based ethics. Even though at first glance the topics seem similar, their different perspectives – anthropocentric versus biocentric – as well as their justifications, principle versus corollary science, form the basis for the difference between the two disciplines. In spite of the lack of consensus in sustainability discourse about possible forms of sustainability ethics, there is, however, agreement that such fundamental principles as responsibility and justice are essential components of it. The definition of the Brundtland Report shows, for example, that the principle of sustainability has at its core the struggle for intra- and intergenerational responsibility and justice. Such approaches that deal with the ethical claims of sustainability are framed by the anthropocentric and Aristotelian question, "How should people live and what is today and tomorrow a 'good' life" (Renn 2007, p. 64–99).

The question about what makes a life a good life is of course by no means a new question that solely belongs to sustainability discourse. On the contrary this question revisits the more than 2000-year-old core question of ethics, which was already asked by Aristotle (384–322 BCE) in his Nicomachean Ethics. The core of Aristotelian ethics is formed by the terms *eudaimonia* (happiness) and *arete* (virtue). Aristotle raises happiness to an ultimate end that all human beings should aspire to and makes it the principle of his ethics, while putting virtue at its side to provide orientation in specific situations calling for a decision (Rapp 2006). From an Aristotelian perspective, the good life consists of the activity of the soul in accordance with *ergon*, which is the function, task or work particular to human beings and represents the best possible state of the soul (Rapp 2006). The excellence or virtuousness (*arete*) of a person is the result of the exercise of their *ergon*. "Human good turns out to be activity of soul in accordance with virtue, and if there are more than one virtue, in accordance with the best and most complete." (Aristotle, EN 1098a 16–20). Aristotle locates this activity in the contemplative life, at the centre of which is found the cultivation of what he calls the theoretical disciplines of, alongside philosophy and theology, for example, astronomy or mathematics (Rapp 2006, p. 73).

In sustainability discourse, this age-old question of what makes the "good life" is given an intertemporal extension beyond the ancient study of virtue in the present day into the future. Even if there are many contemporary answers to the question, it is clear that, as can be seen in Aristotle, the traditional questions and answers of ethics can make an important contribution to a future-oriented discussion of sustainability. By means of exemplary dilemmata from the sustainability debate, namely, the discussion about pension policy in light of demographic change and the question of a just and sustainable distribution of resources, the next section will show exactly what kind of contribution the classic approaches to ethics discussed above will be able to make sustainability discourse.

Question: What are the differences between the environmental and sustainability approaches to ethics?

4 Exemplary Approaches to Dealing with Dilemmata in Sustainability Ethics

4.1 Dilemma 1: Generational Contracts in the Light of Demographic Change

Demographic changes impact social security systems, and the ageing of society exacerbates the question of the duties of the present generation towards future generations. In November 2009 the 12th coordinated population projection of the German Federal Office of Statistics came to the following findings for the time period 2008–2060: "Germany's population is decreasing, its people are getting older and there will be – even if birth rates rise slightly – fewer children born than there are today" (Egeler 2009, p. 8). A consequence of this demographic trend is that the "numerical ratio of potential recipients of benefits of pension insurance schemes compared to the potential contributors to these systems [will] worsen" (Egeler 2009, p. 12). A dwindling number of the working-age population must then in future provide for a growing number of people of pension age. Extensive obligations are being imposed on future generations in comparison to those prior to them (Fig. 15.3).

From a philosophical perspective this situation raises a central question for sustainability ethics, namely, whether it is even possible to impose obligations on future unborn generations, and if so what exactly these might be. This question belongs to a duty and imperative-based ethics and is a core question of sustainability discourse, and it is also not a new one. Above all in the Kantian ethics of duty, it is a central topic, so that Kant can serve as a key reference in structuring the dilemma. In his *Critique of Pure Reason*, published in 1781, he formulated the three key questions of his philosophy (Kant 1973, p. 522f.): "The whole interest of reason, speculative as well as practical, is centred in the three following questions:

1. What can I know?
2. What ought I to do?
3. What may I hope?"

The first question, which metaphysics is to provide the answer for, is directed at determining "the origin, as well as of the extent and limits of our speculative reason"; the second, the province of ethics, builds on the answer to the first and focuses on "transcendental and practical human freedom, that is, a person's capability to freely be causally effective in the world" (Klemme 2009, p. 13); and the third question, to be answered by religion and metaphysics, enquires into the "highest goal we can hope to achieve by means of our pure practical reason" (Klemme 2009, p. 13). In 1793 Kant added a fourth question, one that he thought encompassed all three prior questions: "What is a human being?" (Kant 1969, p. 429).

In sustainability discourse, it is the second question that is at first particularly interesting. However, Kant's imperative-based ethics, at the centre of which is the question of ought and of human duty, is an ethics oriented to the present. Duty is for Kant "an action that is absolutely necessary, that is, it is made absolutely necessary

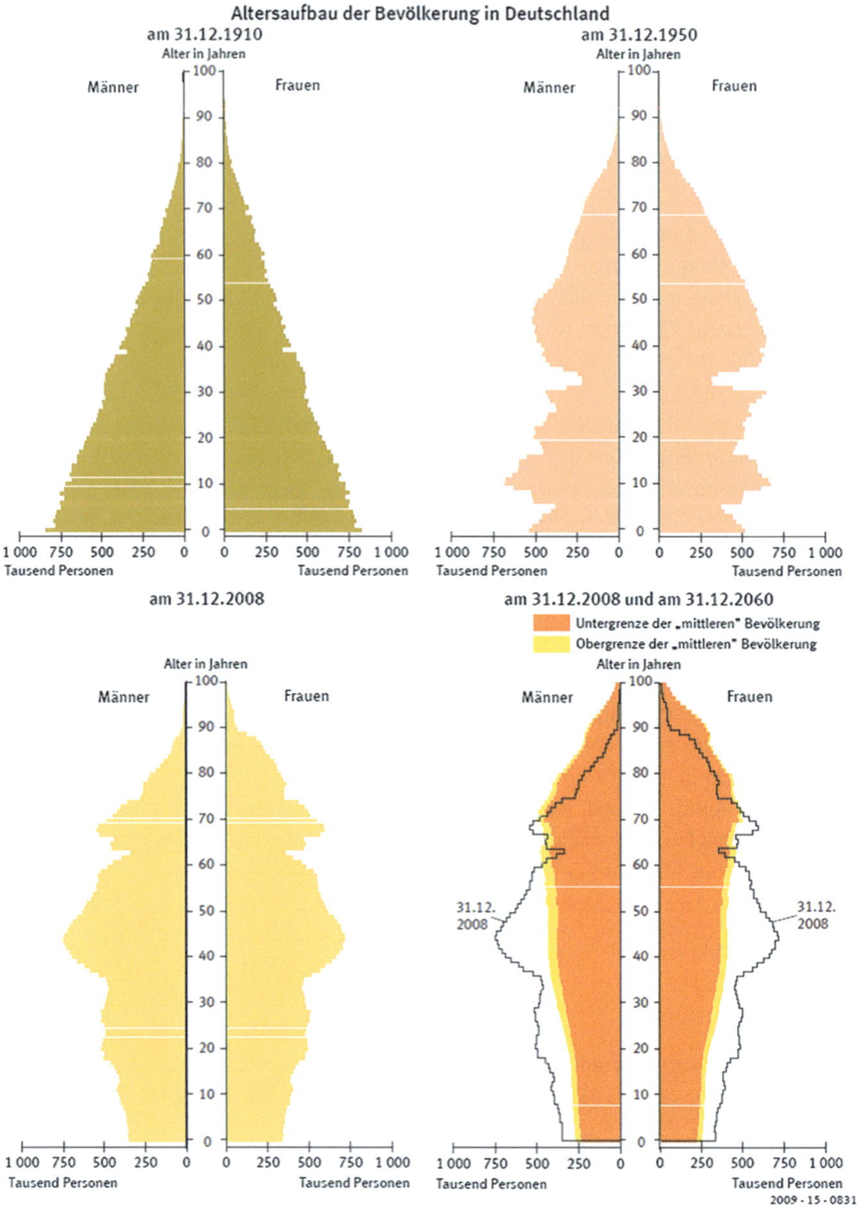

Fig. 15.3 Age structure of the population in Germany (Statistisches Bundesamt 2009)

by reason", and in such a way "'as if' there were a supernatural law" (Eisler 2002, p. 417). The moral necessity behind duty is derived from the freedom of the individual as a rational being and the autonomy of their reason.

The core ethical problem of the principle of sustainability can now be located in the question whether there can be such ethical obligations towards future generations, since future generations at least today from a legal point of view are not yet able to be a claimant because they do not have the legal characteristics of a natural person *ad personam*. In the case of the pension system, the question is thus whether in view of projected demographic trends present generations can already be said to have an ethical duty to act, due to political or individual decisions, towards those generations that in the future – if the present old-age social security system remains unchanged – will have to pay for their own pensions. There is little agreement as to who these Kantian "rational beings" will be in future generations, whose future needs can be so little estimated as the resources that will be available in the future.

From a theological point of view, this dilemma can be ultimately reduced to the New Testament question: "Who is my neighbour?" The answer cannot only include the number of people that an individual actually knows at a given time, such as members of the family, friends or neighbours. Ott argues that ethical judgements can be formulated in the present that involve rights and interests in the future so that an action "can already be impermissible or a norm can already be invalid, even though its consequences and side-effects might first affect persons in the future" (Ott 1996, p. 141). He assumes then that future generations will be similar in relevant characteristics to present ones; would have similar basic needs, interests and preferences to those living today; and would not be willing to accept harms, disadvantages or deficiencies in favour of present generations.

This broad understanding of the concept of duty and moral capability shows again the distance of a purely environmental ethics approach to an exclusively anthropocentric ethics. In order to justify a universalistic position that considers it necessary in principle to morally account for all future persons in actions undertaken in the present, Ott develops six universalistic principles drawn from Kambartel, Habermas, Birnbacher, Singer, Jonas and Apel. In combination these yield the following test questions for the morality of present actions for future generations. "Does this behaviour show consideration for future persons, is it universally generalizable, will all future generations potentially be able to agree with it, does it produce a maximum amount of human happiness over an extended period of time, is it, in the sense of Hans Jonas, compatible, does it contribute to an ideal communication community in the sense of Karl-Otto Apel" (Ott 1996, p. 148). If these questions can be answered positively, then our duties towards future generations will have been adequately accounted for.

Conflicts of ethical duties regarding future generations can thus be structured if, following Kant, sustainability is viewed as a problem of reason on a virtual timeline. Since if we assume that future generations are not dissimilar to present generations in their needs to create a good life, then those who would in the present make it difficult or impossible to meet those needs would be called on not only for reasons of sustainability but would be obligated in a Kantian sense to ensure the *status quo* for coming generations. Whoever breaks a generational contract of his own accord is not merely breaching a contract but, by consciously violating duties towards those who will come after him, is ultimately acting irrationally. In particular in the discussion about demographics and pensions, this dilemma in sustainability ethics is deepened when individuals today seek to profit at the expense of future generations by talking of terminating a "generational contract".

4.2 Dilemma 2: Future-Oriented and Sustainable Distribution of Resources

There must also be a broad ethical discussion about absolute or comparative standards of a maximally equitable distribution of resources on this planet. The ethical dilemma here is that one part of the world is consuming the resources of another part of the world without the two parties meeting each other. This dilemma of global resource distribution leads, after a discussion of the concept of risk, to a further core concept of sustainability and justice and so to the question: "What distribution of goods and opportunities between present and future persons is a just distribution?" This directs attention to intra- and intergenerational justice as a further central principle in sustainability, along with responsibility and a future-oriented management of risk (Kersting 2000). While intragenerative justice postulates equal opportunities regarding access to basic goods, the possibility of satisfying fundamental needs and participation in social decision-making processes, the principle of intergenerative justice refers to a distribution that, in the face of the limited carrying capacity of the ecosystem, will preserve life over the long term.

When answering the question whether absolute or relative standards lead to a more just distribution between present and future generations, John Rawls's (1921–2002) *Theory of Justice* (1971) serves as an example of how ethics can help structure the resulting ethically complex distribution dilemmata sustainable economic activity faces, not only economically but also ecologically and socially. At the core is the question "which principles for institutions that regulate distribution would people agree to in a decision-making process based on fair conditions" (Nida-Rümelin and Ozmen 2007, p. 654). This question is ethically relevant not least because it is prior to the question about the good. The good can only be determined subject to what is just. And the problem of justice in turn refers not only to individual action but also to the social norming of rights and duties in the distribution of goods. People are, according to Rawls, rational beings, and their actions are driven by the search for individual advantage in social cooperation as well as for a greatest possible share in social goods. A solution to this tension between the common good and self-interest is, according to Rawls, found in a concept of justice that all members of society can agree to.

In an intergenerational perspective, Rawls's approach can be extended to the question how it is possible to not only fairly distribute goods among living persons and groups but also among different generations, that is, if we are to act justly how much we should concede future generations from what is currently available. In a similar direction Ott and Döring develop Rawls's approach and outline an "intertemporal extension of John Rawls's theory of justice, in which behind the veil of ignorance the representatives do not know which generation they belong to" (Christen 2011, p. 35). This shift in perspective makes Rawls's theory of justice highly relevant for sustainability discourse. For Rawls a criterion of justice will only be agreed to if natural, social and individual realities are put to one side and the influence of individual preferences and beliefs are curbed. He creates this situation in a thought experiment he names the *original position*, in which the parties do not know their own identity or interests. "Although the decision-making persons or parties do know general facts about psychology or social sciences they do not know who they are; they do not know their gender, age, status, class, race or ethnicity; they do not know which natural talents (such as intelligence or bodily strength) they have nor which social, cultural and religious milieu has shaped them. They also know nothing about their beliefs about good or their psychological inclinations – they decide behind a veil of ignorance" (Nida-Rümelin and Ozmen 2007, p. 656). All the relevant information for determining biased criteria of justice are unavailable to the parties making the decision, assuring their impartiality and thus creating a situation in which "since the differences among the parties are unknown to them, and everyone is equally rational and similarly situated, each is convinced by the same arguments" (Rawls 1976, p. 139). If this thought experiment is supplemented by an intergenerational perspective, the membership of the parties making the decision in a particular generation – whether it is the present or some future one – is hidden behind the veil of ignorance. Under these circumstances, none of the parties can be certain whether the solution they prefer is beneficial for their or for another generation nor whether they will have to suffer the negative consequences of their decisions or first coming generations would. In Rawls's thought experiment, the parties behind a veil of ignorance would agree to two principles of justice.

John Rawls's Principles of Justice

First Principle

Each person is to have an equal right to the most extensive total system of equal basic liberties compatible with a similar system of liberty for all.

Second Principle

Social and economic inequalities are to be arranged so that they are both:

(a) To the greatest benefit of the least advantaged, consistent with the savings principle
(b) Attached to offices and positions open to all under conditions of fair equality of opportunity

First Priority Rule (The Priority of Liberty)

The principles of justice are to be ranked in lexical order and therefore liberty can only be restricted for the sake of liberty.

There are two cases:

(a) A less extensive liberty must strengthen the total system of liberty shared by all.
(b) A less than equal liberty must be acceptable to those with the lesser liberty.

Second Priority Rule (The Priority of Justice over Efficiency and Welfare)

The second principle of justice is lexically prior to the principle of efficiency and to that of maximising the sum of advantages, and fair opportunity is prior to the difference principle. There are two cases:

(a) An inequality of opportunity must enhance the opportunities of those with the lesser opportunity.
(b) An excessive rate of saving must on balance mitigate the burden of those bearing this hardship.

(Rawls 1976, pp. 302–3)

The first principle, which determines the distribution of basic political goods, civil and human rights together with fundamental liberties on a strictly egalitarian basis, is prior to the second and must not be "restricted in favour of the greater efficiency of the economic and social system" (Nida-Rümelin and Ozmen 2007, p. 658). The second principle governs the distribution of basic socioeconomic goods and also makes use of egalitarian distribution as the basis for evaluating possible improvements in their distribution. Unequal distribution is only permissible if it leads to an improvement for all, especially those in the worst off group in a society. In the original position, economic and social relations are evaluated using the efficiency principle so that a situation is considered Pareto optimal if no one can be made better off without making someone else worse off. If generation membership is also included behind the veil of ignorance in the original position, then there is a solution to the demand that no generation should be worse off than another. However, since some efficient distributions go against intuitions of justice, the so-called difference principle is needed to choose among equally efficient, unequal distributions the one that is just to the extent that it contributes to "enhance the opportunities of those with the lesser opportunity" (Rawls 1976, p. 303). As a result any rational person would require as high a minimum as possible for the group with the least opportunities, since he could be a member of this group himself.

This approach leads back to the core of the debate about sustainability and the question, with reference to Kant, as to whether there can be duties towards future generations and whether these – returning to the issue of distributive justice that was the starting point of the case study – also have universal validity. From an intergenerational perspective, each generation would have to have the least possible disadvantage, in the Rawlsian sense, from the decisions and actions of earlier generations if there was to be a just distribution of goods and opportunities. In this context Ott and Döring also ask "whether future generations would have to receive the same amount as present generations have inherited (comparative standard) or whether it would also be just if they were guaranteed a certain minimum amount (absolute standard)" (Christen 2011, p. 35). They argue for a comparative intergenerational standard of distribution. Against this background, there is no longer any reason that people would be satisfied with an absolute minimum standard. The comparative standard is supplemented by an absolute standard, which for Ott and Döring is based on the so-called capability approach of the philosopher Martha Nussbaum, according to which "all human beings should receive the opportunity to exercise certain basic capabilities in order to be able to live a human life" (Christen 2011, p. 36). By means of the absolute standard, it would be possible to ensure that "it will not be permissible for the quality of life to be less than a certain amount, not only now but also over time" (Christen 2011, p. 36). Sustainability can in this sense be reduced to the normatively grounded idea that "regardless of space and time all human beings should be guaranteed an absolute standard without this violating the comparative standard regarding future generations, that is, without future generations being worse off than the present generation" (Christen 2011, p. 36).

Often such debates about specific dilemmata of a just – national as well as global – distribution of goods and resources lead from the sustainability discourse to categorical problems and central topoi of ethics, as, for example, Rawls introduces in his concept of just distribution with recourse to Kant and the utilitarians. What appears to be a purely economic problem about the fair use of natural resources becomes an ethical dilemma that cannot be solved with only the expertise of the World Bank and the IMF. Instead it requires a discussion of a universal ethics, such as Hans Küng and others are already involved in (Küng et al. 2010).

Task: Locate the concept of distributive justice in sustainability discourse. Where could there be a reference in this concept of justice to the approach of John Rawls and to the Kantian concept of duty?
Question: What are the similarities and differences between these two dilemmata?

5 Conclusion

It was Kant who in 1793 simply yet incisively formulated this fundamental question of anthropology as the fourth question of his philosophy: "What is a human being?" (Kant 1969, p. 429 and 1972, p. 25). In the Preface to his Logic, Kant observed that the three prior questions, concerning human knowledge, duty and hope, belong "in principle" to anthropology "because the first three questions are related to the last" (Kant 1972, p. 25). Whoever works on an approach to ethics will first clarify the anthropological and conceptual premises as carefully as possible: Who owes what to whom in the present and in the future? What does it mean to take a risk and how can its consequences be evaluated? What is a just distribution beyond the economic mechanism of distribution? No matter whether an ethics of sustainability is constructed on the basis of Aristotle or Kant, Jonas or Rawls or Marx or Habermas, ethics in general and sustainability ethics in particular are not an addendum, a decoration of business, science, technology or politics; it defines their reach and structures their options. Ethics is not supposed to explain the world to its corollary sciences but is instead an attempt to understand the premises and conditions of human action before it then ethically reflects, structures or criticises more or less moral actions. Sustainability ethics ultimately enquires in an Aristotelian sense into what it is that makes life meaningful and worth living and what makes it a "good" life and so a life that gives a human being their humanity. And it also enquires, following Kant, into what the duty of the individual is and how sustainable action can be rationally justified, not only in the present but over a period of time that far exceeds the life of a single person.

In the end sustainability as an ethical principle describes something similar to what Immanuel Kant describes with the term "reason", though with two particular features. Sustainability, and so also sustainability ethics, projects rational action over time. This temporal aspect has now become – ecologically, socially and economically – urgent and has led to the virulence of debates about sustainability; it can be explained by developments that were not foreseeable for Kant: by industrialisation and globalisation and all of its consequences. Whoever attempts to reconcile the three major factors of production – land, labour and capital – in a discourse about the principles of distributive justice so that economic activity is sustainable over time and no excessive risks are taken will not have a problem accepting sustainability as a value and an intergenerational guideline, even if this description from a philosophical perspective is, at the latest since Heidegger's fundamental criticism of the concept of value as "positivistic substitute for metaphysics" (Heidegger 1977, p. 227), unclear.

Rudolf Schüßler draws attention to a further source of tension in the ethical debate about sustainability. He emphasises that the focus on the relationship between present and future generations rests on an individualistic understanding and that this viewpoint is incompatible with a communitarian social philosophy, which would argue that the compensation of interests and needs across generations is meaningless. Present generations, according to communitarians, have sufficiently fulfilled their duty if they leave the commonwealth, the *polis*, in a well-ordered state (Schüßler 2008, p. 65). It remains an open question what the standard for this well orderedness should be. In this sense sustainability ethics does not only reflect an anthropological image of human beings, their social responsibility or their duties but also the relationship of human beings to each other, to other generations and above all to their natural environment. They do not argue from a purely anthropocentric or biocentric perspective, but they assert the existence of ethical duties beyond geographical and intergenerational borders.

Whoever professes this principle of intergenerational justice and thus would like to grant coming generations similar life chances as those who are now alive will have to, behind the Rawlsian veil of ignorance, make sacrifices, especially regarding the consumption of natural resources. Increasing efficiency and using resources more responsibly without freeing oneself from certain lifestyles habits will not be enough (Renn 2007, p. 95f). A crucial task of future ethics of sustainability will be to structure present as well as future actions in order to emphasise that all action – beyond all systemic limitations and supposed as well as actual constraints – is ultimately based on individual decisions and that ethically responsible action is also dependent on our understanding of what it means to be a human being and on the values of each and every actor, since we must ask ourselves – as did Kant – not in an anonymous collective "What ought we to do?" but as individuals and specifically "What should I do?"

Task: Attempt to define the concept "sustainability ethics" and describe its roots and the controversies it has caused. Exchange your views with other students.

Bibliography

Aristotle (1941) Nicomachean Ethics. 31st printing. Random House, New York

Bayertz K (1994) Praktische Philosophie als angewandte Ethik. In: Bayertz K (ed) Praktische Philosophie. Grundorientierungen angewandter Ethik. Rowohlt, Reinbek, pp 7–47

Birnbacher D (2006) Utilitarismus. In: Düwell M, Hübenthal C, Werner MH (eds) Handbuch Ethik. Metzler, Stuttgart, pp 95–107

Christen M (2011) Nachhaltigkeit als ethische Herausforderung. Inf Philos 2:34–43
Egeler R (2009) Statement auf der Pressekonferenz "Bevölkerungsentwicklung in Deutschland bis 2060" am 18. November 2009 in Berlin. Retrieved on 19 Mar 2012 unter. http://www.destatis.de/jetspeed/portal/cms/Sites/destatis/Internet/DE/Presse/pk/2009/Bevoelkerung/Statement__Egeler__PDF,property=file.pdf
Eisler R (2002) Pflicht. In: Eisler R (ed) Kant-Lexikon. Nachschlagewerk zu Kants sämtlichen Schriften, Briefen und handschriftlichem Nachlaß. Olms Verlag, Hildesheim u.a, pp 410–417
Federal Statistics Office (2009) Germany's population by 2060. Federal Statistics Office, Wiesbaden
Fenner D (2008) Ethik: Wie soll ich handeln? Francke, Tübingen
Fenner D (2010) Einführung in die Angewandte Ethik. Francke, Tübingen
Grober U (2010) Die Entdeckung der Nachhaltigkeit. Kulturgeschichte eines Begriffs. Kunstmann Verlag, München
Haus M (2003) Kommunitarismus. Einführung und Analyse. Westdeutscher Verlag, Wiesbaden
Heidegger M (1977) Nietzsches Wort "Gott ist tot". In: Heidegger M (ed) Holzwege. Klostermann Verlag, Frankfurt a.M, pp 209–267
Höffe O (2008) Einleitung. In: Höffe O (ed) Einführung in die utilitaristische Ethik. Klassische und zeitgenössische Texte. Francke Verlag, Tübingen, pp 7–51
Hübenthal C (2006) Teleologische Ansätze. In: Düwell M, Hübenthal C, Werner MH (eds) Handbuch Ethik. Metzler, Stuttgart/Weimar, pp 61–68
Jonas H (1987a) Das Prinzip Verantwortung. Versuch einer Ethik für die technologische Zivilisation. Suhrkamp, Frankfurt a.M
Jonas H (1987b) Technik, Medizin und Ethik. Zur Praxis des Prinzips Verantwortung. Suhrkamp, Frankfurt a.M
Kambartel F (1995) Prinzip. In J. Mittelstraß (ed) Enzyklopädie Philosophie und Wissenschaftstheorie, Bd. III. Metzler, Stuttgart u.a., p 341f
Kant I (1969) Brief an Carl Friedrich Stäudlin, 4.5.1793. In: Kant's gesammelte Schriften, hrsg. von der Preußischen Akademie der Wissenschaften, Berlin. Reprint of the Berlin und Leipzig edition 1923, vol XI
Kant I (1970) Grundlegung zur Metaphysik der Sitten. In: Kant I (ed) Werke in sechs Bänden. Published by W. Weischedel, vol IV. Wissenschaftliche Buchgesellschaft, Darmstadt
Kant I (1972) Logik. In: Kant's gesammelte Schriften. Published by the Preußischen Akademie der Wissenschaften. Reprint of the Berlin und Leipzig edition 1922, vol IX
Kant I (1973) Kritik der reinen Vernunft. In: Kant's gesammelte Schriften. Published by the Preußischen Akademie der Wissenschaften. Reprint of the Berlin und Leipzig edition 1911, Bd. III
Kersting W (2000) Theorien sozialer Gerechtigkeit. Metzler, Stuttgart
Klemme HF (2009) Immanuel Kant. In: Bohlken E, Thies C (eds) Handbuch Anthropologie. Der Mensch zwischen Natur, Kultur und Technik. Metzler, Stuttgart u.a, pp 11–16
Küng H, Leisinger KM, Wieland J (2010) Globales Wirtschaftsethos: *Konsequenzen und Herausforderungen für die Weltwirtschaft*. Deutscher Taschenbuch Verlag, München
Mieth D (2003) Was ist Nachhaltigkeit? Eine ethische Kursbestimmung. In: Bobbert M, Düwell M, Jax K (eds) Umwelt-Ethik-Recht. Francke, Tübingen/Basel, pp 182–201
Nida-Rümelin J (2005) Theoretische und angewandte Ethik: Paradigmen, Begründungen, Bereiche. In: Nida-Rümelin J (ed) Angewandte Ethik. Die Bereichsethiken und ihre theoretische Fundierung. Ein Handbuch. Kröner, Stuttgart, pp 2–87
Nida-Rümelin J, Ozmen E (2007) John Rawls. Eine Theorie der Gerechtigkeit. In: Brocker M (ed) Geschichte des politischen Denkens. Ein Handbuch. Suhrkamp, Frankfurt a.M, pp 651–666
Ott K (1996) Vom Begründen zum Handeln. Aufsätze zur angewandten Ethik. Narr-Francke-Attempto-Verlag, Tübingen
Ott K (1997) Umweltethik in schwieriger Zeit. In Ott K (ed.) Umweltethik in schwieriger Zeit. Antrittsvorlesung von Prof. Dr. Konrad Ott und Reden von Prof. Dr. Jürgen Kohler u.a. Festschrift anlässlich der Antrittsvorlesung von Dr. Konrad Ott, Professor für Umweltethik an dem Stiftungslehrstuhl der Michael Otto Stiftung an der Ernst-Moritz-Arndt-Universität in Greifswald am 17. Oktober 1997. Michael-Otto-Stiftung für Umweltschutz, Hamburg, pp 51–88

Ott K, Döring R (2008) Theorie und Praxis starker Nachhaltigkeit. Metropolis-Verlag, Marburg
Ott K (2010) Umweltethik zur Einführung. Junius, Hamburg
Patzig G (1971) Ethik ohne Metaphysik. Vandenhoek & Ruprecht, Göttingen
Pieper A (2007) Einführung in die Ethik. Francke, Tübingen/Basel
Pieper A, Thurnherr U (1998) Einleitung. In: Pieper A, Thurnherr U (eds) Angewandte Ethik. Eine Einführung. Beck, München, pp 7–13
Rapp C (2006) Aristoteles. In: Düwell M, Hübenthal C, Werner MH (eds) Handbuch Ethik. Metzler, Stuttgart/Weimar, pp 69–81
Rawls J (1976) A Theory of Justice. Oxford University Press, Oxford
Renn O (2007) Ethische Anforderungen an eine Nachhaltige Entwicklung: Zwischen globalen Zwängen und individuellen Handlungsspielräumen. In: Altner G, Michelsen G (eds) Ethik und Nachhaltigkeit: Grundsatzfragen und Handlungsperspektiven im universitären Agendaprozess. Verlag für akademische Schriften, Frankfurt a.M, pp 64–99
Rogall H (2003) Akteure der nachhaltigen Entwicklung. Der ökologische Reformstau und seine Gründe. Ökom-Verlag, München
Rogall H (2008) Ökologische Ökonomie. Eine Einführung. VS Verlag für Sozialwissenschaften/ GWV Fachverlage GmbH, Wiesbaden
Schüßler R (2008) Nachhaltigkeit und Ethik. In: Kahl W (ed) Nachhaltigkeit als Verbundbegriff. Mohr Siebeck, Tübingen, pp 60–79
Schwemmer O (2005) Anthropologie. In: Mittelstraß J (Hrsg) Enzyklopädie Philosophie und Wissenschaftstheorie, vol I. Metzler, Stuttgart, u.a., pp 153–157.
Statistisches Bundesamt (ed) (2009) Bevölkerung Deutschlands bis 2060. Begleitmaterial zur Pressekonferenz am 18. November 2009 in Berlin. Retrieved on 19 Mar 2012 at: http://www.destatis.de/jetspeed/portal/cms/Sites/destatis/Internet/DE/Presse/pk/2009/Bevoelkerung/pressebroschuere__bevoelkerungsentwicklung2009,property=file.pdf
Thurnherr U (2000) Angewandte Ethik zur Einführung. Junius, Hamburg
Unnerstall H (1999) Rechte zukünftige Generationen. Königshausen u. Neumann, Würzburg
Weber M (1992) Politik als Beruf. Reclam, Stuttgart

Chapter 16
Ocean Space and Sustainability

Jan H. Stel

Abstract The notion of ocean space stands for a holistic, system science approach combined with 4D thinking from the ocean, and the processes within it, towards the land. It is in fact a social-ecological concept that deals with sustainability challenges which are the consequence of the complex interactions between human activities and the marine environment at all scales. Ocean space is a critical player in the Earth System, it's central to climate regulation, the hydrological and carbon cycles and nutrient flows, it balances levels of atmospheric gases, it's a source of raw materials, and a sink for anthropogenic pollutants. On a human scale, it is impressively large. On a planetary scale, however, it's insignificant, although it's an ancient feature of the Earth.

Sustainability in ocean space is still an emerging issue. Since the early seventeenth century the Grotian notion of *Mare Liberum*, has dominated the unsustainable, use of ocean resources. Grotius, main challenge was to warrant freedom of navigation, trade, fisheries and whaling for the Dutch Republic. He was not at all interested in sustainability. In the 1960s Arvid Pardo introduced the principle of the 'Common Heritage of Mankind', which is incorporated in the present international Law of the Sea. It is an ethical and even today, controversial concept.

In this paper the global sustainability framework of the United Nations Convention on the Law of the Sea, and regional European developments with regard to its shared Exclusive Economic Zone, are discussed. It is concluded that for sustainability in ocean space, a more up-to-date and integrated or holistic, approach is urgently needed.

Keywords Earth system science • Anthropocene • Arctic climate change • UNCLOS • Apocalypse • Grotius • EU marine strategy • Ocean space

---

J.H. Stel (✉)
International Centre for Integrated assessment and Sustainable development (ICIS), Maastricht University, P.O. Box 616, 6200 MD Maastricht, The Netherlands
e-mail: janstel@skynet.be

H. Heinrichs et al. (eds.), *Sustainability Science*, DOI 10.1007/978-94-017-7242-6_16

1 Introduction

Perceptions colour our view towards ocean space. Indian Ocean societies and China have viewed the sea as a special place of trade, beyond society and social processes. It was considered an area to be crossed as quickly as possible; not as territory for control, influence or social power (Steinberg 2001). European societies, however, view ocean space in terms of ownership of (territorial zones) or access to resources and the freedom to trade. After World War II, this perception was more and more influenced by new technology that allowed a growing use of ocean resources. As a consequence, Grotius' notion of *Mare Liberum* was increasingly disregarded. This led to a new Law of the Sea treaty, which defines the rights and responsibilities of states with regard to the use of ocean space.

In contrast, the perception of the ancient Polynesian society was unique. Polynesians view the ocean as a multitude of islands connected by short journeys, in a field of crosscurrents, wave patterns, shifting breezes and flotsam, rich in bird and sea life, all laid out under a series of rotating constellations, whose intersection with the horizon easily marks one's place on the trail between islands (Lewis 1978). Their culture was fully adapted to ocean space. They knew how to live, and survive, within the ocean environment. Their sophisticated navigation system was based on observations of stars, ocean swells, flight patterns of birds and other natural signs. They used charts of sticks and shells to record the interference patterns of waves intersecting with islands (McKay and Walmsley 2003). And, as they moved further away from the continents, they developed a portable agricultural system, in which domesticated plants and animals were carried in their canoes for transplantation on the islands they encountered. They lived and survived in an immense, undefined ocean world where they could find their way over the open ocean – the surface of ocean space.

2 What Is Ocean Space?

Oceans are an ancient, 4.4 billion years old, characteristic of the Earth. The word as such is derived from the Ancient Greek 'okeanos', referring to a (3D) body of saline water. Time is the fourth dimension and leads to the notion of ocean space (Stel 2002, 2013). Humans have, just for convenience, divided the world ocean into the Pacific, Atlantic, Indian, Southern (Antarctic) and Arctic oceans. In reality, however, they are only temporary features of a single world ocean. At the dawn of the third millennium, outer space exploration has frequently reemphasised the Earth as a blue dot in the universe. Therefore, exploring and understanding the special colour of our planet, as determined by ocean space, is one of the big challenges of this century.

Ocean space – 1.37 billion km^3 of water covering some 70 % of the Earth's surface – is a different world, which, even today, we barely know. It is a dynamic world with complex currents, waterfalls and cataracts. Just 5 % has been explored. Life is everywhere, from microbes in watery cracks in the deep ocean floor to life in freshwater lakes and streams on the land filled with water temporarily on loan from the ocean. It's a weightless and mostly dark world, like outer space. It's a world alien to us, as a terrestrial species.

From an ocean perspective, phenomena like El Niño and La Niña, the thermohaline circulation, the onset and intensity of the Asian Monsoon, the carbon and water cycles and the release of methane from the (Arctic) ocean floor are shaping life on the land and framing human activities. In truth, processes within ocean space shape and mould our daily lives, our activities, our societies and our history (see Boxes 16.1 and 16.3). Ocean space is the last physical frontier on earth. The main drivers in ocean exploration are new technology (miniaturisation, biomarkers, etc.) and the fast increase in computer power for modelling.

From our human perspective, the oceans seem quite vast, but in regard to the planet as a whole, they are almost as insignificant as we ourselves. There is more water chemically trapped within the Earth's hot interior than there is in ocean space and the atmosphere. Ocean space is a critical player in the Earth system: it controls the climate, the hydrological and carbon cycles and nutrient flows and the gases in our atmosphere, it provides us with raw materials for use and it helps the planet attend to the anthropogenic pollutants, like CO_2, that result from that use. It's hard to understand why ecosystem services, as well as the value of ocean space, are not taken into account when we discuss human activities.

The notion *of ocean space* was coined in the 1960s, and *stands for a system science approach combined with thinking from the ocean, and the processes within it, towards the land.* It includes both human activities that are influenced by ocean space and human activities, like the exploitation of ocean resources and pollution, that affect ocean space itself. It's a concept that joins ideas of sociology and ecology to deal with sustainability challenges resulting from the complex interactions between human activities and the marine environment from the local to global levels. So far, the local to regional scale has been addressed in, for example, the concept of Integrated Coastal Zone Management (ICZM) which advocates a holistic approach for coastal zone management to reach sustainable development. Later, it was widened to the management of regional seas like the Baltic and EEZs (Stel 2006, 2012). [AU: A number of the ideas in these two paragraphs were expressed almost verbatim in the abstract. It is not uncommon for books of this type to use that format, but since that hasn't been the case in the other chapters so far, I felt that these should be re-written slightly so as to not be exact restatements of the same ideas.]

Box 16.1: Apocalypse: Climate Change in the Fourteenth Century

Fig. 16.1 Scene of the Four Horsemen of the Apocalypse Tapestry, Château d'Angers, France. The tapestry was ordered by Duke Louis I of Anjou in 1373, designed by the Flemish painter Hennequin de Bruges or Jan van Bondel and woven by Paris weavers between 1373 and 1389. The four horsemen represent pestilence, war, famine and death and herald the end of the world, according to Christian belief

Subtle changes in thermohaline circulation (THC) in the Northern Atlantic part of ocean space triggered a natural climate variation by bringing warmer seawater into the North Atlantic. The Medieval Warm Period (ca. 950–1300) allowed Vikings to travel far north, colonising Greenland and reaching Canada. It also spurred a cod fishing industry off western Greenland in waters some 4 °C warmer than before. In Western Europe, warm temperatures allowed for a rapidly increasing population, leading to urbanisation, prosperity and pollution on a local to regional level.

All this ended with the onset of the Little Ice Age (ca. 1300–1850), which, most likely, is related to a slowing of the THC. Weather in Europe was colder and wetter, due to a temperature drop of 1 °C. English vineyards disappeared, and fish stocks moved away from the then cold Atlantic waters. North Sea weather became stormier, leading to frequent flooding in the Low Countries. The worst climate-related event during the fourteenth century in NW Europe was the Great Famine between 1315 and 1322. In the spring of 1315, it rained continuously for up to a hundred days. This bad weather led to famine, with mortality rates up to 10 and 18 % in, respectively, Belgium and England.

(continued)

Box 16.1: (continued)

Late medieval societies not only had to cope with climate change but also with alien species causing the Black Death, killing more or less half of Europe's population, and man-made disasters, like the Hundred Years' War. Thus, it's not entirely surprising that some at the time did indeed conclude that the biblical apocalypse was near. But it was not.

3 Ocean Space and Sustainability

Since the early seventeenth century, ocean governance was dominated by the Grotian notion of *Mare Liberum*, the 'Freedom of the Seas'. De Groot, however, was not at all interested in sustainable use of ocean resources (Box 16.2). His main challenge was to warrant freedom of navigation, trade, fishery and whaling for the Dutch Republic (1581–1795). This type of thinking remained standard in use of ocean resources until the 1960s, when Arvid Pardo coined the notion of the Common Heritage of Mankind, a new type of ethical thinking (still controversial to this day) which has been incorporated into the present international Law of the Sea. But, for sustainability in ocean space, a more up-to-date and integrated approach is needed.

3.1 UNCLOS: A Global Framework

The notion of ocean space is derived from the Preamble of the United Nations Convention on the Law of the Sea (UNCLOS 1982). It is closely linked to Arvid Pardo (1914–1999), who became famous for his Draft Ocean Space Treaty, a working paper submitted by Malta to the UN Seabed Committee in 1971. Through the principle of the Common Heritage of Mankind (CHM), Pardo considered ocean space and its resources to be a global common that could not be owned by states. His principle forms a contrast with Grotius' *Mare Liberum*, which creates an open access regime and allows for its laissez-faire use.

Pardo, as well as Mann Borgese (1918–2002), advocated a sustainable use of ocean resources, its conservation and the transfer of knowledge and funds (capacity building; Stel 1990, 1994) to developing countries. The CHM concept comprises four building blocks: economic development, environmental protection, peace building and ethics for the sharing of the benefits. Basically, they are the three pil-

Box 16.2: Santa Catarina: A Tipping Point in Ocean Governance

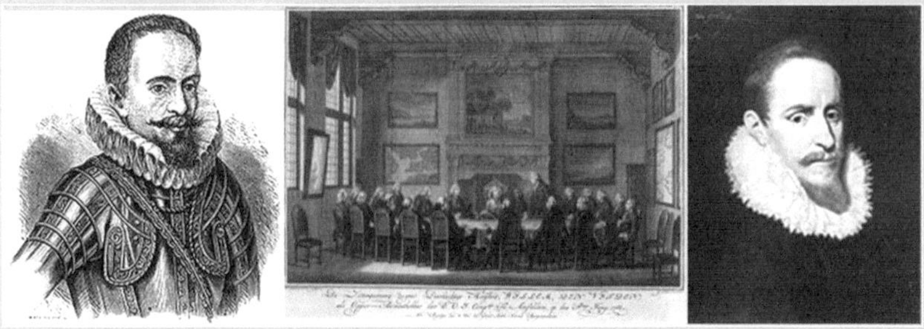

Fig. 16.2 Jacob van Heemskerck, the Gentleman XVII of the VOC, and Hugo de Groot, players in a lawsuit that changed the world through the introduction of the notion of *Mare Liberum*

The capture of the 1.500 tonnes, Portuguese carrack *Santa Catarina* on 25 February 1603 by Admiral Jacob van Heemskerck off the coast of Malacca turned out to be a tipping point in international maritime law.

Although the 750 passengers, among whom a hundred women were allowed to leave peacefully, the ship and its cargo of Chinese silk, musk and Ming porcelain were kept as a prize: a valuable jackpot. When auctioned in Amsterdam in the fall of 1604, the profit was around 3.35 million Dutch guilders. In today's currency, this would be an estimated €54 million.

To the Dutch and certainly most of the shareholders of the United Dutch East India Company (VOC), van Heemskerck was a hero (Fig. 16.2). To the Portuguese, he was a pirate, and they reclaimed the ship and its valuable cargo. The Gentleman XVII hired Hugo de Groot or Grotius, a young brilliant lawyer. In his defence, de Groot wrote *De Jure Praedae*, which was largely based on van Heemskerck's own reasoning – revenge for the mistreatment of Dutch merchants in the East Indies by the Portuguese – to attack the carrack (Van Ittersum 2003, 2010). On 4 September 1604, the VOC formally confiscated the *Santa Catarina*. The decision of the Amsterdam Admiralty Court was widely publicised to gain national and international support.

Grotius also tipped the international scale in maritime law by introducing the notion of *Mare Liberum*, the principle of the 'Freedom of the Seas'. For this, he still is referred to as the 'father of international law' (Kröner 2011). However, Grotius' main aim was the right of free trade in Asia and the Americas for the Dutch Republic. By this, he was one of the founders of Dutch colonial rule (Boschberg 2006; Van Ittersum 2010). Even today, Grotius' notion is facilitating a whole range of unsustainable human activities in ocean space.

lars of sustainability. They were, however, strongly opposed by the USA. Nevertheless, the CHM concept is partly incorporated in the final text of part of UNCLOS. In retrospect, the visionary Pardo and Mann Borgese were frontrunners in a transition towards sustainability in ocean governance, a transition still to come.

UNCLOS led to the greatest 'land grab' in human history through the introduction of the concept of the exclusive economic zone (EEZ) (Fig. 16.3). This is a marine zone of 200 nautical miles in which states have the right to exploit marine resources in a sustainable way. Technically, it does not include the state's territorial waters. The EEZ's inner boundary follows the borders of the state's territorial waters (usually 12 nautical miles from the coast). The present enclosure through the EEZs covers approximately 142 million km^2, an area almost as large as the land surface, and covering 40 % of the world's oceans. They contain 90 % of marine resources.

Another new element of UNCLOS was the establishment of the International Seabed Authority for the exploitation of non-marine, ocean resources outside the EEZs. The open waters of the High Seas, however, are still a global common, where the 'tragedy of the commons' in deep sea fishery is part of daily life (Ostrom et al. 2000) The concept of *ocean states – a hierarchy of states based upon the size of the EEZ* – is also an effect of UNCLOS (Stel 2002, 2012). The European Union, with its 28 member states, has a shared EEZ of some 25 million km^2. As such, it's by far the largest in the world. The ocean-land ratio for the EU is about 5:1. Based on this ratio, one could consider the marine domain as the most important feature of the EU-28. As a terrestrial species, however, we tend to focus on the land instead of the sea (Steffen et al. 2011). Moreover, this ratio is also blurring the real situation, as most of the shared EEZ is situated outside Europe and relates to former colonies. From a national perspective, the USA has the world's largest EEZ, followed by France, Australia and Russia.

UNCLOS, ratified by 165 states and the European Union per 19 February 2013, governs all aspects of ocean space. This includes the delimitation of maritime boundaries, environmental regulations, scientific research, commerce and the settlement of international disputes involving marine issues. With Chapter 17 of Agenda 21, which

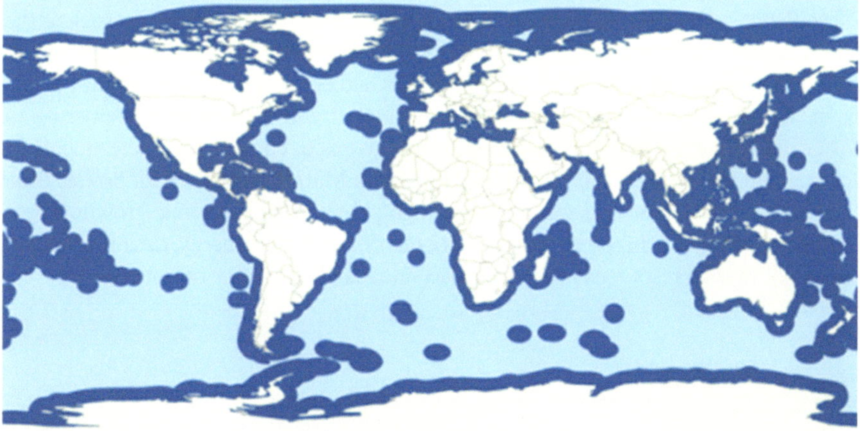

Fig. 16.3 The exclusive economic zones of the world

resulted from the United Nations Conference on Environment and Development (UNCED 1992), as well as the outcome of its successors, Johannesburg 2002 and Rio+20 in 2012, and the Convention on Biological Diversity (CBD 1993), it sketches the contours of a new and holistic vision of governance. In this perception, an ecosystem approach and stakeholder participation are key building elements. The 2012 UN initiative 'Oceans Compact' builds on the outcome of the Rio+20 conference. It aims at 'Healthy Oceans for Prosperity' or sustainability in ocean space.

3.2 Regional European Union

Europe, through the European Union (regional level), is, with the USA and Australia (national level), taking the lead in developing integrated approaches towards ocean space sustainability. In Europe, this transition has been dominated by the introduction of the Integrated Maritime Policy (IMP) in 2007 and the Marine Strategy Framework Directive (MFSD), which came into force in June 2008. The latter is the environmental pillar of the IMP, which was developed through extensive stakeholder consultation. The 23 EU member states with marine territories are obliged to protect the marine environment, to achieve a 'good marine environmental status' by 2020 and to protect the resource base for marine-related economic and social activities. A *good marine environmental status* is defined as *an ecologically diverse, dynamic, healthy and productive ocean space*.

Member states have to develop marine strategies that serve as action plans for applying an ecosystem-based approach to the management of their human activities. These strategies must be based on marine regions (Greater North Sea, Baltic) or subregions (Adriatic Sea) (Fig. 16.4) and should address the 11 Marine Strategy Framework Directive descriptors, like biological diversity, nonindigenous species, eutrophication, contaminants, marine litter, energy and noise (EU 2012).

The IMP covers a number of cross-cutting policy objectives in areas like marine data and knowledge, maritime spatial planning, Blue Growth, integrated maritime surveillance and coordination at the level of regional seas or marine basins such as the North Sea and the Baltic. Blue Growth refers to long-term economic growth based on different maritime sectors. Thus, it is about jobs and economic growth, and might just be the new magic word for continuing unsustainability in ocean space. At any rate, to foster the future exploitation of Europe's ocean space, considerable investments in science and technology will be needed (European Marine Board 2013). So far, Blue Growth has focused on five sectors with a high potential for economic growth. These sectors are short sea shipping, coastal tourism, offshore wind energy, desalination and the use of marine resources in the pharmaceutical and cosmetics industries.

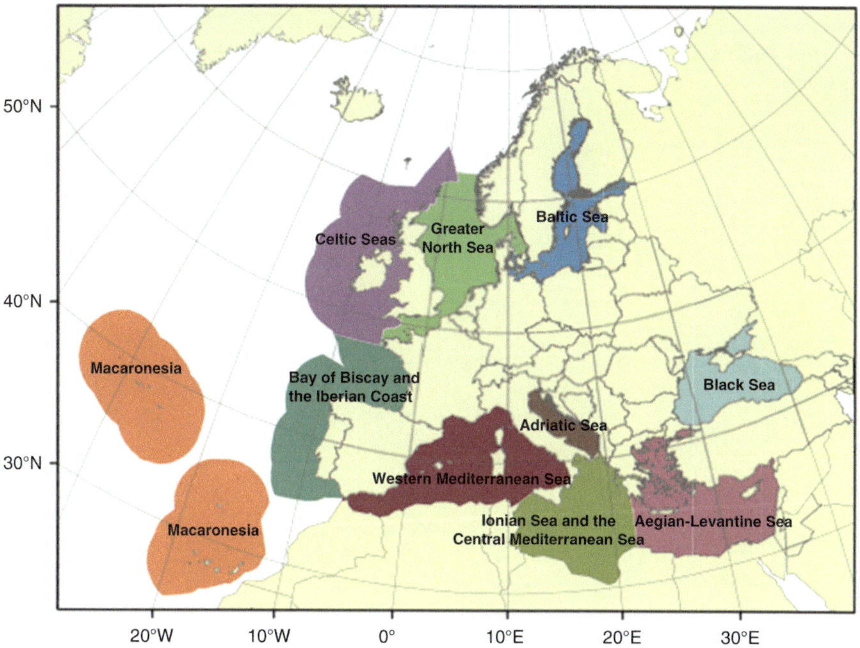

Fig. 16.4 Regions and subregions of the EU Marine Strategy Framework Directive

4 Conclusions

In 1967, when Arvid Pardo addressed the United Nations General Assembly about the new Law of the Sea, disciplinarity was the common research mode. Then, just as today, most oceanographers were trained in one of the traditional sciences, like physics, chemistry, biology and geology, or in a related field of engineering, meteorology, etc. (Pinet 2009). In dedicated research institutes or university departments, these disciplinary boundaries mostly blur though multi- and interdisciplinary research efforts. But even today, and despite the recognition of anthropogenic forcing in many modern environmental issues, social sciences are mostly not a part of or affiliated with these ocean research institutes. This is hampering the development of sustainability in ocean sciences.

Since Pardo, sustainability issues in ocean space have evolved in ways that never could have been imagined when UNCLOS was negotiated during the 1970s. Firstly, the world population grew from 3.4 to an estimated 7.2 billion by the end of 2013. Doubling the population also caused a tremendous growth in human activities, welfare and consumption. This leads to an ecological footprint that is increasingly overshooting the carrying or bio-capacity of the Earth (WWF 2012) with more than 50 % in 2012 and a biodiversity loss of 38 % between 1970 and 2008. Secondly, new technology allowed us to explore outer space, ocean space, the deep Earth, the micro- and nanoworld, etc. It dramatically changed society and lead to globalisation

Box 16.3: CO2 Pollution: Are We Ready for an Ice-Free Arctic

Fig. 16.5 Canadian research vessel in the Arctic in 2012

The Arctic is critical to our understanding of the global dimensions of anthropogenic climate change. It is the canary in a coal mine. In the old days, coal miners brought these small birds with them into the mines to detect odourless and colourless, but rather dangerous, pockets of methane or carbon monoxide. As long as the bird kept singing, the miners knew their air supply was safe. A dead canary, however, signalled an immediate evacuation. They were used in British coal mines until the late 1980s, when technology took over. Likewise, there are a selected number of signals in the Arctic that convey change and danger in the near future.

On 16 September 2012, Arctic summer ice cover reached its lowest level since instrumental records began. At just 3.4 million km^2, it follows an alarming decadal trend. Many scientists are now predicting an ice-free Arctic within a few years or decades at best. The environmental and societal implications are enormous, and as the ice is disappearing faster than predicted, we are largely unprepared. How will this, for instance, impact the European and North American weather system? We simply do not know. So, one could conclude that we are not at all ready for an ice-free Arctic.

and the post-industrial information society. It goes without saying that this new technology is also constantly reshaping ocean space research.

The management challenges of ocean space are changing rapidly, because of the increasing demand for resources, as well as the negative impact of human activities through CO_2 and other types of pollution, ocean acidification, dead zones and algal blooms. Since UNCLOS, fishing fleets have grown larger and more efficient, leading to overfishing that threatens 85 % of the world's fish stocks (FAO 2012). New technology will also allow deep sea oil and gas exploration and deep sea mining in the very near future. Finally, new scientific insights have paved the way for the development of pharmaceutical and cosmetic uses of marine genetic resources. So, ongoing unsustainable use of ocean resources might lurk just around the corner. This is one of the main challenges for sustainability science in the near future.

New standards of environmental planning and decision-making have been developed over recent decades and are, as a consequence, not (yet) dealt with in UNCLOS. These new standards are, for instance, the precautionary principle, the ecosystem approach and the ecosystem services. On the other hand, new tools like marine-protected areas, maritime spatial planning, strategic environmental assessments, environmental impact assessments and marine bioregional plans have been developed to protect ocean space, its resources and its biodiversity. Some of them are incorporated in new regional or national approaches, like the European IMP, and national management plans like those of Australia and the USA. But the need for sustainability in ocean space, based upon an internationally agreed-upon holistic view and vision, is urgent (Stel 2010). Ocean space is also a crucial element of the biosphere and delivers ecosystem services that dwarf traditional economic returns (Costanza et al. 1997, 2007).

Questions

1. How did climate change affect early fourteenth-century Europe?
2. How did technological advances affect everyday life?
3. What is ocean space?
4. Why did Grotius coin the notion of Mare Liberum? What was the effect for the Dutch Republic and other European countries and why has this notion become a leading principle in ocean law?
5. How did Arvid Pardo and Elisabeth Mann Borgese introduce sustainability in the present UN Law of the Sea?
6. Why should UNCLOS be actualised? What are the new elements and what are the threats?
7. Are we ready or not for an ice-free Arctic?
8. What are the main sustainability challenges in ocean space?

Further Reading

Borgese EM (1998) The oceanic circle. Governing the seas as a global resource. United Nations University Press, Tokyo/New York/Paris, 240 pp

Davor V, Schei PJ (eds) (2011) The world ocean in globalisation. Martinus Nijhoff Publishers, Leiden/Boston, 549 pp

Earle S (2009) The world is blue: how our fate and the ocean's are one. National Geographic Books, Washington, DC, 303 pp

Gelpke N, Visbeck M (eds) (2010) World Ocean Review 2010: living with the oceans. Mare, Bremen, 232 pp. (Downloadable at: http://worldoceanreview.com/)
Pinet PR (2009) Invitation to oceanography. Jones and Bartlett Publishers, Inc., Ontario, 626 pp
Stel JH (2013) Ocean space and the anthropocene, new notions in geosciences? An essay. In: Mulder EWA, Jagt JWM, Schulp AS (eds) The Sunday's child of Dutch earth sciences – a tribute to Bert Boekschoten on the occasion of his 80th birthday. Neth J Geosci – Geologie en Mijnbouw 92(2/3):194–211

Websites

GMES	http://www.gmes.info/
ICZM	http://ec.europa.eu/environment/iczm/
IMP	http://ec.europa.eu/maritimeaffairs/
IOI	http://www.ioinst.org/
NASA	http://www.nasa.gov/
NASA-Aollo Project	http://www.nasa.gov/mission_pages/apollo/
NOAA	http://www.noaa.gov/
Oceans Compact	http://www.un.org/depts/los/ocean_compact/oceans_compact.htm
Rio+20	http://www.uncsd2012.org/
UNCLOS	http://www.unclos.com/

References

Boschberg P (2006) Hugo Grotius' theory of trans-oceanic trade regulation: revisiting *Mare Liberum* (1609). History and theory of international law series, Working Paper Rev Aug 2006. Cite as: Institute for International Law and justice, IILJ, New York, 60 pp
Costanza R, D'Arge R, De Groot R, Farber S, Grosso M, Hannon B (1997) The value of the world's ecosystem services and natural capital. Nature 387:253–260
Costanza R, Graumlich L, Steffen W (eds) (2007) Integrated history and future of people on Earth, Dahlem workshop report 96. MIT Press, Cambridge, 495 pp
European Commission (2012) Links between the Marine Strategy Framework Directive (MSFD 2008/56/EC) and the Nature Directives (Birds Directive 2009/147/EEC (BD) and Habitats Directive 92/43/EEC (HD)), 31 pp
European Marine Board (2013) Navigating the future IV. Position paper 22. European Marine Board, Ostend, 202 pp
FAO (2012) The state of the world fisheries and aquaculture. FAO, Rome, 209 pp
Kröner C (2011) Crossing the *Mare Liberum:* The settlement of disputes in an interconnected world. Fundación Universitaria San Pablo CEU, Madrid, 25 pp
Lewis D (1978) The voyaging stars: secrets of the Pacific island navigators. William Collins, Sydney
McKay B, Walmsley A (2003) Maori time: notions of space, time and building form in the South Pacific. In: John Franz (ed) IDEA 2003, Journal of the Interior Design Educators Association, pp 85–95
Ostrom E, Dietz T, Dolsak N, Stern PC, Stonich S, Weber EU (2000) The drama of the commons. National Academy Press, Washington, DC, 534 pp
Pinet PR (2009) Invitation to oceanography. Jones and Bartlett Publishers, Inc, Ontario, 626 pp

Steffen W, Persson Å, Deutsch L, Zalasiewicz JK, Williams M, Richardson K, Crumley C, Crutzen P, Folke C, Gordon L, Molina M, Ramanathan V, Rockström J, Scheffer M, Schnellhuber HJ, Svedin U (2011) The anthropocene: from global change to planetary stewardship. Ambio 40:739–761

Steinberg PE (2001) The social construction of the ocean. Cambridge University Press, Cambridge

Stel JH (1990) Partners in science. In: Proceedings 23' annual conference of the Law of the Sea Institute, Hononulu, pp 535–553

Stel JH (1994) Partners in marine science; facing our global responsibility. In: Oceanology International 94, The Global Ocean, conference proceedings, vol 3, section 2, pp 1–10

Stel JH (2002) Mare Nostrum – Mare Liberum – Mare sit Aeternum, duurzaam gebruik van de oceanische ruimte. Inaugural speech, Maastricht University, The Netherlands

Stel JH (2006) Governance of Europe's exclusive economic zones, a vision. In: Dahlin H, Flemming NC, Marchand P, Petersson SE (eds) European operational oceanography: present and future. Proceedings of the fourth international conference on EuroGOOS. European Communities, pp 302–311

Stel JH (2010) Multilevel governance of ocean space, a challenge for the informed society of tomorrow. In: Proceedings of international conference 50 years of Education and Awareness Raising for Shaping the Future of the Oceans and Coasts, St. Petersburg 27–30 April 2010. Russian State Hydrometeorological University, St. Petersburg, pp 98–115

Stel JH (2012) De oceaan anders bekeken. ACCO, Leuven

Stel JH (2013) Ocean space and the anthropocene, new notions in geosciences? – An essay. In: Mulder EWA, Jagt JWM, Schulp AS (eds) The Sunday's child of Dutch earth sciences – a tribute to Bert Boekschoten on the occasion of his 80th birthday. Neth J Geosci – Geologie en Mijnbouw 92(2/3): 167- AANVULLEN

Van Ittersum MJ (2003) Hugo Grotius in context: Van Heemskerck's capture of the *Santa Catarina* and its justification in *De Jure Praedae* (1604–1606). Asian J Soc Sci 31(3):511–548

Van Ittersum MJ (2010) The long goodbye: Hugo Grotius and the justification of Dutch expansion overseas (1604–1645). Hist Euro Ideas 36:386–411

WWF (2012) Living planet report 2010: biodiversity, biocapacity and development. WWF, Gland, 162 pp

Chapter 17
Sustainable Landscape Development

Michael Stauffacher and Pius Krütli

Abstract Sustainable landscape development is situated at the centre of sustainable development, covers the urban and rural areas, spans multiple scales, raises problems of justice, is multi-sectorial and can thus only be understood and managed through holistic approaches. The scientific field of sustainable landscape development is located at the interfaces of several disciplines, namely, landscape ecology, urban and landscape planning and rural and regional sustainable development. A bridging concept between the more natural and the more social scientifically oriented landscape perspectives is that of landscape functions and services. Sustainability science can contribute to this research with its coupled system perspective on the socioecological dimensions of landscapes. Further, sustainability science understood as transdisciplinary collaborative process of science and society offers also guidance on how to tackle the normative character of sustainable landscape development.

Keywords Functions of landscapes • Landscape services • Urban and rural planning • Socioecological systems • Transdisciplinary research • Resilience

Before reading this chapter, please reflect on your own as to what you expect to read here. Doing so, you can focus on the term "landscape", as you should already be familiar with sustainable development from the other chapters. Thus, take a sheet of paper and write down in your own words what is your understanding of the term "landscape".
After you have done so, review your notes and try distinguishing when you have referred to (i) the bio-physical environment (natural, ecological, etc.), (ii) the social side (human, individual, institutional, etc.) of the phenomenon and (iii) the interface, the interactional aspects of both (socio-ecological, human-environmental, etc.). Further, have you focussed on rural areas or have you also considered urban areas?

Now, you are prepared to read this chapter. After each chapter, you should stop and write down in your own words what you have learned so far.

M. Stauffacher (✉) • P. Krütli
ETH Zurich, USYS Transdisciplinarity Laboratory (TdLab), Sonneggstrasse 33, SOL F5, 8092 Zurich, Switzerland
e-mail: michael.stauffacher@env.ethz.ch; pius.kruetli@env.ethz.ch

H. Heinrichs et al. (eds.), *Sustainability Science*, DOI 10.1007/978-94-017-7242-6_17

1 What Is a 'Landscape'?

The meaning of the term landscape is multifaceted (Selman 2010). One can roughly distinguish between a natural scientific, biophysical definition (ecocentric) and a definition driven by aesthetic and sociocultural criteria (anthropocentric). Whilst the first postulates an objectively given and often functionally defined area, the latter conceives of landscape primarily as a subjectively defined and purely mental object (Kirchhoff et al. 2013). According to Kirchhoff and colleagues, the second meaning can be traced back to the historical roots of the terms 'Landschaft' in German and 'landschap' in Dutch, designating 'a painting using central perspective, in which an area of land is represented as an aesthetically whole' (Kirchhoff et al. 2013, p. 38). A pragmatist's definition at the interface of both perspectives is offered by the European Landscape Convention (see Box 17.1), stressing public perception as well as physical properties. Selman (2012) introduces long lists of physical and social drivers and understands landscape development as a coupled process of both. Thus, landscapes can best be conceptualised as socioecological systems (Ostrom 2009; Walker et al. 2004) or human–environmental systems (Turner et al. 2003; Scholz 2011).

Box 17.1: European Landscape Convention: Definitions (Council of Europe 2000)

- 'Landscape' means an area, as perceived by people, the character of which is the result of the action and interaction of natural and/or human factors.
- 'Landscape protection' means actions to conserve and maintain the significant or characteristic features of a landscape, justified by its heritage value derived from its natural configuration and/or from human activity.
- 'Landscape management' means action, from a perspective of sustainable development, to ensure the regular upkeep of a landscape, so as to guide and harmonise changes which are brought about by social, economic and environmental processes.
- 'Landscape planning' means strong forward-looking action to enhance, restore or create landscapes.

2 What Is 'Sustainable Landscape Development'?

Parts of the discussion around sustainable landscape development are strongly influenced by the tradition of landscape preservation (i.e. protection from any use), which in the USA led, in the nineteenth century, to the creation of national parks based on an understanding of wilderness (Selman 2010) as reaction to the ubiquitous urbanisation and industrialisation processes. The discussions are still 'dominated by a simplistic dualism between 'traditional' landscapes and modern' (Widgren 2012, p. 105). The latter disregards that so-called traditional landscapes are themselves the product of human history and ever changing (see the various works by Marc Antrop 2005; 2006). The European Landscape Convention (see

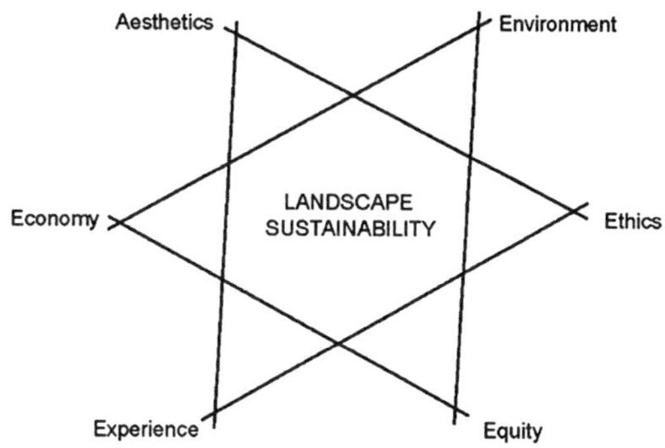

Fig. 17.1 The six Es of sustainable landscape development: environment, economic, equity, aesthetics, experience and ethics (Musacchio 2009, p. 998)

Box 17.1) became, in recent times, at least in Europe, a core reference for a much broader understanding of sustainable landscape development.

To define sustainable landscape development, Paul Selman distinguishes between different core functions of landscapes: (i) environmental sustainability concerned with spatial patterns and especially habitat fragmentation; (ii) economic sustainability in which the landscapes offer jobs (e.g. in tourism) and yield (e.g. food); (iii) social sustainability securing a just allocation of access and risks; (iv) political sustainability emphasising public participation in related political decision processes; and, finally, (v) aesthetic sustainability, because visual perception is always essential for landscapes (Selman 2009, 2010, 2012). In a similar vein, Laura Musacchio distinguishes between the six Es of sustainable landscape development: environment, economy, equity, aesthetics, ethics and experience (see Fig. 17.1). Musacchio thus adds specifically the experiential character, as landscapes have to be experienced and cannot solely be studied abstractly. She further proposes a conceptual framework for research and practice, with a special emphasis on the recognition of the coupled character of human and natural systems prevalent in landscapes, the importance of landscapes in urban areas and the multiple scales involved, from the local to the global, because various global drivers, like climate change, urbanisation and globalisation of value chains, actually impact local landscapes (Musacchio 2009).

Sustainable landscape development is thus, in fact, at the centre of sustainable development (see as well Selman 2010, p. 397): landscapes are universal, covering not only rural areas but urban ones as well (Wu 2010); they are dynamic (Antrop 2006); they are hierarchical, spanning multiple scales (Musacchio 2009); they can restrict access to certain groups and expose some groups to hazards, i.e. raise problems of justice (Walker 2011); and finally, they are multi-sectorial and can only be understood and managed through holistic approaches (Kirchhoff et al. 2013).

3 How Is 'Sustainable Landscape Development' Scientifically Tackled?

The field of sustainable landscape development is located at the interface of several disciplines, to name just the three most essential: (i) landscape ecology, (ii) urban and landscape planning and (iii) rural and regional sustainable development. The first deals with questions about 'how landscape structure affects the functioning of landscapes' (Wiens 2013) and their ecosystem services, i.e. emphasis on analysis of the biophysical processes; the second aims at 'human and ecological communities that are resilient, sustainable, and less vulnerable to disturbance events' (Gobster 2011, p. 315), i.e. emphasis on planning and management; and the third focuses on (economic) development, including a critical discourse around sustainable development itself (Marsden and Sonnino 2008), and addresses, for instance, the role of power and knowledge (Bruckmeier and Tovey 2008). Classical landscape ecology focuses both on spatial patterns and ecological processes and largely excludes people or sees them primarily as causing landscape change.

A proposal for a bridging concept between the more natural and the more social scientifically oriented landscape perspectives has been presented by Jianguo Wu: 'landscape sustainability is the capacity of a landscape to consistently provide long-term, landscape-specific ecosystem services essential for maintaining and improving human wellbeing' (Wu 2013, p. 1013). Termorshuizen and Opdam (2009) in a similar vein introduce the concept of 'landscape services' (see Fig. 17.2). 'Landscape

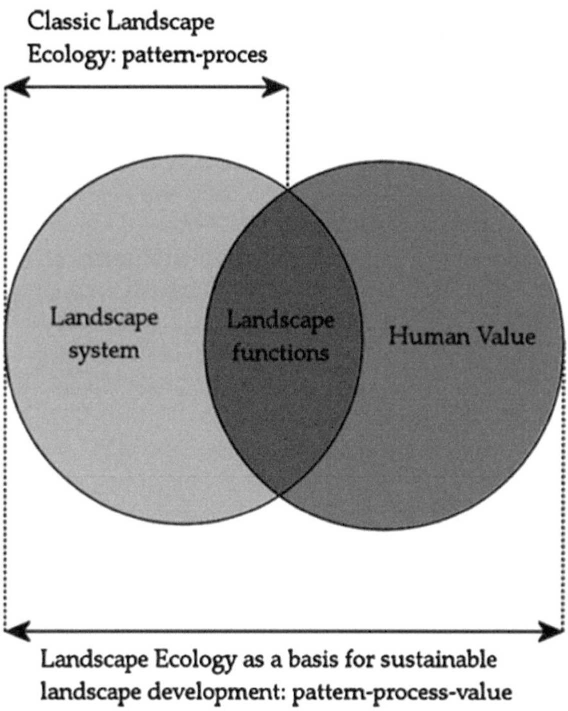

Fig. 17.2 Landscape functions as a bridging concept between the physical landscape system and the human values attributed to them (Termorshuizen and Opdam 2009, p. 1040)

services' translate landscape functions into 'services' valued by people. In contrast to the very popular concept of 'ecosystem services', Termorshuizen and Opdam claim that the landscape level is actually better suited to local planning than the generally much larger scale of ecosystems. And they further stress that the term landscape is as well used by social scientists and legitimate for local people. These perspectives share the idea of multifunctionality (de Groot 2006). Production, regulation and habitat functions are generally distinguished and focused empirically; the concept is thus 'fundamentally ecocentric, having a primary concern for the functioning of the earth systems' (Selman 2009, p. 47). In contrast, the anthropogenic dimension needs to be emphasised (Bolliger et al. 2011). Cultural services are often mentioned in this regard, but their role remains vague and disputed (Daniel et al. 2012; Kirchhoff 2012).

4 What Can Sustainability Science Contribute to Sustainable Landscape Development?

Sustainability science has to make reference to and integrate the above-presented different disciplinary perspectives from a socioecological or human–environmental system approach and given its transformational character to move beyond mere analysis and address as well practical application in real-world contexts (two examples of such approaches are presented in Boxes 17.2 and 17.3) – see Chap. 3 in this book.

Box 17.2: Human–Environmental Systems (HES)-Based Transdisciplinary Processes

The aim of the so-called transdisciplinary case study design is a process of mutual learning between science and people from outside academia for the development of orientations for sustainable development. We use a case study on sustainable landscape development in the Swiss canton of Appenzell Ausserrhoden (AR) to sketch this approach (Stauffacher et al. 2008; Stauffacher and Scholz 2011). All the presented steps are pursued in transdisciplinary collaboration between science and society, following a functional-dynamic approach, i.e. the purpose and the form of collaboration is carefully designed (Krütli et al. 2010). Further, seven postulates constituting the HES framework (Seidl et al. 2013) help in structuring the complexities of the case (Fig. 17.3).

(continued)

Box 17.2: (continued)

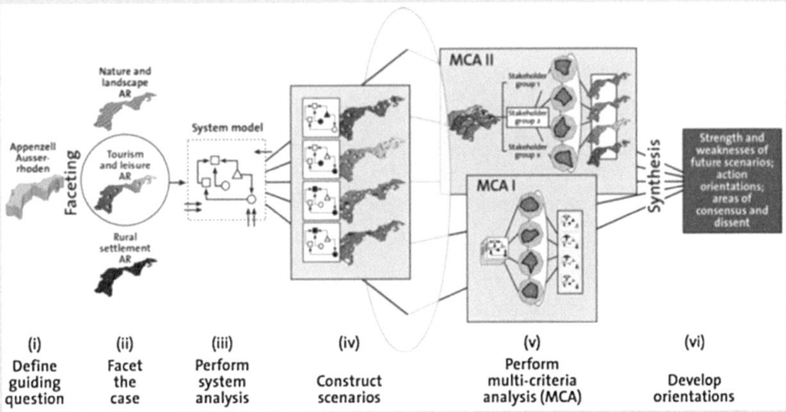

Fig. 17.3 The transdisciplinary case study design to tackle sustainable landscape development (Adapted from Scholz et al. 2006, p. 238)

- Define a guiding question
- The system (temporally and spatially) and the targeted sustainability transition are defined. In the AR case study, the guiding question reads as follows: 'How can the ecological quality of landscape in Appenzell Ausserrhoden be preserved or improved and, at the same time, the added value be sustained or even increased?'
- Facet the case
- Specific subsystems are conceptually defined to investigate the case and answer the guiding question. 'Nature and landscape', 'tourism and leisure' and 'rural settlement' were selected for the AR case study as facets; the topic 'local industries' was postponed for a subsequent case study.
- Perform system analysis
- A semi-quantitative system model is developed to describe the current state and future development of the case. In the AR case study, the history and dynamics of the region were investigated using document analysis, interviews with key stakeholders and analysis of relevant data from the statistical office. For the facet 'tourism and leisure', a massive decline in overnight stays and daily tourism was observed and many driving factors were found.
- Construct scenarios
- Based on the results of the preceding steps, different scenarios are constructed. The scenarios serve as a basis for the assessment. Three to four scenarios were constructed for each of the facets in the AR case study

(continued)

Box 17.2: (continued)

combining intuitive and analytical scenario construction (Wiek et al. 2006). For the facet 'tourism and leisure', scenarios were constructed which contrasted daily with overnight stay tourism and landscape and cultural heritage oriented towards highly intensified forms of tourism.

- Perform multi-criteria analysis (MCA)
- Two different approaches are combined: assessments referring to science-based arguments (MCA I) and obtaining individual preference information from different stakeholder groups (MCA II). A small set of nine evaluation criteria was defined for each facet in the AR case study, covering ecological, economical and social aspects. For the facet 'tourism and leisure', for instance, the number of working places in tourism, energy use and aesthetics was chosen. The analysis helped show the promising economic performance of overnight stays and the general acceptance of all stakeholder groups with respect to a tourism based on the local cultural heritage of the farmers.
- Discuss the results and develop orientations
- Insights from the previous steps serve as a basis for developing strategic orientations guiding the sustainability transition. Overall, the AR case study concluded that landscape as the main capital of the canton should be conserved but used, for agriculture, tourism and also for housing. For the facet 'tourism and leisure', it was concluded that a family holiday village with clear links to the local heritage and that marketed local farming products would be a promising option for future development. This was, in fact, latter successfully implemented in the poorest village of the canton and is still attracting families from all over Switzerland and abroad.[1]

[1] http://www.reka.ch/en/rekaholidays/rekaholidayvillages/seiten/unraesch.aspx

Both perspectives presented in Boxes 17.2 and 17.3 share a common weakness, which asks for future improvements: the societal level remains shallow. As a result, essential dimensions of societies like 'power, class, gender and ethnicity' and stratification and their consequences are largely neglected (Widgren 2012, p. 104). Thus, it is necessary to develop a more pronounced (environmental) sociological perspective on landscapes. The broad research field of environmental justice (Schlosberg 2007; Walker 2011) would certainly offer an initially promising route to follow, as social sustainability (Selman 2012) or ethics/equity (Musacchio 2009) are key for sustainable landscape development.

Box 17.3: Four-Step Framework of Resilience Management in Socioecological Systems

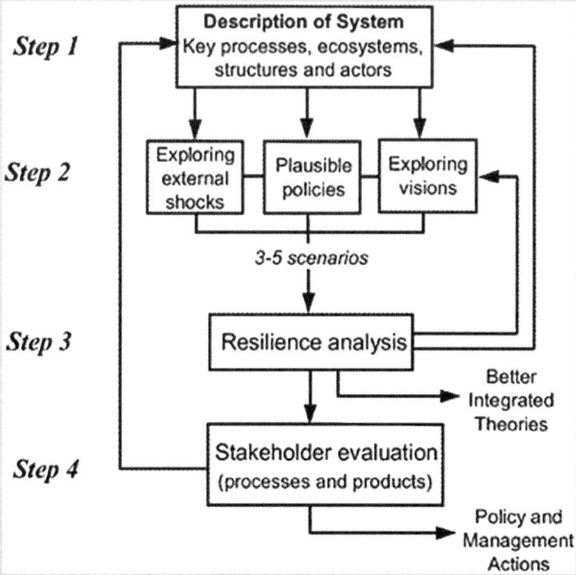

Fig. 17.4 A framework for the analysis of resilience in socioecological systems (Walker et al. 2002)

Resilience management aims at preventing 'the system from moving to undesired system configurations in the face of external stresses and disturbance', as well as nurturing and preserving 'the elements that enable the system to renew and reorganize itself following a massive change' (Walker et al. 2002). All steps briefly introduced here are implemented in interaction between scientists and various stakeholders. For more details, please refer to Resilience Alliance (2007) and Walker et al. (2002) (Fig. 17.4).

- Step 1. Resilience[2] of what?
  Development of a conceptual model of the system (including its historical profile). It should cover all of what is known and deemed important by the stakeholders and what determines them.

[2] Resilience is 'the potential of a system to remain in a particular configuration and to maintain its feedbacks and functions, and involves the ability of the system to reorganize following disturbance driven change' (B. Walker et al. 2002).

(continued)

Box 17.3: (continued)

- Step 2. Resilience to what? Visions and scenarios
  Collection and examination of external factors (major policy drivers, as well as action of stakeholders) and development of different possible future scenarios to which the system needs to be resilient.
- Step 3. Resilience analysis
  In this step, the results from the two preceding steps are integrated and the interactions between these are explored. This should allow for identifying drivers and processes that have an impact on the important characteristics of the system.
- Step 4. Resilience management (evaluation and implications)
  Finally, from the whole process and its results, a concrete action plan to increase the resilience of the system is derived.

In conclusion, sustainability science could contribute to research on sustainable landscape development with its coupled system perspective on human–environmental/socioecological dimensions of landscapes. At the same time, sustainability science, understood as a transdisciplinary collaborative process of science and society, also offers guidance on how to tackle the normative character of sustainability transitions. Conversely, the broad literature of sustainable landscape development can enrich sustainability science with concrete expertise, for instance, in landscape ecology (spatial patterns and ecological processes) and landscape aesthetics (cultural heritage).

- ***Task***: *Review your notes from the beginning and the short summaries you wrote after each chapter and reflect on the following (you would preferably do this in a group with 2–3 students): Would you revise your initial notes on 'landscape'? If yes, how and why?*

Further Reading

Antrop M (2006) Sustainable landscapes: contradiction, fiction or utopia? Landsc Urban Plan 75:187–197

Council of Europe (2000) The European landscape convention. Retrieved 3 June 2013, from http://conventions.coe.int/Treaty/en/Treaties/Html/176.htm

Musacchio LR (2009) The scientific basis for the design of landscape sustainability: a conceptual framework for translational landscape research and practice of designed landscapes and the six Es of landscape sustainability. Landsc Ecol 24:993–1013

Selman P (2012) Sustainable landscape planning. The reconnection agenda. Routledge, London/New York

Termorshuizen JW, Opdam P (2009) Landscape services as a bridge between landscape ecology and sustainable development. Landsc Ecol 24:1037–1052

Important Weblinks

http://go.worldbank.org/S2G5CQ4KD0
http://www.coe.int/t/dg4/cultureheritage/heritage/landscape/
http://www.resalliance.org/index.php/resilience_assessment
http://www.tdlab.usys.ethz.ch/
http://www.tdlab.usys.ethz.ch/education/tdcs/former/cs2001.html

References

Antrop M (2005) Why landscapes of the past are important for the future. Landsc Urban Plan 70:21–34

Antrop M (2006) Sustainable landscapes: contradiction, fiction or utopia? Landsc Urban Plan 75:187–197

Bolliger J, Bättig M, Gallati J, Kläy A, Stauffacher M, Kienast F (2011) Landscape multifunctionality – a powerful concept to identify effects of environmental change. Reg Environ Chang 11:203–206. doi:10.1007/s10113-010-0185-6

Bruckmeier K, Tovey H (2008) Knowledge in sustainable rural development: from forms of knowledge to knowledge processes. Sociol Rural 48(3):313–329

Council of Europe (2000) The European landscape convention. Retrieved 3 June 2013, from http://conventions.coe.int/Treaty/en/Treaties/Html/176.htm

Daniel TC, Muhar A, Arnberger A, Aznar O, Boyd JW, Kai MA, Chan RC, Elmqvist T, Flint CG, Gobster PH, Grêt-Regamey A, Lave R, Muhar S, Penker M, Ribe RG, Schauppenlehner T, Sikor T, Soloviy I, Spierenburg M, Taczanowska K, Tam J, von der Dunk A (2012) Contributions of cultural services to the ecosystem services agenda. Proc Natl Acad Sci U S A 103(23):8812–8819

de Groot R (2006) Function-analysis and valuation as a tool to assess land use conflicts in planning for sustainable, multi-functional landscapes. Landsc Urban Plan 75:175–186

Gobster PH (2011) Landscape and urban planning at 100: looking back moving forward. Landsc Urban Plan 100:315–317

Kirchhoff T (2012) Pivotal cultural values of nature cannot be integrated into the ecosystem services framework. Proc Natl Acad Sci U S A 109(46)

Kirchhoff T, Trepl L, Vicenzotti V (2013) What is landscape ecology? An analysis and evaluation of six different conceptions. Landsc Res 38(1):33–51

Krütli P, Stauffacher M, Flüeler T, Scholz RW (2010) Functional-dynamic public participation in technological decision making: site selection processes of nuclear waste repositories. J Risk Res 13(7):861–875. doi:10.1080/13669871003703252

Marsden T, Sonnino R (2008) Rural development and regional state: denying multifunctional agriculture in the UK. J Rural Stud 24:422–431

Musacchio LR (2009) The scientific basis for the design of landscape sustainability: a conceptual framework for translational landscape research and practice of designed landscapes and the six Es of landscape sustainability. Landsc Ecol 24:993–1013

Ostrom E (2009) A general framework for analyzing sustainability of social-ecological systems. Science 325(5939):419–422

Resilience Alliance (2007) Assessing and managing resilience in social-ecological systems: a workbook for scientists. Retrieved 3 June 2013, from http://www.resalliance.org/srv/file.php/210

Schlosberg D (2007) Defining environmental justice. Theories, movements, and nature. Oxford University Press, New York

Scholz RW (2011) Environmental literacy in science and society: from knowledge to decisions. Cambridge University Press, Cambridge

Scholz RW, Lang D, Walter AI, Wiek A, Stauffacher M (2006) Transdisciplinary case studies as a means of sustainability learning: historical framework and theory. Int J Sustain High Educ 7(3):226–251

Seidl R, Brand F, Stauffacher M, Krütli P, Le Q, Spörri A, Meylan G, Moser C, González M, Scholz R (2013) Science with society in the anthropocene. Ambio 42(1):5–12. doi:10.1007/s13280-012-0363-5

Selman P (2009) Planning for landscape multifunctionality. Sustain Sci Pract Policy 5(2):45–52

Selman P (2010) Landscape planning – preservation, conservation and sustainable development. Town Plan Rev 81(4):381–406

Selman P (2012) Sustainable landscape planning. The reconnection agenda. Routledge, London/New York

Stauffacher M, Flüeler T, Krütli P, Scholz RW (2008) Analytic and dynamic approach to collaborative landscape planning: a transdisciplinary case study in a Swiss pre-alpine region. Syst Pract Action Res 21(6):409–422. doi:10.1007/s11213-008-9107-7

Stauffacher M, Scholz RW (2011) Applying the HES-framework. Case 2: transdisciplinary case studies for regional and organizational transitions. In: Scholz RW (ed) Environmental literacy in science and society: from knowledge to decisions. Cambridge University Press, Cambridge, pp 473–482

Termorshuizen JW, Opdam P (2009) Landscape services as a bridge between landscape ecology and sustainable development. Landsc Ecol 24:1037–1052

Turner BL, Matson PA, McCarthy JJ, Corell RW, Christensen L, Eckley N, Hovelsrud-Broda GK, Kasperson JX, Kasperson RE, Luers A, Martello ML, Mathiesen S, Naylor R, Polsky C, Pulsipher A, Schiller A, Selin H, Tyler N (2003) Illustrating the coupled human–environment system for vulnerability analysis: three case studies. Proc Natl Acad Sci 100(14):8080–8085

Walker B, Carpenter S, Anderies A, Abel N, Cumming C, Janssen M, Lebel L, Norberg J, Peterson GD, Pritchard R (2002) Resilience management in social-ecological systems: a working hypothesis for a participatory approach. Conserv Ecol 6(1):14, http://www.consecol.org/vol6/iss1/art14

Walker BH, Holling CS, Carpenter SC, Kinzig AP (2004) Resilience, adaptability and transformability. Ecol Soc 9(2):5, http://www.ecologyandsociety.org/vol9/iss2/art5/

Walker G (2011) Environmental justice: concepts, evidence, and politics. Routledge, London

Widgren M (2012) Resilience thinking vs. political ecology: understanding the dynamics of small-scale, labour-intensive farming landscapes. In: Plieninger T, Bieling C (eds) Resilience and the cultural landscape: understanding and managing change in human-shaped environments. Cambridge University Press, Cambridge, pp 95–110

Wiek AH, Binder CR, Scholz RW (2006) Functions of scenarios in transition processes. Futures 38(7):740–766

Wiens JA (2013) Is landscape sustainability a useful concept in a changing world? Landsc Ecol 28(6):1047–1052. doi:10.1007/s10980-012-9801-9

Wu J (2010) Urban sustainability: an inevitable goal of landscape research. Landsc Ecol 25:1–4

Wu J (2013) Landscape sustainability science: ecosystem services and human well-being in changing landscapes. Landsc Ecol 28(6):999–1023

Chapter 18
Sustainable Development and Material Flows

Beatrice John, Andreas Möller, and Annika Weiser

Abstract A major target of strategies toward a more sustainable resource use must be to find ways of remaining within the planetary boundaries, not only by reducing overall resource use but also through keeping within the system what we are already using. This makes it necessary to take a systemic perspective and look at the whole life cycle of joint product systems, raw material inputs, and respective emissions. Knowing and understanding the dynamics of material stocks and flows may be a first step toward managing them. In the context of society, this approach is known as socioeconomic metabolism and is increasingly applied especially in regional and urban contexts. Here, we introduce material flow analysis as a possible method for constructing and evaluating material and energy flows to gain an insight into the flows of specific substances within the anthropogenic system. We show the main characteristics and applications as well as possible limitations of such a modeling approach and conclude with implications for a further development of such methods to enable a shift from analysis to assessment and strategy building that reflects sustainability principles and goes beyond efficiency.

Keywords Material flow analysis • Socioeconomic metabolism • Industrial ecology • Life cycle assessment • Stocks and flows

B. John • A. Weiser (✉)
Institute of Ethics and Transdisciplinary Sustainability Research, Leuphana University Lüneburg, Scharnhorststraße 1, 21335 Lüneburg, Germany
e-mail: beatrice.john@leuphana.de; weiser@leuphana.de

A. Möller
Institute for Environmental and Sustainability Communication, Leuphana University Lüneburg, Scharnhorststraße 1, 21335 Lüneburg, Germany
e-mail: amoeller@uni.leuphana.de

H. Heinrichs et al. (eds.), *Sustainability Science*, DOI 10.1007/978-94-017-7242-6_18

1 Society's Dependency on Materials: Avoiding End-of-Pipe Solutions

Issues of current and future raw material use and the resource needs are vital for sustainable development (Graedel and van der Voet 2010). Either the resources are actually vital, like water, or they are built into devices that make life easier or contribute to certain aspects of sustainable development (such as clean energy production devices). Take the case of smartphones and mobile computing. Far exceeding all estimations made on their future demand, worldwide mobile phone subscriptions have increased from 33.9 % in 2005 to an estimated 96.2 % in 2013 (Fig. 18.1) (ITU 2013). They have become essential components of our everyday life and fulfill vital functions, especially in developing countries, for example, in banking and the organization of health care in remote areas (cf. the VillageReach program in Malawi (VillageReach 2014)).

But this also makes us highly dependent on the availability of raw materials. Before being used or built into a device, they must be mined and refined, and measures must be taken once they are put out of use. There are, for example, more than 40 chemical elements built into a single phone (Wäger and Lang 2010), partly in such small amounts that it is next to impossible to recycle them.

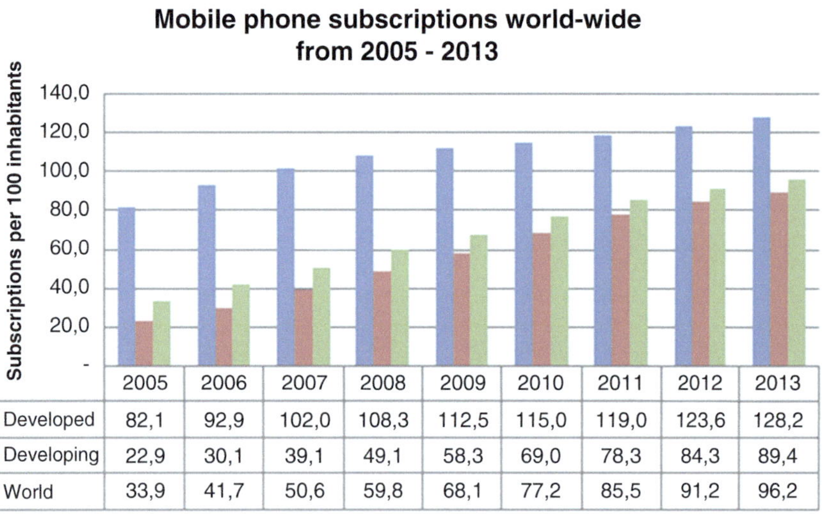

	2005	2006	2007	2008	2009	2010	2011	2012	2013
Developed	82,1	92,9	102,0	108,3	112,5	115,0	119,0	123,6	128,2
Developing	22,9	30,1	39,1	49,1	58,3	69,0	78,3	84,3	89,4
World	33,9	41,7	50,6	59,8	68,1	77,2	85,5	91,2	96,2

Fig. 18.1 The development of mobile phone subscriptions from 2005 to 2013 (Source: ITU ICT Facts & Figures 2013; numbers for 2012 and 2013 are ICT estimates)

The aim, therefore, must be not only to reduce the overall amount of resources we take from the Earth but also to keep those we are already using within the system. In order to minimize the amount of valuable resources being disposed of in landfills or lost through dissipation (understood as "the 'dilution' of materials into the technosphere or ecosphere in such a way that their recovery is made difficult or impossible" (Wäger et al. 2012)), we need to aim for an optimization of the according processes and interfaces between the various phases. In other words, we must look at the whole life cycle of a product and its raw materials. Ideally, this leads us to point at which (i) products that have reached their end-of-life stage become a stock (a *mine*, if you like) for new products and (ii) products that cannot be reused and recycled are compostable and therefore not producing waste and emissions.

2 Material Flows, the Socioeconomic Metabolism Concept and Industrial Ecology

In order to achieve a level of resource consumption that remains within the planetary boundaries, it is necessary to take a systemic perspective. This includes shifting the focus from resource efficiency (i.e., from a rather *static perspective*) to the dynamics of flows and stocks (*dynamic perspective*). Knowing and understanding these material flows is an essential step toward managing them. It also allows us to identify and take into account side effects of the respective material use, not only throughout a product's value chain but also in the broader context of human-environmental interactions.

This way of looking at material flows in the context of society can be subsumed under the concept of socioeconomic metabolism. Originally derived from the biological and ecological perspective on an organism, the metabolism concept is widely used to comprehend energetic and material stocks and flows; their reactions, processes, and interrelations of and between entities; and different forms of inputs and outputs (Fischer-Kowalski 1998). Various disciplines, such as geology and anthropology, have developed this concept further, increasingly shifting its application to human society. Especially with the rise of environmental movements and society's growing critique of economic growth in the 1960s, the metabolism of highly developed societies became a major issue and required a stronger collaboration to develop approaches that are capable of analyzing human-environmental interactions by radically crossing disciplinary boundaries of social, natural, and human sciences.

Box 18.1: Industrial Ecology

Industrial ecology - as an academic concept and a practical tool for policy-makers (Gibbs and Deutz 2007) - arose in the 1990s from the idea of creating an industrial circular economy, in which industrial waste may serve as a source for others, also coining the term "industrial ecosystem" (Frosch et al. 1989).

Industrial ecology employs a metaphor that makes the idea tangible but also a little fuzzy (Graedel 1996).

The question of the practical utility of the idea of industrial ecology is consequently also a major point of critique (Gibbs and Deutz 2007). The "industry" part refers to its focus on improving industrial processes, which are a major cause of environmental disturbance, making companies the main addressees. The "ecology" part shows (i) the concept's origin, i.e., taking natural ecosystems as a model for the design of industrial activities, as well as (ii) its intention, namely, to keep all human (industrial) action within the ecological frame that enables such action and to achieve effects of "industrial symbiosis" comparable to those found in nature (Lifset and Graedel 2002). "Industrial symbiosis engages traditionally separate industries in a collective approach to competitive advantage involving physical exchange of materials, energy, water, and/or by-products. The keys to industrial symbiosis are collaboration and the synergistic possibilities offered by geographic proximity." (Chertow 2000)

Two prominent examples of eco-industrial parks, which apply IE principles, are:

- Kalundborg in Denmark (Lowe 1997): See online Case Kalundborg
- Ulsan in South Korea (Behera et al. 2012): See online Cases in South Korea

These historical developments, as well as current urbanization trends, require us to look into the functioning of our cities, i.e., human activities such as transportation, communication, living, and working (Baccini and Brunner 2012). More broadly, Kennedy et al. (2007) define it as "the sumtotal of the technical and socio-economic processes that occur in cities, resulting in growth, production of energy, and elimination of waste" (Kennedy et al. 2007).

Becoming more prevalent over the course of the last few decades, Wolman began a first holistic approach in accounting the urban metabolism of a hypothetical American city, revealing the first systemic insights on the impacts of production and consumption in 1965. Since then, a number of urban metabolism studies with varying foci have been conducted (Kennedy et al. 2011).

Comprehensive holistic overviews of urban metabolisms are not very common. Frequently, studies focus on isolated flows, depending on availability and quality of data. The example of Duvigneaud and Denayeyer-De Smet (1977) (see Fig. 18.2)

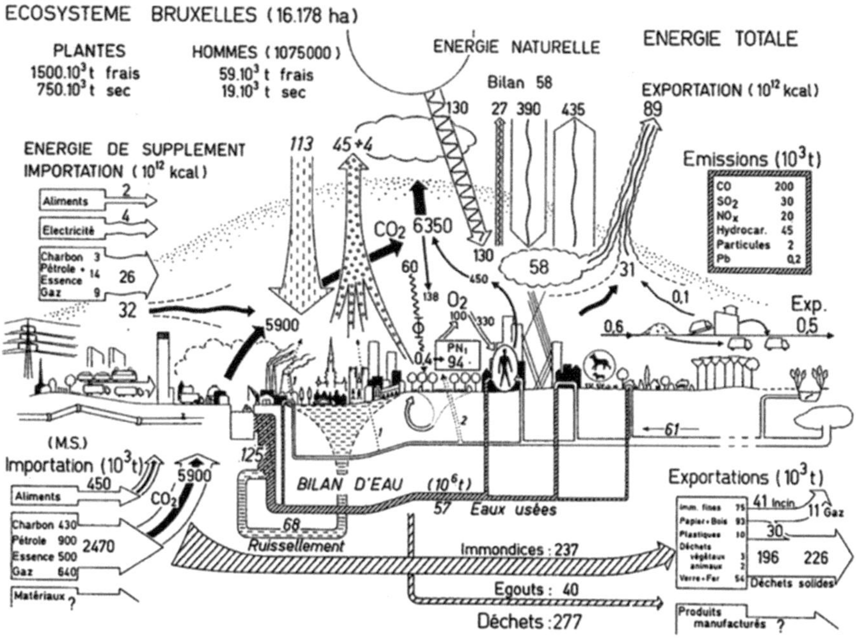

Fig. 18.2 Informal Sankey diagram of the material and energy flow of Brussels (Duvigneaud and Denayeyer-De Smet 1977)

illustrates one of the first analyses on energy flows within Brussels, adding the natural energy balance to the anthropogenic energy inputs. An integrative perspective is given through the assessment of carbon cycling with storage, input, and export, as it ascribes systemic flows to different sectors of buildings, transportation, humans, and vegetation from which carbon emissions originate (Kellett et al. 2013).

The focus on the spatial scale, the boundaries of the metabolism, is basic in order to define the significance, importance, and relative contribution of flows for the system and its relation to others, e.g., to global boundaries. Case studies are conducted from the household level (Cohen et al. 2005) to the neighborhood level (Kellett et al. 2013; Codoban and Kennedy 2008; Berg and Nycander 1997), and to the city level. They also relate across those scales to the respective hinterlands and regions. The hinterland fulfills a twofold role: as sink for urban waste and as source for resources and material. The work of Lenzen and Peters (2009) follows localized household consumption demands throughout Australia and reveals the upstream impacts on the hinterland of greenhouse gas emissions, water usage, and labor provision.

Given that input and output of materials happen with a certain time delay, the temporal scale also plays an important role in assessing resource flows in cities. Very young, fast-growing cities and older, slow-growing or shrinking cities differ in pace and amount of intake and output of materials and waste. This manifests, for example, in the building stock: When point in time for retrofitting of existing building stock arrives, dynamically, waste and bound resources (such as copper) are discharged and made available for future endeavors, but also require waste disposal, and consequently add on to the disadvantage of future generations. For example, the German Federal Environment Agency (Umweltbundesamt) approaches this issue with the project "Anthropogenic Stocks," with which buildings, roads, and other facilities are inventoried to estimate future flows of material like concrete, steel, or copper. Such a model requires a dynamic material flow analysis (UBA 2012).

In order to strengthen the concept of a sustainable metabolism, it is crucial to consider the foregoing aspects in relation to the normative dimensions of the sustainability framework, including inter- and intragenerational components, as well as geographical inequalities in regions of the Global South versus the Global North. A transformative approach to reshaping resource flows into sustainable metabolisms requires broadening the conceptual basis of a current state analysis and including a sustainability appraisal.

2.1 *Material Flow Analysis*

One of the tools that help to better understand material flows and to assess the society's metabolism is material flow analysis (MFA). "Material flow analysis refers to the analysis of the throughput of process chains comprising extraction or harvest, chemical transformation, manufacturing, consumption, recycling and disposal of materials." (Bringezu and Moriguchi 2002).

The most important purpose of material flow analysis nowadays is efficiency analysis, such as in life cycle assessment (LCA), a special material analysis approach (Guinée 2002), which analyzes the impacts of products and services on the natural environment.

2.1.1 Material Flow Analysis as a Modeling Procedure

At its core, material flow analysis is a modeling process: the process of constructing and evaluating material and energy flow models. The purposes of those models are mainly insights into the flows of specific substances into the anthropogenic system, and the impacts of production, usage, and disposal of products and services on the environment, as well as the design of new production and supply networks. Regarding the metabolism concept, the modeling instrument analyzes the effects of societal institutions and processes on material and energy stocks and flows and therefore on the natural environment.

Four main modeling steps generally apply: (i) defining the goals and framing the system that is to be analyzed, (ii) informal analysis of the process chain, (iii) modeling and calculation, and (iv) evaluation of the model (cf. Bringezu and Moriguchi 2002). Core modeling steps are an informal analysis of supply chains, product life cycle, and the transformation of the conceptual model into a formal model that allows for calculating the dynamics of flows, stocks, and other metabolic indicators.

From a mathematical perspective, the step of constructing formal material flow models is the construction of a graph $G=(V, E)$ that consists of a finite set of nodes V (in case of MFA processes) and a set of links or edges E between the nodes where the flows take place. These graphs are called flow sheets or networks. The second step is to specify the processes by defining the relationships between the input and the output flows of each process in a way that a modeling expert or computer can evaluate them to calculate formerly unknown flows within the system. A third step is to specify already known "manual" flows, for instance, the planned product output per year (reference flows in life cycle assessment, cost objects in future-oriented cost accounting), feed streams on the input side (chemical engineering), or other parameters of the model.

Figure 18.3 shows such a model with two nested loops: The production system consists of two different chemical processes (chemical reactor 1 and chemical reactor 2). The second chemical process uses a by-product of the first process as its input. The purpose of MFA is to determine all relevant flows and the process levels of the unit processes.

The modeling steps result in a system of nonlinear equations (Westerberg et al. 1979, p. 14), which (1) specify the relationships within the unit processes, (2) link the processes (connecting equations) to one another, and (3) link the manual flows or design specifications (Chen and Stadtherr 1985).

The aim of the calculation step, mainly performed by computers today, is to solve the system of nonlinear equations and to know all material and energy flows, which occur in the material flow model. The algorithms are called solvers. Different solvers can be distinguished. If we want to find future steady states of the material and energy flow system (steady-state modeling), the solver has to calculate time-independent flow rates (flows per time unit). The main problem of calculating steady

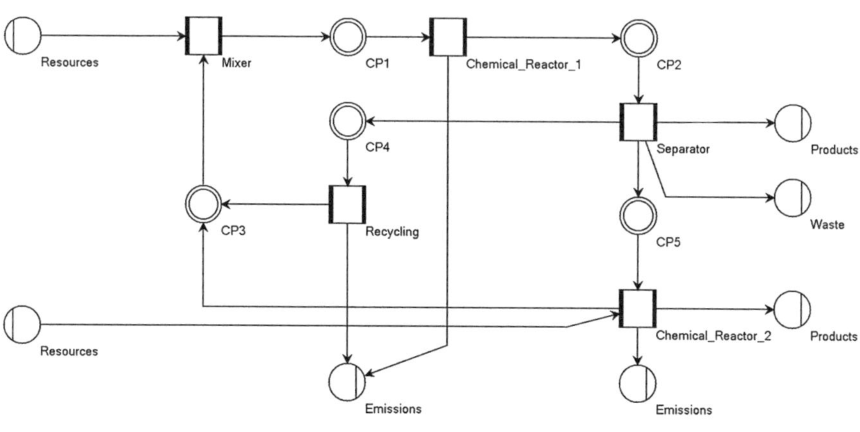

Fig. 18.3 Material flow model of two different chemical processes, including recycle flows

states is dealing with loops. Loops are a characteristic of metabolisms, but they inhibit direct sequential calculation. Different approaches are developed to deal with loops and recycle flows: the sequential modular approach, the simultaneous modular approach, and the equation-based approach (Westerberg et al. 1979; Westerber and Piela 1994).

Another class of solutions emphasizes the dynamics of material and energy flow systems, in which the flow rates are rather time-dependent functions. Consequently, the solver has to deal with a system of (ordinary) differential equations, as in system dynamics (Hannon and Ruth 1994). This kind of analysis is called a dynamic MFA. Here, steady-state approaches cannot be applied to solve a system of (ordinary) differential equations (ODEs) as initial value problems (Finlayson 2012, p. 356). Continuous simulation and integration methods like Euler–Cauchy and Runge–Kutta 2/4 come into play.

Different evaluation methods exist, including very simple approaches like a direct evaluation through input/output balance or sophisticated impact assessment methods to estimate effects like climate change or ozone depletion. Substance flows can be visualized by splitting flows into their components (e.g., phosphorus) and with aid of Sankey diagrams.

Direct evaluation methods do not include efficiency analyses like LCA, but they may be performed as a subsequent modeling step, consisting of (1) a period-oriented material flow analysis with the purpose of providing a material and energy flow model for a given system (region (urban metabolism), supply chain, production site, etc.) and (2) a subsequent efficiency analysis, for which the period-oriented material model provides the data.

2.2 *Focus and Limitations of Different Types of MFA*

Because LCA is in line with generalized action orientations in our economic system (efficiency (Taylor 1911), key performance indicators, etc.) and can be combined with cost accounting, it is a very attractive instrument in the field of material flow analysis, just like carbon or water footprinting (Weidema et al. 2008).

However, such efficiency analyses do not cover all relevant metabolism core indicators (Fischer-Kowalski & Hüttler 1999), for instance, absolute stocks and flows and their dynamics. In addition to life cycle thinking, several modeling images are discussed for MFA. One of them is "verbund." The term is used in the chemical industry as one of the most important design metaphors. The idea of the verbund, which has existed in the chemical industry for more than 100 years, becomes more and more important in other parts of our socio-technical metabolism, for instance, as an approach to regional material flow analysis. Regions are interpreted as a verbund, including production, consumption, agriculture, services, waste management, and energy supply. Chemical processes feed other chemical processes, so that ideally, no waste streams and emissions occur, making consistency and symbiosis important design principles, as new chemical processes and products must fit into the verbund.

Also, LCAs do not support insights into the interplay between stocks and flows. Relevant challenges in the field of sustainability result from the fact that, in modern societies, this interplay of stocks and flows has been changed dramatically: scarcity of raw materials, waste disposal, artificial substances in the natural environment, the concentration of carbon dioxide in the atmosphere, etc. Therefore, metaphors like verbund can be applied to broaden the horizon.

Nevertheless, the main focus of all MFA instruments is on material flows and stocks. In modern societies, material flows and stocks are extended by immaterial flows and stocks: computer-based information systems and networks. It is very difficult to model the effects of these flows on the natural environment, for instance, the carbon footprint of single emails. Direct and indirect impacts must be distinguished, and the indirect impacts of already existing infrastructures like computers and routers are very high. Often, typical examples in the field of MFA and LCA are steel, concrete, and packaging, but not services of modern computer-based communication and information infrastructures.

3 Toward Sustainable Material Flows: Further Developments and the Role of MFA for Sustainable Development

While theoretical and conceptual thoughts on material flows and their value for sustainable development are fairly strongly developed, the operationalization of sustainability principles within their tools and practical application are still rather undeveloped in parts: In order to achieve material flows that may be considered "sustainable," we will have to shift from analysis to assessment and strategy building (Wiek et al. 2012); specifically, this means to link analytical results (what are the material flows in a given region) to sustainability principles that go beyond efficiency (what are *sustainable* material flows in this region).

At first sight, today's smartphones are an example of dematerialization: They are 1 billion times more powerful as, while being only a fraction of the size of, the first computers in the 1940s, such as Konrad Zuse's Z1, which weighed about 500 kg and consumed electric energy of 1 kWh. Still, the total material input has increased due to the fact that there are many more smartphones and computers in the world now than back then (known as *rebound effect*). Therefore, we have to consider total amounts, as well as developments over time, in order to achieve material use that is consistent with our planetary boundaries.

Consequently, when going into the direction of sufficiency principles, the necessity to shift from analysis to assessment clearly comes to the fore: more precisely, instead of focusing on improving existing processes (analysis), sustainability assessments would consider whether the process is actually "good." This advancement demands more interdisciplinary work, such as tapping into social and cultural norms in order to understand and integrate questions of lifestyle and behavior change, or taking into account the possibility of applying low-tech solutions.

So far, strategies focusing purely on optimization of material flows might disregard a resilience perspective. Improvement of flows and infrastructure from an efficiency perspective should be in close interrelation with stability and resilience of the system at hand: Shocks affecting the system, such as extreme climate change events in urban areas, can severely damage dependent systems operating without any redundancies and diversity. This may result in lack of service provision, social segregation, and security issues and eventually threaten the well-being of residents.

Taking all these aspects into account will afford an increased amount of planning and management. This may include material flow management, which also replaces approaches for end-of-life waste disposal, i.e., aiming at waste reduction instead of reuse or mere disposal (Lang et al. 2006), and may be supported through national legislation and international agreements. But it also calls for improved procedures for the process of finding the best possible solutions considering regional characteristics, norms, and needs, including transdisciplinary approaches. A practical example for such a process is the BRIDGE project (see http://bridge-fp7.eu). Across several transdisciplinary case studies with Firenze (Italy), Helsinki (Finland), and Gilwice (Poland), the BRIDGE Decision Support System (DSS) aims to connect the analytical tool with a sustainability appraisal and evaluation design for urban planning interventions on the basis of both material flow data and socioeconomic criteria (Chrysoulakis et al. 2013). These are first steps toward a transformational approach and metamorphosis of the urban metabolism to broaden its conceptual basis away from a purely analytical tool but with serious chances for interdisciplinary and transdisciplinary research.

Further Reading

Ayres RU, Ayres LW (eds) (2002) A handbook of industrial ecology. Edward Elgar Publications, Cheltenham/Northampton

Baccini P, Brunner PH (2012) Metabolism of the anthroposphere: analysis, evaluation, design. MIT Press, Cambridge

Fischer-Kowalski M (1998a) Society's metabolism – the intellectual history of material flow analysis, part I, 1860–1970. J Ind Ecol 2(1):61–78

Fischer-Kowalski M, Hüttler W (1999a) Society's metabolism – the intellectual history of material flow analysis, Part II, 1970–1998. J Ind Ecol 2(4):107–136

Graedel TE, van der Voet E (eds) (2010) Linkages of sustainability. The MIT Press, Cambridge

Journal of Industrial Ecology (2012) Special Issue: Sustain Urban Syst 16(6): 775–715

References

Baccini P, Brunner PH (2012) Metabolism of the anthroposphere: analysis, evaluation, design. MIT Press, Cambridge

Behera SK et al (2012) Evolution of "designed" industrial symbiosis networks in the Ulsan Eco-industrial Park: "research and development into business" as the enabling framework. J Clean Prod 29–30:103–112

Berg PG, Nycander G (1997) Sustainable neighbourhoods - a qualitative model for resource management in communities. Landsc Urban Plann 39(2):117–135
Bringezu S, Moriguchi Y (2002) Material flow analysis. In: Ayres RU, Ayres LW (eds) A handbook of industrial ecology. Edward Elgar Publications, Cheltenham/Northampton, pp 79–90
Broyden CG (1965) A class of methods for solving nonlinear simultaneous equations. Math Comput 19:577–593
Chen H-S, Stadtherr MA (1985) A simultaneous-modular approach to process flowsheeting and optimization, part I: theory and implementation. AIChE J 31(11):1843–1856
Chertow MR (2000) Industrial symbiosis: literature and taxonomy. Annu Rev Energy Environ 25:313–37
Chrysoulakis N et al (2013) Sustainable urban metabolism as a link between bio-physical sciences and urban planning: the BRIDGE project. Landsc Urban Plann 112:100–117
Codoban N, Kennedy CA (2008) Metabolism of neighborhoods. J Urban Plann Dev 134(1):21–31
Cohen C, Lenzen M, Schaeffer R (2005) Energy requirements of households in Brazil. Energy Policy 33:555–562
Duvigneaud P, Denayeyer-De Smet S (1977) L'Ecosystéme Urbain Bruxellois, in Productivité en Belgique. In: Duvigneaud P, Kestemont P (eds) Traveaux de la Section Belge du Programme Biologique International, Bruxelles, pp 581–597
Erkman S (2002) The recent history of industrial ecology. In: Ayres RU, Ayres LW (eds) A handbook of industrial ecology. Edward Elgar Publications, Cheltenham/Northampton, pp 27–35
Finlayson BA (2012) Introduction to chemical engineering computing, 2nd edn. Wiley, Hoboken
Fischer-Kowalski M (1998) Society's metabolism – the intellectual history of material flow analysis, part I, 1860–1970. J Ind Ecol 2(1):61–78
Fischer-Kowalski M, Hüttler W (1999) Society's metabolism – the intellectual history of material flow analysis, part II, 1970–1998. J Ind Ecol 2(4):107–136
Frosch RA, Gallopoulos NE (1989) Strategies for manufacturing. Sci Am 189(3):144–152
Gibbs D, Deutz P (2007) Reflections on implementing industrial ecology through eco-industrial park development. J Clean Prod 15(17):1683–1695
Graedel TE (1996) On the concept of industrial ecology. Ann Rev Energy Environ 21:69–98
Graedel TE, van der Voet E (2010) Linkages of sustainability. MIT Press, Cambridge/London
Graedel TE et al (2012) Methodology of metal criticality determination. Environ Sci Technol 46(2):1063–1070
Guinée JB (Final ed) (2002) Handbook on life cycle assessment – operational guide to the ISO standards. Kluwer Academic Publishers, Dordrecht/Boston/London
Hannon B, Ruth M (1994) Dynamic modeling. Springer, Berlin/Heidelberg/New York
Heijungs R (1994) A generic method for the identification of options for cleaner products. Ecol Econ 10:69–81
Heijungs R, Suh S (2002) The computational structure of life cycle assessment. Kluwer Academic Publishers, Dordrecht
ITU (International Telecommunications Union) Statistics (2013) ICT facts and figures. Available at: http://www.itu.int/en/ITU-D/Statistics/Pages/stat/default.aspx. Accessed 21 Apr 2014
Kellett R, Christen A, Coops NC, van der Laan M, Crawford B, Tooke TR, Olchovski I (2013) A systems approach to carbon cycling and emissions modeling at an urban neighborhood scale. Landsc Urban Plan 110:48–58
Kennedy C, Cuddihy J, Engel-Yan J (2007) The changing metabolism of cities. J Ind Ecol 11(2):43–59
Kennedy C, Pincetl S, Bunje P (2011) The study of urban metabolism and its applications to urban planning and design. Environ Poll 159(8–9):1965–73
Lang DJ, Binder CR, Scholz RW, Schleiß K, Stäubli B (2006) Impact factors and regulatory mechanisms for material flow management: Integrating stakeholder and scientific perspectives The case of bio-waste delivery. Resour Conserv Recy 47:101–132
Lenzen M, Peters GM (2009) How city dwellers affect their resource hinterland. J Ind Ecol 14(1):73–90

Lifset R, Graedel TE (2002) Industrial ecology: goals and definitions. In: Ayres RU, Ayres LW (eds) A handbook of industrial ecology. Edward Elgar Publications, Cheltenham/Northampton, pp 3–15
Lowe EA (1997) Creating by-product resource exchanges: strategies for eco-industrial parks. J Clean Prod 5(1–2):57–65
Möller A (2000) Grundlagen stoffstrombasierter Betrieblicher Umweltinformationssysteme. Projekt Verlag, Bochum (in German)
Möller A (2004) Continuous simulation in material flow networks. In: Proceedings of the iEMSs 2004 "Complexity and Integrated Resource Management", Osnabrueck
Möller A (2005) Dynamic material flow analysis in the life cycle assessment tool chain. In: Geldermann J, Treitz M, Schollenberger H, Rentz O (eds) Challenges for industrial production. Universitätsverlag, Karlsruhe
Reisig M (1985) Petri nets – an introduction. Springer, Berlin/Heidelberg/New York
Riebel P (1994) Core features of the 'Einzelkosten- und Deckungsbeitragsrechnung'. Euro Account Rev 3(3):515–546
Taylor FW (1911) The principles of scientific management. Harper & Brothers Publications, New York/London
UBA 2012 What matters (2012): annual report of the Federal Environment Agency. Available at: https://www.umweltbundesamt.de/sites/default/files/medien/publikation/long/4296.pdf
VillageReach (2014) http://villagereach.org/
Wäger PA, Lang DJ (2010) Seltene Metalle – Rohstoffe für Zukunftstechnologien. Available at: http://www.satw.ch/publikationen/satwinfo/SelteneMetalle.pdf
Wäger PA et al (2012) Towards a more sustainable use of scarce metals - a review of intervention options along the metals life cycle. GAIA 21(4):300–309
Weidema B, Thrane M, Christensen P, Schmidt J, Loekke S (2008) Carbon footprint – a catalyst for life cycle assessment? J Ind Ecol 12(1):3–6
Westerberg AW, Piela PC (1994) Equational-based process modeling. Technical report. Department of Chemical Engineering and the Engineering Design Research Center - Carnegie Mellon University, Pittsburgh
Westerberg AW, Hutchinson HP, Motard RL, Winter P (1979) Process flowsheeting. Cambridge University Press, London/New York/Melbourne
Wiek A, Ness B, Brand FS, Schweizer-Ries P, Farioli F (2012) From complex systems analysis to transformational change: a comparative appraisal of sustainability science projects. Sustain Sci 7(Suppl 1):5–24
Wolman A (1965) The metabolism of cities. Sci Am 213(3):179–190

Chapter 19
Sustainable Energy Systems

Stefan Lechtenböhmer and Lars J. Nilsson

Abstract Sustainable energy systems are crucial to all three dimensions of sustainable development and thus central for mitigating climate change and achieving sustainable economic and social development. In order to play this role, current unsustainable energy systems need to undergo a major transition. This chapter first sketches core structural features of anthropogenic energy systems and discusses their relevance for addressing global challenges. Then, the main elements and strategies to make energy systems more sustainable as well as examples on political and societal challenges of this transformation are given.

Keywords Energy systems • Climate change • Millennium development goals • Limits to fossil energy resources • Sustainable energy scenarios • Energy transition governance

1 Introduction

Energy use and energy systems are deeply woven into the modern economy and our daily lives. Reliance on and demand for modern energy carriers is increasing, especially in countries with high economic growth rates, and most of the energy comes from fossil fuels. Despite this growth, billions of people still lack reliable access to modern energy services.

Energy systems are closely linked to the three dimensions of sustainable development, and thus, energy "is central to addressing major challenges of the 21st Century, challenges like climate change, economic and social development,

S. Lechtenböhmer (✉)
Wuppertal Institut für Klima Umwelt Energie, Döppersberg 19, 42103 Wuppertal, Germany
e-mail: stefan.lechtenboehmer@wupperinst.org

L.J. Nilsson
Adjunct Professor, Department for Technology and Society/Environment and Energy Systems, Lund University, Box 118, SE- 221 00 Lund, Sweden
e-mail: lars_j.nilsson@miljo.lth.se

H. Heinrichs et al. (eds.), *Sustainability Science*, DOI 10.1007/978-94-017-7242-6_19

human well-being, sustainable development, and global security" as the Global Energy Assessment (GEA 2012) states in its first sentence.

Future energy systems can be considerably more sustainable. The natural flows of renewable energy sources are several thousand times greater than present and projected human needs. Many technologies to utilise renewable energy sustainably are already available, and more are being developed (see IPCC 2011). Thus, the challenge is not mainly about technologies or energy resources. The key challenge is rather to change the current direction of development and realise a societal transformation where (a) the sustainability of energy systems is increased **and** (b) clean energy for sustainable development is provided.

The conversion of the energy system from being part of the problem to become part of the solution involves a major transition. For this transition to happen, we need future visions, scenarios and roadmaps as well as policies and governance for implementing sustainability transitions and strategies.

In the following, we first sketch some structural features of anthropogenic energy systems and then discuss how and why they are crucial for addressing global challenges. After that, we describe the main elements and strategies needed to make energy systems more sustainable and show that implementation will need several decades. This makes it necessary to have long-term visions and strategies. Finally, we give some examples on political and societal challenges for implementation.

2 Types, Structures and Characteristics of Energy Systems

Our current and unsustainable energy systems are dominated by fossil fuels as energy sources which today supply almost 80 % of all energy globally. They are also characterised by losses of more than two thirds along the supply chains from primary energy source to final energy services delivered (Fig. 19.1 and BP 2013). Energy systems can be looked upon as national or global energy systems, including all energy sources and conversion routes as depicted in Fig. 19.1. They can also be looked upon in smaller scales and scopes, e.g. the energy system of a single town (Fig. 19.2) or down to the heating system of a building (cp. Everett et al. 2012, 1ff).

Global and national energy systems are large, complex and capital intensive although their ultimate purpose is to provide relatively simple energy services such as heating, cooking, motive power and powering the Internet. Energy is also needed as an input in the production of important basic materials (e.g. metals, cement, paper, glass, insulation materials, etc.). It is important to remember that energy (e.g. a kWh or litre of fuel) is not an end in itself but the means to an end, namely, the delivery of end-use energy services that people and society need. That distinction shifts our focus to the demand for energy services and how they can be efficiently delivered, rather than how supply can continue to expand (cp. Lovins 2004, 2011; IEA 2012, 267ff).

Energy systems involve expensive and long-lived infrastructures in production, transports (e.g. power grids) and end uses (e.g. buildings). These characteristics cre-

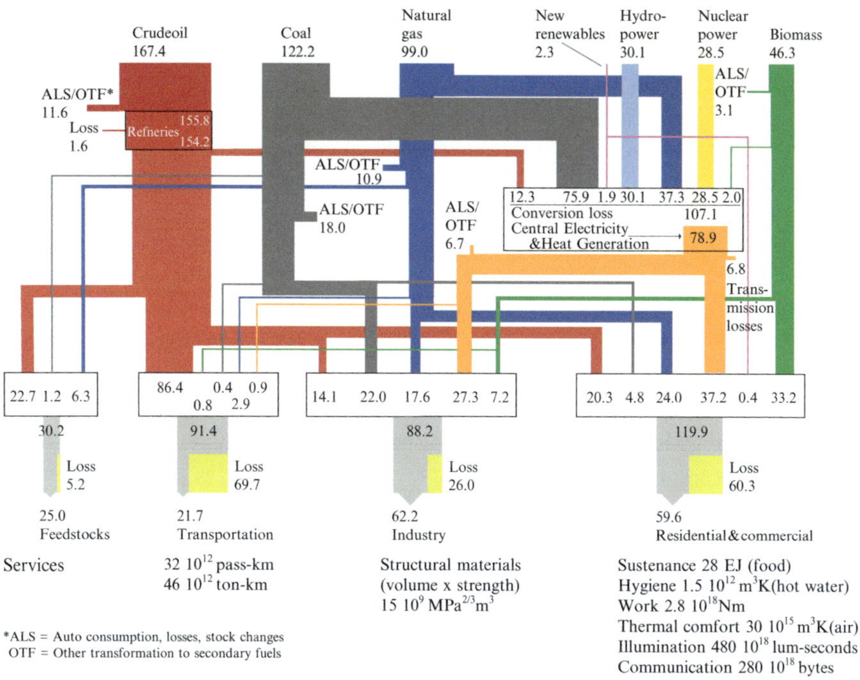

Fig. 19.1 Global energy flows (in EJ) from primary to useful energy by primary resource input, energy carrier (fuels) and end-use sector applications in 2005 (Source: GEA 2012, 45. Reproduced with permission from the International Institute for Applied Systems Analysis (IIASA))

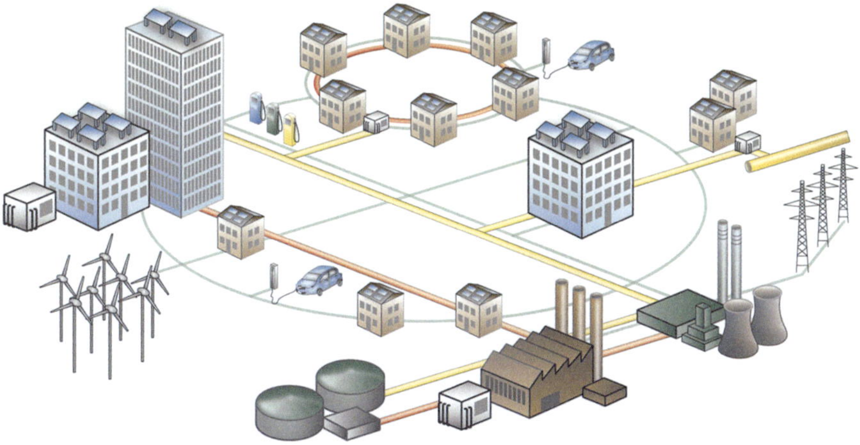

Fig. 19.2 Scheme of an urban energy system (Source: Wuppertal Institute, VisLab 2011)

ate path dependencies that can be barriers to changes towards increasing sustainability as they create technology lock-ins and vested economic interests of investors and other stakeholders. For example, building a new coal-fired power plant determines high CO_2 emissions of power generation for decades. But existing infrastructures may also enable transitions to sustainable energy systems, e.g. through strong power grids that can balance variable renewable electricity generation over wider areas or district heating systems that are flexible with respect to heat sources used.

The critical importance of energy for society, including large environmental, economic and social implications, makes it an important political issue. The overall energy policy goals, or pillars of energy policy, are shared by most countries. These include energy security (safe and uninterrupted supply), economic efficiency and affordability and low environmental impacts (i.e. climate change, acidification or respiratory health problems resulting from air pollution) (see, e.g. IEA 2013, 9). It may be noted that these general energy policy goals mirror the three dimensions of sustainability.

Please discuss the following questions:
- *What does the term "energy system" mean and how can sustainability be understood in the context of energy systems?*
- *Why are energy systems so difficult to change?*

3 The Major Global Challenges and Their Linkages to Energy

As noted, energy is strongly linked to pressing global problems such as poverty eradication and socioeconomic development, degradation of the environment, health, resource depletion and geopolitical conflicts. Energy can be a major cause of, or aggravate, such problems, but energy is also a central means for addressing these challenges.

3.1 Energy, Development and Poverty

Energy use is very unequally distributed worldwide. North America, Japan and Europe have very high per capita consumptions, while particularly sub-Saharan Africa and South Asia dispose of very low amounts of energy per capita, often below 20 GJ per person per year, which is equivalent to 1.2 l of gasoline per day, much of which is typically consumed in industrial and tertiary sector operations (Fig. 19.3).

Poverty and the linked problems of poor health, access to marketed goods and services and outlooks for personal progress and better lives are the sad reality for more than three billion people worldwide (GEA 2012). For most of them this

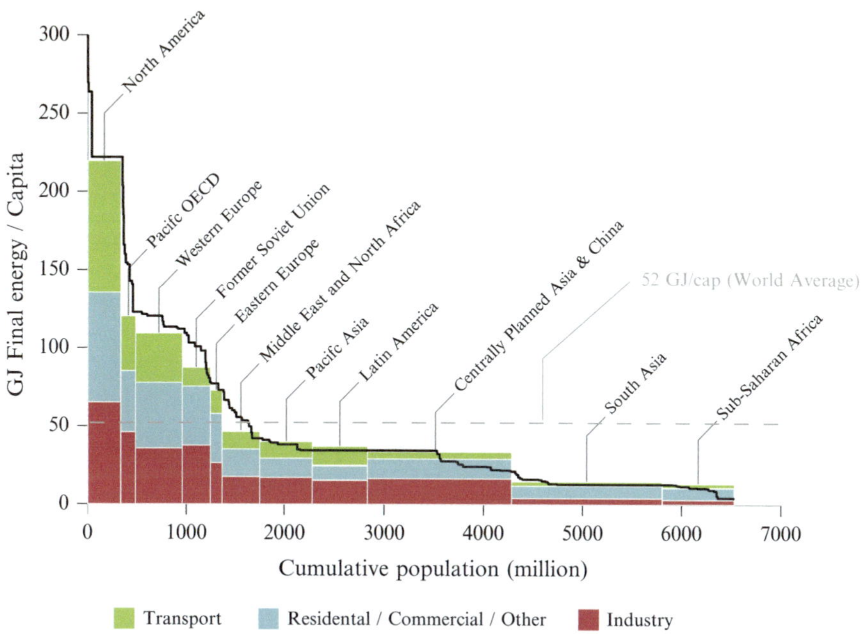

Fig. 19.3 Final energy (GJ) per capita versus cumulative population for 11 world regions sorted by declining per capita energy use (with final energy use disaggregated by sector and total, *colour bars*) and final energy per capita for 137 countries in 2005 (*black, solid line*). *Dashed horizontal line* shows the average final energy per capita, which indicates that approximately 1.5 billion people are above and 5.5 billion below that level (Source: GEA 2012, 36. Reproduced with permission from the International Institute for Applied Systems Analysis (IIASA))

includes a lack of access to modern energy services, which often affects them threefold:

- First, it cuts them off from various options for (economic) development. For example, lack of (electric) lighting reduces the chance to use evening hours for learning or productive occupations which could generate additional income. Farmers who have access to electricity and thus refrigeration can better store and process their crops. This can enable them to adding value to their agricultural production.
- Second, cooking (and heating) with traditional wood fuels, cow dung or coal is inconvenient as well as a major source of indoor pollution and related respiratory diseases.
- Third and adding to the problem, the lack of access to modern energy causes high costs for poor people. The costs per unit of energy service from kerosene in cans or charcoal are often higher than those of "modern" energies. Thus, poor people spend high shares of their income on the purchasing of energy, and in many countries they, women in particular, spend several hours a day only to collect fuels.

A very concrete example is solar lanterns (with PV cells, batteries and LED lamps) that can provide high-quality efficient lighting and simultaneously reduce reliance on purchased fuel for kerosene lamps.

In this context the UN has launched the "sustainable energy for all" initiative (http://www.se4all.org/) which aims at delivering accessible, efficient and renewable energy to as many people as possible as a major initiative to achieve the eight Millennium Development Goals (IEA 2011). Energy will also be important in the context of future Sustainable Development Goals (Chap. 22).

3.2 *Climate Change and the Environment*

Energy use today contributes to about 80 % of CO_2 and 30 % of CH_4 emissions, which are together the main greenhouse gases causing global warming and related climate change. Cumulative historic CO_2 emissions have tripled from 1970 (~420 $GtCO_2$ since 1750) to 2010 (~1300 $GtCO_2$ since 1750) (IPCC 2014). Energy systems also emit many other pollutants (e.g. SO_x, NO_x, heavy metals, particulate matter and organic compounds), and they are relevant for land use and degradation, e.g. for deforestation, for mining of energy resources or for energy infrastructures, as well as for water consumption through, e.g. use of cooling water for power plants and industrial installations (see Fig. 19.4).

Energy use is thus the major single factor that drives the emissions of anthropogenic greenhouse gases and emissions continue to grow. Despite all incremental improvements through higher shares of renewable energy use and improved technologies, the environmental burden from energy use continues to increase at high speed.

3.3 *Limits to (Growing) Fossil Energy Supply and the Role of Unconventional Resources*

Global energy consumption is constantly rising, but fossil fuel resources, notably oil, are limited. This is an obvious fact that has led to a debate on whether the world will soon reach a peak in global oil production and subsequently also peaks of natural gas, uranium and coal production. However, fossil fuel resources are more than sufficient to fuel dangerous climate change well beyond the target to limit average global warming below 2 °C.

While it is undeniable that fossil resources are limited – as the world and the amount it contains are finite – the estimates of which fossil resources are (economically) recoverable and which are not are relatively uncertain and changing over time. Increasing fossil fuel prices and technology developments (e.g. spurred

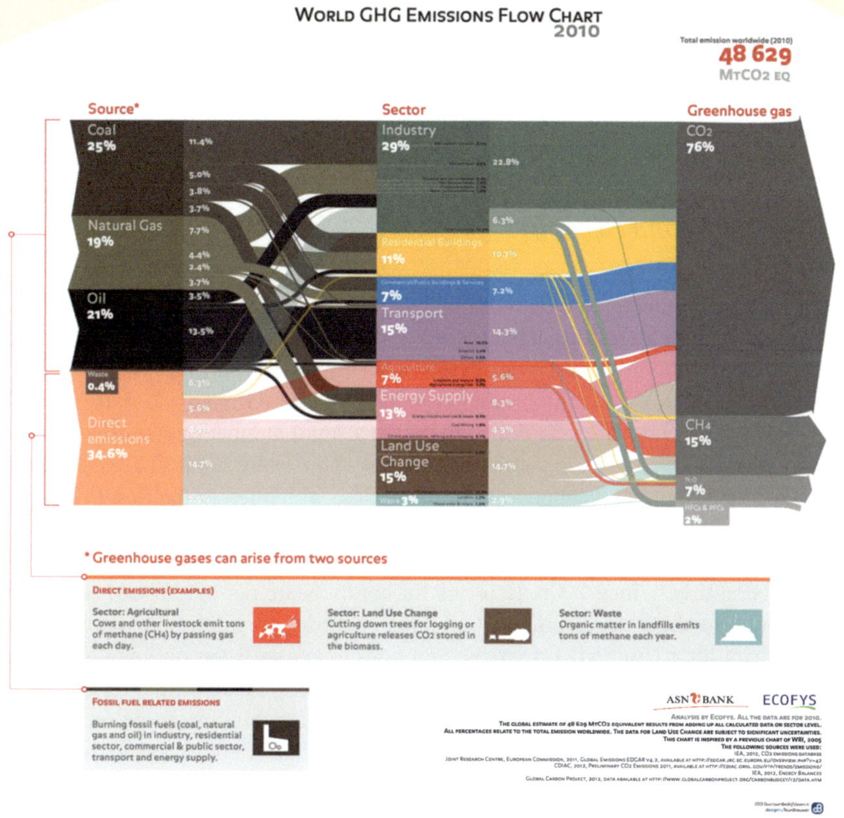

Fig. 19.4 Global greenhouse gas emissions in 2010 by source (estimated)

by high price expectations) make it profitable to extract fossil resources that were previously un-economical to exploit. Such factors postpone the point at which global fossil energy production will de facto not increase anymore (this is assuming that extraction will not be limited by environmental concerns).

The recent development of unconventional gas production in the USA is a good example of this effect. Declining domestic gas production resulted in increasing gas prices and expectations of significantly increasing import dependence. At the same time, drilling techniques including the technology of hydraulic fracturing ("fracking") to better exploit existing unconventional gas were developed. Together with reduced environmental restrictions and a couple of other favourable factors, this made domestic unconventional gas production economic and led to the "shale gas boom" in the years 2010–2013. This boom significantly reduced gas (and subsequently other energy) prices and made the USA the largest producer of gas worldwide (cp. Boersma and Johnson 2013).

However, unconventional energy resources come at a price. Often their extraction is not only more expensive but also more harmful to the environment than that of conventional fossil resources. One example is shale gas which is highly debated due to environmental threats to groundwater resources but also possible large greenhouse gas emissions. Another is the oil extraction from tar sands, mainly in Canada, which among other things consumes huge amounts of energy and water (Lechtenböhmer and Altmann 2013).

3.4 *Energy and Security*

Society is highly dependent on a reliable supply of energy for providing necessary energy services (e.g. heating during cold winters or for food production, transport and cooking). Energy security is therefore an important political driver in many countries, and it is also an important geopolitical factor (GEA 2012; chapter 5). Oil reserves, in particular, are very concentrated in a small number of countries, most of them in the Middle East, and sometimes not very politically stable. The access to oil was used as a political weapon, e.g. in the 1970s, which led to the first (1973) and second (1979) oil crises. The dependence on Russian gas in the EU, Ukraine and other countries has been used by Russia to exert political pressure in various conflicts several times since 2006. Reasons for conflicts are not only about access and power but also about the revenues. Producing countries are interested in maximising profits from their (limited) resources, and consumers are keen on having a steady supply at competitive and stable prices. Thus, producers and consumers are mutually dependent. This is an important factor to consider when thinking about security.

Risks and conflicts are also linked to transport routes. For example, more than 40 % of global oil trade is going to the narrow straits of Hormuz and Malacca making them particularly vulnerable to all sorts of political or even terrorist interventions. In order to reduce vulnerability to interruptions in supply, most countries keep strategic petroleum reserves in the event of crisis. A different set of security issues are associated with nuclear energy, not least since fissile materials can also be used in weapons. For this reason, together with accident risks and the radioactivity of spent fuel, the nuclear enterprise is subject to extensive safeguards and security arrangements.

Less reliance on imported energy through shifting to renewable energy sources that are geographically closer and by improving energy end-use efficiency is generally considered as good for energy security (GEA 2012; chapter 5).

Please discuss:

- *In what ways does energy matter with regard to causing or addressing the major global challenges (e.g. climate change, economic and social development, human well-being, sustainable development and global security)?*

4 The "Energy Equation"

Impacts from societal energy use can be framed with the following simplified equations which are formulated here with regard to greenhouse gas emissions but can be used also for other environmental impacts[1]:

Population	X	Welfare / Capita	X	Activities / Welfare [2]	X	Energy / Activity	X	GHG / Energy	=	GHG emission
				Energy service intensity		*Energy intensity*		*Emission intensity (decarbonisation)*		*Impact*
				Sufficiency		Efficiency		Consistency		

By pointing out the overall drivers of GHG emissions on a macroeconomic level, this equation also provides for a rough structure of the three main elements to reduce GHG emissions and thus to increase sustainability of the energy system: while population and per capita welfare are not typically subject to energy policy, the energy intensity of a nation and its economy and the GHG emission intensity of the respective energy systems are the most important (overall) drivers to reduce energy use and its impacts on the environment, economy and society. Finally, the amount of activities (such as distances travelled or building space heated) per unit of wealth is another element that can be influenced in order to reduce emissions per capita.

In terms of strategies, these elements are often termed as follows:

Energy service intensity (sufficiency)	In the context of developed societies, this mainly means to use less energy services, e.g. travel less by car or have smaller flats and still enjoy a high standard of living and happiness. The reduction of energy service demand successively reduces energy demand and GHG emissions. For developing countries, however, sufficiency often means access to an increased level of energy services.	Activities (energy services)/capita

(continued)

[1] Such functions, which are often defined slightly different, are known as IPAT, or ImPACT, equations (see Waggoner and Ausubel 2002) or as Kaya identity (Kaya 1990; Kaya and Yokobori 1997).

[2] Both, activities and welfare, are often expressed in terms of GDP, which of course is to many respects a simplification. At least above certain levels, welfare is not a direct function of income levels nor is the income a very appropriate measure for the various activities which are supplied by the use of energy.

(Energy) efficiency	Increasing the efficiency of cars, machines and household appliances as well as reducing the energy consumption of buildings leads to a reduced energy use without changes in the demand of energy services (see box on rebound effects). This is done by technical improvements of technology which can be instrumented by various policies and measures.	Energy/activity (GDP)
Emission intensity (consistency)	This means reducing the GHG emissions per unit of energy consumed, e.g. by increasing the share of renewable energy generation and by increasing the efficiency of the energy supply system (e.g. modern power plants and the use of waste heat from power plants). Consistency refers to the idea that the emission intensity should be consistent with the goal of reduced GHG emissions.	GHG/energy

There are basically four *technical options* for reducing carbon dioxide emissions from energy systems: (i) energy efficiency, and in the field of consistency, there are (ii) renewable energy, (iii) nuclear energy and (iv) fossil fuels with carbon capture and storage. From the point of view of sustainability or their ability to meet energy policy goals, they score differently.

Energy efficiency is generally considered a robust strategy to improve on all the energy policy goals and contribute to sustainable development. The same is true for renewable energy although some forms and conversion technologies are still expensive compared to other supply options. But it should also be warned that the use of renewable energy (notably bioenergy and hydropower), can be highly unsustainable with large environmental and social impacts.

Nuclear energy has essentially zero carbon dioxide emissions but is associated with another set of major concerns: nuclear weapon proliferation, accidents and safety and waste handling issues. Fossil fuels are generally considered as unsustainable, but their climate impact can be considerably reduced through deploying technologies for capturing the CO_2 and storing it in underground geological formations. However, fossil fuels remain unsustainable since CO_2 storage capacity as well as fuel resources are limited (see Everett et al. 2012). The sustainability of these four technical options depends strongly on *how* they are implemented and *how* the concept of sustainability is interpreted or defined.

- What are the key elements and strategies to create more sustainable energy systems?

5 Scenarios and Pathways Towards Future Sustainable Energy Systems

The Rebound Effect in Energy Policy

Despite widespread acknowledgement of energy efficiency as a core strategy for sustainable energy systems, it is often claimed that rebound effects will absorb a large part of the energy savings or even lead to higher energy consumption in the long run.

There are however different effects that are mixed under the term rebound. Direct rebound effects which are directly and short-term linked to energy efficiency measures (e.g. heating a well-insulated building to a slightly higher temperature or the more frequent use of a fuel-efficient car) are estimated in several studies for OECD countries to be between 0 and 30 % and expected to decline. This means that more than 70 % of the energy savings achieved through energy efficiency remain.

The effect of other, often more long-term and structural, so-called indirect and macroeconomic rebound effects as well as their attribution to energy efficiency policy is, however, more unclear. While orthodox economics believe them to be small, ecological economists have argued that they are likely to be large or even "backfire" (Sorrell 2010). We, however, are convinced that given the narrow absolute global limits of natural resources, using them as efficiently as possible is a necessary strategy to supply everybody with adequate amounts of energy services.

Over the last years many outlooks or scenarios have been developed by various stakeholders ranging from Greenpeace (2012) and the WWF (Deng et al. 2012) to the International Energy Agency (IEA) and others which try to quantitatively describe the future development towards a global energy system that is compatible with the target to prevent dangerous climate change. For the energy system this means that global greenhouse gas emissions from the energy system (and other sources) have to be cut by at least 50 % by the mid of the century compared to emissions in 1990.

Figure 19.5 shows a result from a scenario analysis commissioned by the WWF (Deng et al. 2012) which can be used as an example. It reveals that in a baseline scenario – if no intensified climate and environmental goals are pursued globally – energy use and related GHG emissions will constantly increase over the next decades (dotted line in Fig. 19.5) putting global climate at high warming risks and causing a range of other problems.

Contrasted to this is a 100 % renewable energy scenario with two major strategies towards a more sustainable system – (i) higher end-use efficiency, energy savings and electrification ("sufficiency" and "efficiency") and (ii) the substitution of fossil fuels by renewable sources (decarbonisation or "consistency"):

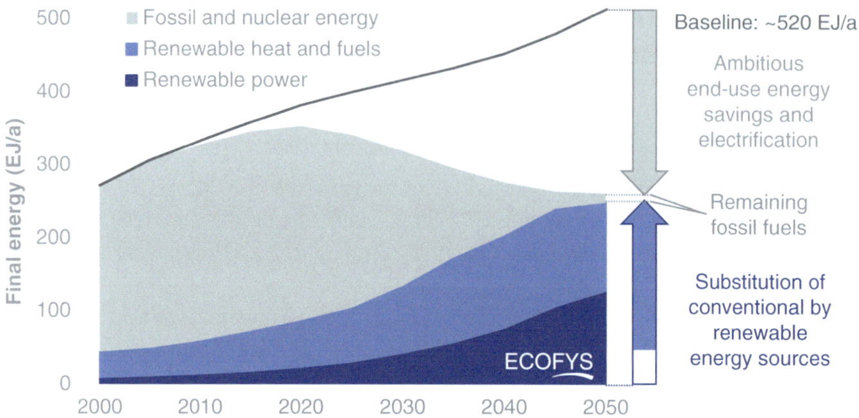

Fig. 19.5 Key developments of the transition to a fully sustainable energy system (Source: Deng et al. 2012, 119)

- About 50 % of the energy demand expected by 2050 under baseline conditions can be avoided by exploiting the large existing potentials for energy efficiency in households, commercial and industrial sectors as well as in transportation and in the energy sector itself.
- The remaining energy demand is almost fully decarbonised by the expansion of renewable energies. The main share of the remaining future energy demand will be delivered by renewable electricity from wind energy, PV cells and other solar energy, hydropower and other sources, while transport and heating fuels will come from biomass, geothermal energy and other renewable sources.

Together, this makes it possible to almost completely phase out fossil fuels and nuclear energy globally.

As can be seen in Fig. 19.5, the use of fossil and nuclear energy will not decline before 2020. The reason for this is that all technical strategies towards a sustainable energy system need some time to gain momentum.

As energy systems consist in part of very long-lived infrastructures and long-lived goods and equipment, it takes time to make the complete shift to improved technologies such as plug-in hybrid-electric cars (few decades), LED lighting (some years) and efficient buildings (many decades). Furthermore, saving almost 50 % of the energy and supplying the rest from renewable sources will require the development of new and improved technologies. The costs and, or, reliability and performance of many technologies still have to be improved, and production lines have to be established.

Thus, in order to realise such ambitious strategies as depicted in many of the scenarios, most of the low-carbon technologies will be needed soon, and to avoid delays in the realisation of the scenarios, the remaining time for successful R&D is already short. Therefore, it is necessary also to revise the priorities of (public) R&D funds which were in the past mainly focussing on nuclear energy, fossil fuels and other aspects but would need to strongly emphasise energy efficiency and renewable energies which make up for the bulk of future sustainable energy technologies that are needed for the transformation towards a sustainable energy system.

This transformation of global, national and local energy systems is, however, not only a technical question. It implies also major social, institutional and economic changes as well as local acceptance. It is also linked to social norms and individual consumption patterns ("sufficiency"). The implementation of energy-efficient technologies and solutions often takes place at the local level. Also, with renewable energies such as wind and solar energy, production becomes more decentralised and local converting the role of energy consumers to so-called prosumers which are simultaneously producers (of, e.g. renewable electricity) and consumers. In this context particularly community levels are important. Consequently, there are numerous scenarios, targets and plans in many cities and communities which try to figure out the local challenges and potentials for action. Cities have (re)acquired their energy supply companies, and, for example, people in rural areas have started renewable energy cooperatives (DGRV 2013).

Local Green Energy and Climate Initiatives
Visions and policies for sustainable energy systems are not limited to national policy-making levels. Since the adoption of the local Agenda 21 at the UN Rio conference 1992, many local initiatives have emerged, and respective networks such as ICLEI (www.iclei.org) and C40 (www.c40cities.org/) comprise thousands of municipalities worldwide including many of the largest megacities (Bulkeley 2010).

- *Please discuss your ideas on how our future energy systems could or should look like.*

6 Policy and Governance for Transitions to Sustainable Energy Systems

Our overview shows that making the transition to sustainable energy systems is not so much an energy resource or technology problem. But it requires investments in energy efficiency, renewable energy technologies, infrastructure, etc. Much, or even all, of these investments can be recouped through lower fuel and other operating costs. Thus, the net effect on total economic costs for energy services can be small.

The main challenge for sustainable energy systems is to get political support and design policies and policy instruments (e.g. regulations, economic incentives and market designs) that stimulate investors, producers, planners and consumers to make the right choices. Important barriers to this include goal conflicts (e.g. when an energy or climate policy is perceived to be in conflict with other societal goals such as jobs and growth) and conflicts of vested interest due to distributional effects. Most policy-induced changes result, more or less, in winners and losers. In our case, fossil fuel producers, car manufacturers and petrochemical companies are examples of likely losers unless they can reinvent themselves and their business. Potential winners are future companies engaged in energy efficiency and renewable energy. Some of these do not exist yet and therefore have little influence in ongoing power struggles and policy processes.

Thus, the challenge is mainly how a transition can be governed through integrated policy strategies that facilitate the shift to new practices, end-use patterns, technologies, infrastructures and primary energy resources. This presents challenges in all areas, but future-oriented energy studies indicate, in particular, four potential governance challenges (aside from those associated with CCS and nuclear energy that we do not expand on here). These include handling the increased pressure on bio-resources and land use, sustainable mobility and transport, electricity system infrastructure development and the decarbonisation of industry.

Biomass is and will remain an important primary energy resource, but energy and climate policies that are designed to make fossil fuels more expensive will also generate a higher willingness to pay for bioenergy. As a result the pressure on these resources increases, and it comes into conflicts with, for example, biodiversity and food production. It can also lead to even greater CO_2 emissions than from fossil fuels if forests are cleared and organic soils are cultivated with high soil carbon emissions as a result. The policy response to this should not be a ban on bioenergy, but it requires good governance and policy that regulates land use and ensures that bioenergy and biofuels are produced sustainably.

Sustainable mobility requires broad policy strategies where one part is the development of new vehicle and fuel technologies based on biofuels, electricity, hydrogen or even hydrocarbons based on electricity. An important challenge here is handling uncertainty concerning how successful technology development will be. It also requires a range of strategies to influence transport demand and the choices of transport modes (where public transport, bicycling and walking are preferred). See Chap. ## on Sustainable Transport.

Scenarios consistently show that electricity will be an increasingly important energy carrier in the future. Wind and solar energy can be readily converted into electricity. Electricity does not contain carbon atoms, and it is a versatile and efficient energy carrier. Electrification is also a key option for reducing fossil fuel use in transport and industry. However, investments in grid infrastructure and redesign of markets are needed in order to handle the variable output of renewable sources through, for example, more flexible demand, transmission across longer distances, energy storage technologies and greater integration with heating and cooling systems.

Finally, several basic material industries, including iron and steel and cement, use large amounts of fossil fuels in their processes. Such basic materials are important for making the transition in other sectors. In cement, roughly half of the CO_2

emissions result from converting calcium carbonate ($CaCO_3$) into calcium oxide (CaO), so using non-fossil fuels will only partly reduce the emissions. Although there are technical options to make industry carbon neutral, it can be quite costly. Higher costs will be difficult to pass through to consumers if competitors in other parts of the world do not meet the same environmental restrictions. Thus, part of the challenge is to facilitate technology shifts while at the same time avoiding that production and emissions move elsewhere.

Conclusion: Energy Systems Can Become Increasingly Sustainable
Looking back, our energy systems have undergone several transitions in history, but we often tend to take present systems as given. It is probably safe to assume that if a transition can be made into sustainable energy systems, with some sufficiency in the use of energy services, future generations will not consider it strange or astounding that the provision of energy services in 2050 or 2100 is, by some measure, sustainable.

Please discuss the following questions:
- *How can sustainable energy systems be implemented?*
- *What needs to be done to achieve sustainable energy systems?*

Recommended Literature for Further Reading

Bhattacharyya SC (2011) Energy economics, concepts, issues, markets and governance. Springer, London/Dordrecht/Heidelberg/New York. doi:10.1007/978-0-85729-268-1
BP (2013) Statistical review of world energy, June 2013 bp.com/statisticalreview
Deng YY et al (2012) Transition to a fully sustainable global energy system. Energy Strat Rev 1(2):109-121. http://dx.doi.org/10.1016/j.esr.2012.07.003
Everett B, Boyle G, Peake S, Ramage J (eds) (2012) Energy systems and sustainability, power for a sustainable future, 2nd edn. Oxford University Press, Oxford. ISBN 978 0 19 959374 3
GEA (2012) Global energy assessment: toward a sustainable future. Cambridge University Press/ International Institute for Applied Systems Analysis, Cambridge/New York/Luxemburg
Greenpeace, GWEC, EREC (2012) Energy [r]evolution. A sustainable world energy outlook. Report 4th edition 2012 world energy scenario

References

Boersma T, Johnson C (2013) Twenty years of US experience – lessons learned for Europe. In: Musialski C, Altmann M, Lechtenböhmer S, Zittel W (eds) Shale gas in Europe – a multidisciplinary analysis with a focus on European specificities. Claeys & Casteels, Leuven/Deventer, pp 9–34
BP (2013) Statistical review of world energy, June 2013. bp.com/statisticalreview

Bulkeley H (2010) Cities and the governing of climate change. Annu Rev Environ Resour 35:229–253. doi:10.1146/annurev-environ-072809-101747

Deng YY et al (2012) Transition to a fully sustainable global energy system. Energy Strat Rev 1(2):109–121. http://dx.doi.org/10.1016/j.esr.2012.07.003

DGRV – German Cooperative and Raiffeisen Confederation (2013) Energy Cooperatives, Findings of survey conducted by the DGRV and its member associations, Spring 2013. Accessed 18 Dec 2013. http://www.dgrv.de/weben.nsf/272e312c8017e736c1256e31005cedff/41cb30f29102b88dc1257a1a00443010/$FILE/Study%20results%20survey%202013.pdf

Greenpeace, GWEC, EREC (2012) Energy [r]evolution. A sustainable world energy outlook. Report 4th edition 2012 world energy scenario

IEA (2011) World energy outlook 2011. Energy for all, financing access for the poor. OECD/IEA, Paris

IEA (2012) World energy outlook 2012. Part B focus on energy efficiency. OECD/IEA, Paris

IEA (2013) Energy policy highlights. OECD/IEA, Paris. http://www.iea.org/publications/freepublications/publication/Energy_Policy_Highlights_2013.pdf

IPCC (2011). IPCC special report on Renewable Energy Sources and Climate Change Mitigation. Prepared by Working Group III of the Intergovernmental Panel on Climate Change [Edenhofer O, Pichs-Madruga R, Sokona Y, Seyboth K, Matschoss P, Kadner S, Zwickel T, Eickemeier P, Hansen G, Schlömer S, von Stechow C (eds)]. Cambridge University Press, Cambridge/New York, 1075 pp

IPCC (2014) Climate Change 2014: Mitigation of Climate Change. Contribution of Working Group III to the Fifth Assessment Report of the Intergovernmental Panel on Climate Change [Edenhofer O, Pichs-Madruga R, Sokona Y, Farahani E, Kadner S, Seyboth K, Adler A, Baum I, Brunner S, Eickemeier P, Kriemann B, Savolainen J, Schlömer S, von Stechow C, Zwickel T, Minx JC (eds)]. Cambridge University Press, Cambridge/New York

Kaya Y (1990) Impact of carbon dioxide emission control on GNP growth: interpretation of proposed scenarios. Paper presented to the IPCC Energy and Industry Subgroup. Response Strategies Working Group, Paris

Kaya Y, Yokobori K (1997) Environment, energy, and economy: strategies for sustainability. United Nations University Press, Tokyo/New York/Paris

Lechtenböhmer S, Altmann M (2013) The greenhouse gas balance. Int Shale Gas Oil J 1(0):24–29

Lovins AB (2004) Energy efficiency, taxonomic overview. Encyclopedia Energy 2: 383–401, 6 vols. Elsevier, San Diego/Oxford. www.rmi.org/sitepages/pid171.php#E04-02

Lovins AB (2011) Reinventing fire: bold business solutions for the new energy era. Chelsea Green Publishing, White River Junction. ISBN 978-1-60358-371-8

Sorrell S (2010) Energy, economic growth and environmental sustainability: five propositions. Sustainability 2(6):1784–1809

Waggoner PE, Ausubel JH (2002) A framework for sustainability science: a renovated IPAT identity. Proc Natl Acad Sci U S A 99(12):7860–7865

Chapter 20
Sustainability and Health

Maud M.T.E. Huynen and Pim Martens

Abstract Achieving good health should be an integral part of the current discussions about sustainable development. It is increasingly recognized that health research (and policy) requires a systems approach and the past decades have witnessed an emerging recognition of the multidimensional and multilevel causation of population health. An ever growing number of health researchers argue that the health of a population can – or must – be viewed within the broader system of health determinants. Consequently, in our effort to assess the health impacts of global (environmental) change, we have to be aware of the limitations of the traditional reductionist approach.

Stressing the need for a system-based approach toward health, this chapter discusses and illustrates a conceptual model describing the broader context and multicausality of our health. We apply this framework to a widely discussed health impact of climate change, namely, the emergence of malaria in the African highlands. This clearly demonstrates that malaria in East Africa's highlands presents an interesting case study for understanding the importance of the system's interactions between climate and non-climate factors in shaping human vulnerability to the adverse health impacts of global warming. Climate change is believed to primarily affect the intrinsic malaria transmission potential, but this relationship interacts with other factors and developments that affect disease dynamics as well.

However, trying to conceptually describe the system involved is only one of the first steps in applying a system-based approach toward health. Hence, we briefly elaborate on some example tools from the sustainability science toolkit (modeling, scenario analyses, and participatory methods) that are available and conceivable in order to advance further systems research in the field of health and sustainable development. The chapter concludes with a discussion of possible barriers to adopting a sustainability science approach toward health, in an effort to explain the slow progress made so far.

Keywords Climate change • Health • Highland malaria • System-based approach • Sustainability science

---

M.M.T.E. Huynen (✉)
International Centre for Integrated assessment & Sustainable development (ICIS), Maastricht University, P.O. Box, 616, 6200 MD Maastricht, The Netherlands
e-mail: M.Huynen@maastrichtuniversity.nl

P. Martens
Maastricht University, Maastricht, The Netherlands
e-mail: p.martens@icis.unimaas.nl

H. Heinrichs et al. (eds.), *Sustainability Science*, DOI 10.1007/978-94-017-7242-6_20

1 Introduction

Achieving good health has become an accepted international goal, and our (future) health should be an integral part of the current discussions about sustainable development. The Brundtland Commission (Brundtland 1987) argued that "the satisfaction of human needs and aspirations is the major objective of development" and "sustainable development requires meeting the basic needs of all and extending to all the opportunity to satisfy their aspirations for a better life." There have been several attempts to identify what our basic needs actually encompass. Well-known theories are, for example, the ones developed by Maslow (1954, 1968) and Max-Neef (1991), and in both approaches maintaining or improving our physical (and mental) health is seen as a crucial element.

The relationship between sustainable development and population health works two ways. As the world around us is becoming progressively interconnected and complex, human health is increasingly perceived as the integrated outcome of its ecological, social-cultural, economic, and institutional determinants. Due to its multidimensional causality, good health is often seen as an outcome of sustainable development. McMichael (2006; McMichael et al. 2000) argues that health can be seen as an important high-level integrating index that reflects the state – and, in the long term, the sustainability – of our natural and socioeconomic environment. The increasing widespread and long-term risks to population health are, therefore, at the heart of non-sustainability. Wilcox and Colwell (2005), for example, agree that no issue could be a more fundamental measure of sustainability than public health. The other way around, however, a healthy population is also necessary to achieve sustainable development. As Brundtland (2002), Director-General Emeritus of the World Health Organization (WHO), puts it: "a healthy life is an outcome of sustainable development, as well as a powerful and undervalued means of achieving it." The WHO Commission on Macroeconomics and Health (2001), for example, concluded that good health is a central input to poverty reduction and socioeconomic development.

The past decades have witnessed a growing recognition of the multidimensional and multilevel causation of population health. An ever growing number of health researchers (Wilcox and Colwell 2005; Pearce and Merletti 2006; Albrecht et al. 1998; Colwell 2004; McMichael 2005) argue that the health of a population can – or must – be viewed within the broader system of health determinants. Populations are not simply the collection of individuals, but are shaped by, and shape, the systematic context in which they operate (Pearce and Merletti 2006). Risk factors for disease do not operate in isolation but occur in a particular population context. Upstream forces play an important role in global health research (Sreenivasan and Benatar 2006). These upstream or *contextual factors* may have large impacts, but their effects are nonlinear and less predictable (Philippe and Mansi 1998). As our attention moves upstream in the causal chain of health determinants, there is an increasing interest in multilevel – and systems – approaches (Pearce and Merletti 2006; McMichael 1995, 1999; Pearce 2004). Various terms have been used to describe such broader approaches to population health, such as eco-epidemiology (Martens 1998; Susser and Susser 1996; Ladd and Soskolne 2008; Soskolne and Broemling 2002), ecological perspective on health (McLaren and Hawe 2005), social-ecological systems perspective on health (McMichael 1999), ecosystem approach to public health (Arya et al. 2009), ecological public health (Morris 2010), and bio-complexity approach to health (Wilcox and Colwell 2005; Colwell 2004). As Soskolne et al. (2007) state, we "must embrace greater complexity" as "the traditionally used, reductionist, linear approaches are inferior for understanding the interactive webs that are critical for sustainable development and for the health and well-being of future generations." Similarly, the WHO argues that systems thinking works to reveal the underlying characteristics and relationships of systems (de Savigny and Adam 2009).

Stressing the need for a *system-based approach toward health*, this chapter first discusses a conceptual model describing the multi-causality within the health system. This will be further illustrated by a description of the climate and non-climate drivers behind the observed emergence of malaria in the African highlands. Accordingly, we will briefly elaborate on some example tools from the sustainability science toolkit that are available and conceivable in order to advance further systems research in the field of health and sustainable development. Finally, the chapter concludes with a discussion of possible barriers to adopting a sustainability science approach toward health, in an effort to explain the slow progress made so far.

Q: Reflect on the notion that "health is an integrated index of how sustainable we are managing our natural, social, and economic resources."

2 A Systems Approach Toward Population Health

In order to illustrate the broader context and multi-causality of our health, Huynen et al. (2005; Huynen 2008) developed a *conceptual framework for population health* (Fig. 20.1). Their model combines the nature of health determinants and their level of causality into a basic framework that conceptualizes *the multi-causality of population health*.

In order to differentiate between determinants of a different nature, the customary distinction between institutional, sociocultural, economic, and environmental determinants is made. These determinants operate at different hierarchical levels of causality. The chain of events leading to a specific health outcome includes both proximal and distal causes—proximal factors act directly to cause disease or health gains, while distal determinants are set further back in the causal chain and act via intermediate causes. In addition, contextual determinants play an important role. These can be seen as the upstream macro-level conditions shaping the distal and proximate health determinants; they form the context within which the distal and proximate factors operate and develop. Determinants with different positions in the

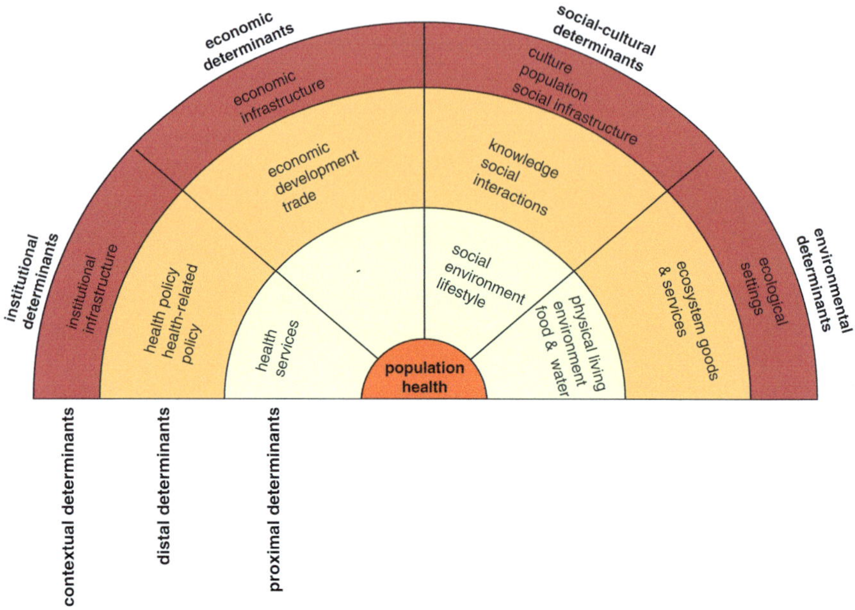

Fig. 20.1 Multi-nature and multi-level framework for population health, developed by Huynen et al. (2005)

causal chain probably also differ in their temporal dimensions. Individual-level proximal health risks can be altered relatively quickly, for example, by a change in personal behavior; for disease rates in whole populations to change requires slower and more structural changes in contextual factors, often over the course of a few decades (Huynen 2008).

Figure 20.1 shows the wide-ranging overview of the health determinants that can fit within this framework. The way different factors and developments within the system interact is critical to how the whole system works and, subsequently, how vulnerable populations are to a particular health risk.

A key example of a global health challenge to sustainable development is the health impact of climate change, and one of the first steps in applying a system-based approach toward climate change and health entails describing the system involved. Box 20.1 discusses the climate and non-climate drivers behind a well-recognized health impact of climate change, namely, the emergence of highland malaria in the East African highland. Accordingly, Table 20.1 applies the above framework (in Fig. 20.1) to this case study.

Box 20.1 and Table 20.1 clearly demonstrate that malaria in East Africa's highlands presents an interesting case study for understanding the importance of the interactions between climate and non-climate factors in shaping human vulnerability to the adverse health impacts of global warming (Huynen et al. 2013). A 2011 report by the Africa Initiative (Tesi 2011) also stressed the multi-causality of malaria; although climate change has been associated with the observed malaria

Box 20.1: Case Study—The Emergence of Highland Malaria in the East African Highland (Based on Huynen et al. 2013)

Climate (change) is believed to be an important factor in the dynamics of malaria transmission (Martens et al. 1999; Chaves and Koenraadt 2010; IPCC 2007, 2014). Temperature affects mosquito survival as well as parasite development. Additionally, mosquito survival is also affected by changes in humidity, while developments in rainfall (patterns) can affect the number of suitable breeding sites. As a result, the past two decades have witnessed considerable debate about the importance of climate change in driving the observed changes in malaria distribution and transmission in highland regions. A review study by Chaves and Koenraadt (2010) concluded that the linkage between climate change and malaria in the highlands of Africa is rather robust. The same publication stressed, however, that overemphasizing the role of climate as the autonomous main driver of highland malaria does not account for the clear multifactorial causality of disease transmission (Chaves and Koenraadt 2010).

In an elaborate literature review, Cohen et al. (2012) identified the following suggested causes of past malaria resurgence events: weakening of control activities (e.g., due to funding constraints, poor execution, purposeful cessation), technical problems (e.g., vector resistance, drug resistance), human or mosquito movement, development/industry changes (including land use change), socioeconomic weakening, climate/weather, and war. Malaria is also closely linked to poverty; poorer communities have a higher disease risk due to, for example, lower (financial) access to health services, poorer nutritional status, lower education levels, poor sanitation, and inadequate housing (Ricci 2012). Although the above listing is probably far from exhaustive, it clearly illustrates that climate change is just one of many processes that affect infectious disease risk (Morse 1995; Cohen 2000; Sutherst 2004; McMichael 2004; IPCC 2014. Hence, the assessment of climate change impacts on malaria is challenged by the complex interactions between climate and non-climate factors. We will explore this in more detail by looking at the various drivers of malaria emergence in the East African highlands.

The highlands are a fragile ecosystem under great pressure from increasing populations, deforestation, and increased farming (McMichael 2003). East African highlands are one of the most populated regions in Africa, and their population growth rates are among the highest in the world. As a result, the regions are also faced with high rates of poverty. Poverty and demographic pressures have spurred massive land use and land cover changes (including massive deforestation) for agricultural practices (Himeidan and Kweka 2012). The upland communities are often remote from regional health centers, and health services are patchy making the surveillance and control of malaria difficult. It is increasingly acknowledged that the risk of highland malaria moving to higher altitudes depends on the interplay between climate change and, for example, land use change, population growth, population movement,

(continued)

Box 20.1: (continued)
agricultural practice (e.g., pesticide use, irrigation systems), cessation of malaria control activities, drug resistance, and socioeconomic status.

Malaria invasion of the African highlands has been associated with the migration of people from the lower areas to the higher altitudes (Lindsay and Martens 1998), introducing the malaria parasite into highland regions. The limited immunity of people living at higher altitudes could have played a role. Furthermore, the massive deforestation in East Africa has shown to be associated with changes in the local climate. As such, both the land use changes and global warming may act together in causing the observed regional change in the local climate of the East African highlands (Himeidan and Kweka 2012). Changes in crop choice can also play a role, as demonstrated by the invasion of malaria in the Bure highlands of Ethiopia due to the fact that the mosquito vector thrived on feeding on maize pollen, just shortly after this crop was introduced (Ye-Ebiyo et al. 2000; Kebede et al. 2005). Irrigation activities and forest clearing have been associated with increases in vector densities due to, for example, enhancing mosquito breeding sites (Himeidan and Kweka 2012). Susceptibility to the increasing mosquito densities and associated malaria risk is further complicated by the high poverty rates in the East Africa highlands. Fortunately, the highlands have experienced a reduction in malaria prevalence since the early 2000s, due to ongoing malaria interventions (Chaves and Koenraadt 2010; Himeidan and Kweka 2012; Stern et al. 2011). However, the sustainability of these interventions may be questioned (Himeidan and Kweka 2012). African countries mostly rely on external donors, and global funding levels for malaria are in an increasingly precarious state (Pigott et al. 2012); weakening of malaria control programs has been an important driver of observed past causes of malaria resurgence (Cohen et al. 2012). Recently, Artzy-Randrup et al. (2010) hypothesized that the influence of climate change on malaria also interacts with the spread of drug resistance through altered levels of transmission intensity.

invasion in African highlands, other factors are involved as well in accelerating this process. The report argues that climatic factors (increases in temperature, rainfall, and humidity) act as primary factors, because as long as the disease transmission is constrained by climatic factors, the disease will automatically be limited as well. The secondary factors, such as drug resistance, agricultural development, population growth, migration, conflicts, and land use change, can accelerate this process put in motion by climatic factors. Similarly, Chaves and Koenraadt (2010) emphasize that "a multidimensional array of underlying factors is likely to be at play here, most of which may be sensitive to climatic change." Hence, although climate change is believed to primarily affect the intrinsic malaria transmission potential (Cohen et al. 2012; Tesi 2011), it interacts with other factors and developments that affect disease dynamics as well. Most of them are expected to be affected by climate

Table 20.1 The emergence of highland malaria in Africa: example system variables

Causal level of health determination	Institutional	Economic	Sociocultural	Environmental
Contextual	Public health infrastructure, including a number of health-care centers in highland areas	Economic infrastructure	High population growth and density resulting in demographic pressures	Climate change, ecosystem change
Distal	Health policy including efforts to reduce malaria, agricultural policies	Slow economic development, agricultural sector developments	Population movement, high poverty rates	Substantial land use/cover change, agricultural irrigation, altered local climate regulation
Proximal	Pre-2000: lack of (access to) health care and control/ surveillance activities Post-2000: increasing malaria interventions and control	–	Lack of immunity to malaria in highlands, wrong use of antibiotics or bed nets, drug resistance	Changes in local climate including temperature rise, increase in mosquito breeding sites

change, such as agriculture, food security, migration, and poverty (IPCC 2007; McMichael et al. 2012). Hence, it is increasingly recognized that research and policy in the field of climate change and health requires a systems approach (Huynen et al. 2013), building on insights from sustainability science.

Q: In what world regions will vector-borne disease, like malaria, be most sensitive to climatic changes?

3 Adopting a Systems Approach to Health: Sustainability Science Tools

The idea that problem framing using conceptual models may be used to address complex (policy) challenges is not new (Morris 2010), and the previous section has put the infectious disease risks associated with climate change within a broader systems context. Although problem framing in order to wrap your head around all relevant variables within the climate-health system is an important step forward, it might represent only the tip of the iceberg. Within this system there are dynamic processes and feedback loops, resulting in emergent system properties (i.e., sum more than its parts), points of bifurcation, and possible tipping points.

So how must we address such a broad issue, encompassing debated relationships between multiple interacting factors operating at different positions in the causal chain? Building on insights from Mode-2 science (Gibbons et al. 1994), post-normal science (Ravetz 1999; Funtowicz and Ravetz 1993, 1994), and sustainability science (Kates et al. 2001; Martens 2006), a systems approach toward health should account for a number of shared research principles such as transdisciplinarity, participation of nonscientific stakeholders, co-production of knowledge, recognition of uncertainty and system's complexity, and the quest for an exploratory science instead of a predictive one. This challenges epidemiologists, as well as scientists and practitioners in other relevant disciplines, to extend their conventional methodological boundaries. To date, however, an unprecedented gap is apparent between paradigm and practice. Yet innovative methods and tools are emerging in other fields, providing examples of those available and conceivable in order to advance further systems research in the field of health and sustainable development (Soskolne et al. 2009):

- *Modeling the health system*: In modeling population health, traditional epidemiological approaches usually use regression techniques to explore the relations between health determinants and health outcomes (Soskolne et al. 2009; Galea et al. 2010). However, these usually provide only limited insight into the dynamics behind changing health patterns; the fundamental limitation of these statistical techniques in addressing interacting, dynamic, discontinuous, or changing relationships within the system remains (Galea et al. 2010). Hence, there is an increasing interest in adopting complex system dynamic simulation models in health research (e.g., Galea et al. 2010; Sterman 2006; Trochim et al. 2006;

Mendez 2010) that allow for causal influence at multiple levels, the interaction among system variables, dynamic feedback, nonlinearity, and discontinuities. As explained by the Mendez (2010) system, modeling in public health can be seen as "a formal expression of our thoughts about the mechanisms that drive a real phenomenon [...]. Models can provide a common framework to exchange ideas, crystallize our thoughts, highlight what we know and what we still need to find out, and experiment with possible solutions." In this respect, Galea et al. (2010) argue that epidemiologists and other health scientists can learn from other fields that have been applying such simulation approaches, such as systems biology, ecology and environmental sciences, and organizational science.

- *Scenario analysis of future health*: A system-based approach implies a lower emphasis on prediction but an accompanying greater emphasis on understanding the processes involved, acknowledging (inherent) uncertainties, and exploring alternative health futures. In sustainability science, scenario analysis is used as a tool to assist in the understanding of possible future developments of complex systems. Scenarios can be defined as descriptions of journeys to possible futures that reflect different assumptions about how current trends will unfold, how critical uncertainties will play out, and what new factors will come into play (UNEP 2002). In other words, scenarios are plausible but simplified descriptions of how the future may develop, according to a coherent and internally consistent set of assumptions about key driving forces and relationships (Swart et al. 2004). UNEP (2007), for example, provides an interesting guideline for developing scenarios. Looking at the main global-scale scenario studies, it can be concluded, however, that the health dimension is largely missing (Huynen 2008; Martens and Huynen 2003).
- *Transdisciplinary/participatory methods*: The omnipresence of uncertainty in complex systems allows for different valid views on the essence and functioning of these systems. The use of participatory/transdisciplinary methods is more exclusively linked to the emerging paradigm of post-normal science. As such, the involvement of actors from outside academia into the research process is also seen as a key component of sustainability science; it facilitates the integration of the best available knowledge and the co-production of knowledge, the identification and reconciliation of values and preferences, as well as creation of ownership for problems and solutions. Transdisciplinary, community-based, interactive, or participatory approaches have been suggested in order to meet these goals (Lang et al. 2012). Van Asselt and Rijkens-Klomp (2002) indicated, for example, that a multitude of participatory methods (e.g., focus groups, participatory modeling, scientist-stakeholder workshops, scenario analysis, and policy exercises) could be used to help assessors in structuring and eliciting tacit knowledge about and identifying perspectives on the complex issue being studied in the face of uncertainty.

Q: In addressing the complexity of "sustainability and health," which of the above methods is most useful? In what context?

4 Conclusion: The Need for Overcoming Barriers

To conclude, there is a growing acknowledgment of the multidimensional and multilevel causation of (global) health and the importance of a system-based approach, building on insights from sustainability science (Martens et al. 2011). Consequently, in our effort to assess the health impacts of global (environmental) change, we have to be aware of the limitations of the traditional reductionist approach; population health cannot be disassembled to their constituent elements and then reassembled in order to develop an understanding of the system as a whole. For example, this chapter shows that many of the factors within the climate-health system will interact with each other in ways that, as yet, may not be fully understood. Additionally, the outcomes of these interactions will vary across geographical locations but also across different disease outcomes (IPCC 2007; Cohen 2000; Sutherst 2004). We need to be moving away from discussion about the relative importance of climate change compared to other stressors, toward approaches that take possible synergies between different developments into account. As climate and non-climate factors work together, climate change cannot be seen as "a standalone risk factor" but rather as an amplifier of existing health risks (Costello et al. 2009). In order to avoid an escalation of health risk synergies, there is a need to better understand the multifaceted and complex linkages involved (Canfalonieri and McMichael 2006).

However, over the past several decades, questions of closely related cause-and-effect relationships have dominated epidemiological practice. Linear, reductionist approaches to research questions – focusing on proximate cause-and-effect relationships – have characterized much of what epidemiology has contributed to public health in the second half of the twentieth century (Soskolne et al. 2009). As a result, however, the exploration of long-term and complex risks to human health seems far removed from the tidy examples that abound in textbooks of epidemiology and public health research. There is a need to broaden the traditional view on disease causation in order to account for a multilevel understanding of disease etiology and the interrelations among these multiple health determinants (Galea et al. 2010). Such system thinking challenges the epidemiological concern with studying single causes of disease in isolation; by training, epidemiologists and public health researchers are less accustomed to studying causes within a systems context or addressing far longer time frames than current boundaries of the health sciences and the formal health sector (Martens and Huynen 2003).

A sustainability science approach to public health also implies recognizing that there is no single discipline or single operational method for systems thinking (Leishow and Milstein 2006). Such interdisciplinarity demands from health researchers to be particularly open to (learn from) the contributions of other traditions and approaches. Moving even beyond research collaborations among and above disciplinary boundaries, transdisciplinarity requires the involvement of and collaborations with nonacademic stakeholders from business, policymaking, and/or civil society. However, scientists taking a more conventional research perspective, such as traditional epidemiologists and health researchers, might question the reliability, validity, and other epistemological and methodological aspects of this type of research (Lang et al. 2012). From a more practical perspective, transdisciplinary research is a relatively new field, still in need of further enhancement in order to overcome its teething problems. Lang et al. (2012) recently published a very elaborate overview of the main challenges (and possible coping strategies) in conducting transdisciplinary research, including difficulties concerning design principles (e.g., lack of joint problem framing, selection of stakeholders/team members), methodological issues (e.g., conflicting methodological standards, discontinuous participation), and problems in the application of co-created knowledge (e.g., lack of transferability of results). They conclude that further developing the practice of transdisciplinary research requires "continuous structural changes in the academic system in order to build capacity for transdisciplinarity among students and researchers." The identified (practical) research challenges, as well as their conclusions about the need for capacity building, seem equally valid for conducting transdisciplinary research regarding the field of health and sustainable development.

Furthermore, the use of complex systems dynamic modeling approaches demands a shift from singling out a single cause as main research objective to a focus on understanding interactions and interrelations between various causal factors operating at multiple levels in order to gain insights into how these relationships (and feedbacks) contribute to the emergence of disease patterns within a population (Galea et al. 2010). These models need to be parameterized with observational (epidemiological) data, but this data needs to be applied in a creative way combining information from disparate sources and allowing for assumptions to be made in order to create simulation models in face of imperfect data and uncertainty about parameter values, relationships, and future developments. Accounting for system's complexity and uncertainty will also require a conceptual shift for epidemiology and public health – from statistical association models focused on observed effect estimates to simulations of complex dynamic systems of health determination in which we test scenarios under different conditions (Galea et al. 2010). Thinking critically about "what-if scenarios" entails moving from a predictive science in search for eliminating uncertainty to an exploratory science in the face of (inherent) uncertainties.

Hence, as stressed by Galea et al. (2010), unfamiliarity with methods and limited training in their implementation are probably enough reasons to delay epidemiologists' adaptation of systems approaches. Sterman (2006) even states that "faced with overwhelming complexity of the real world, time pressure, and limited cognitive capabilities, we are forced to fall back on rote procedures, habits, rules of thumb, and simple mental models." But – although health scientists might feel very comfortable with more reductionist approaches and we are, consequently, very slow adopters of systems thinking – we have to face the reality that we are dealing with complex real life health risks that we need to understand and address in the face of many sustainable development challenges.

Further Reading

Huynen MMTE, Martens P, Hilderink HBM (2005) The health impacts of globalisation: a conceptual framework. Global Health 1: article number 14 (12 pages). http://www.globalizationandhealth.com/content/1/1/14

Lee K, Collin J (2005) Global change and health. Open University Press, Maidenhead. ISBN 0335218482

Rayner G, Lang T (2012) Ecological public health: reshaping the conditions for good health. Earthscan, New York. ISBN 13: 978–1844078325

WHO/WMO (2012) Atlas of health and climate. http://www.who.int/globalchange/publications/atlas/report/en/index.html

Websites

http://www.who.int/globalchange/en/
http://www.who.int/social_determinants/en/

References

Albrecht G, Freeman S, Higginbothman N (1998) Complexity and human health: the case for a transdisciplinary paradigm. Cult Med Psychiatry 22:55–92

Artzy-Randrup Y, Alonso D, Pascual M (2010) Transmission intensity and drug resistance in malaria population dynamics: implications for climate change. PLoS One 5, e13588

Arya N et al (2009) Time for an ecosystem approach to public health? Lessons from two infectious disease outbreaks in Canada. Glob Public Health 4:31–49

Brundtland GH (ed) (1987) Our common future: The World Commission on Environment and Development. Oxford University Press, Oxford

Brundtland GH (2002) World summit on sustainable development: importance of health in economic development makes it a priority. BMJ 325:399–400

Canfalonieri U, McMichael AJ (eds) (2006) Global environmental change and human health: science plan and implementation strategy. Earth System Science Partnership (ESSP report no. 4)

Chaves L, Koenraadt C (2010) Climate change and highland malaria: fresh air for a hot debate. Q Rev Biol 85:27–55

Cohen M (2000) Changing patterns of infectious disease. Nature 406(6797):762–767

Cohen J et al (2012) Malaria resurgence: a systematic review and assessment of its causes. Malar J 11:122

Colwell RR (2004) Biocomplexity and a new public health domain. Ecohealth 1:6–7

Costello A et al (2009) Managing the health effects of climate change. Lancet 373:1693–1733

de Savigny D, Adam T (2009) Systems thinking for health systems strengthening. Alliance for Health Policy and Systems Research. World Health Organization, Geneva

Funtowicz SO, Ravetz JR (1993) Science for the post-normal age. Futures 25:735–755

Funtowicz SO, Ravetz JR (1994) Uncertainty, complexity and post-normal science. Environ Toxicol Chem 13:1881–1885

Galea S, Riddle M, Kaplan G (2010) Causal thinking and complex system approaches in epidemiology. Int J Epidemiol 39:97–106

Gibbons M et al (1994) The new production of knowledge: the dynamics of science and research in contemporary societies. Sage, London

Himeidan Y, Kweka E (2012) Malaria in East African highlands during the past 30 years: impact of environmental changes. Front Physiol 3:315
Huynen MMTE (2008) Future health in a globalising world. Universitaire Pers Maastricht, Maastricht
Huynen MMTE, Martens P, Hilderink HBM (2005) The health impacts of globalisation: a conceptual framework. Global Health 1: article number 14 (12 pages)
Huynen M, Martens P, Akin S (2013) Climate change: an amplifier of existing health risks in developing countries. Environ Dev Sustain 15(6):1425–1442. doi:10.1007/s10668-013-9450-4
IPCC (2007) Climate change 2007: impacts, adaptation and vulnerability. Cambridge University Press, Cambridge
IPCC (2014) Climate change 2014: impacts, adaptation and vulnerability; Working Group II contribution to the IPCC fifth assessment report. Cambridge University Press, Cambridge/New York
Kates R et al (2001) Sustainability science. Science 292:641–642
Kebede A, McCann J, Ye-Ebiyo Y (2005) New evidence of the effects of agro-ecological change on malaria transmission. Am J Trop Med Hyg 73:676–680
Ladd BD, Soskolne CL (2008) A toolkit for ecoepidemiological enquiry under global ecological change. In: Soskolne CL et al (eds) Sustaining life on earth: environmental and human health through global governance. Lexington Books, Lanham
Lang D et al (2012) Transdisciplinary research in sustainability science: practice, principles, and challenges. Sustain Sci 7(1):25–43
Leishow S, Milstein B (2006) Systems thinking and modeling for public health practice. Am J Public Health 96(3):403–405
Lindsay S, Martens W (1998) Malaria in the African highlands: past, present and future. Bull World Health Organ 76:33–45
Martens W (1998) Health impacts of climate change and ozone depletion: an ecoepidemiologic modeling approach. Environ Health Perspect 106(Suppl 1):241–251
Martens P (2006) Sustainability: science or fiction. Sustain Sci Pract Policy 2:36–41
Martens P, Huynen MMTE (2003) A future without health? Health dimension in global scenario studies. Bull World Health Organ 81:896–901
Martens P et al (1999) Climate change and future populations at risk of malaria. Glob Environ Chang 9:S89–S107
Martens P et al (2011) Globalisation and human health: complexity, links and research gaps. IHDP Update 1:5–7
Maslow AH (1954) Motivation and personality. Harper & Row, New York
Maslow AH (1968) Toward a psychology of being. Van Nostrand Reinhold, New York
Max-Neef MA (1991) Human scale development: conception, application and further reflections. The Apex Press, New York
McLaren L, Hawe P (2005) Ecological perspectives in health research. J Epidemiol Community Health 59:6–14
McMichael AJ (1995) The health of persons, populations and planets: epidemiology comes full circle. Epidemiology 6:633–636
McMichael AJ (1999) Prisoners of the proximate: loosening the constraints on epidemiology in an age of change. Am J Epidemiol 10:887–897
McMichael AJ (2003) Global climate change: will it affect vector-borne infectious diseases? Intern Med J 33:554–555
McMichael AJ (2004) Environmental and social influences on emerging infectious diseases: past, present and future. Philos Trans R Soc 359(1447):1049–1058
McMichael AJ (2005) Detecting the health effects of environmental change: scientific and political challenge. Ecohealth 2:1–3
McMichael AJ (2006) Population health as a primary criterion of sustainability. Ecohealth 3:182–186

McMichael AJ, Smith KR, Corvalan CF (2000) The sustainability transition: a new challenge. Bull World Health Organ 78:1067
McMichael C, Barnett J, McMichael A (2012) An ill wind? Climate change, migration, and health. Environ Health Perspect 120:646–654
Mendez D (2010) A systems approach to a complex problem. Am J Public Health 100(7):1160
Morris G (2010) Ecological public health and climate change policy. Perspect Public Health 130:34–40
Morse S (1995) Factors in the emergence of infectious diseases. Emerg Infect Dis 1(1):7–15
Pearce N (2004) The globalization of epidemiology: introductory remarks. Int J Epidemiol 33:1–5
Pearce N, Merletti F (2006) Complexity, simplicity, and epidemiology. Int J Epidemiol 35:515–519
Philippe P, Mansi O (1998) Nonlinearity in the epidemiology of complex health and disease processes. Theor Med Bioeth 19:591–607
Pigott D et al (2012) Funding for malaria control 2006–2010: a comprehensive global assessment. Malar J 11:246
Ravetz J (1999) What is post-normal science. Futures 31:647–653
Ricci F (2012) Social implications of malaria and their relationships with poverty. Mediterr J Hematol Infect Dis 4, e2012048
Soskolne C, Broemling N (2002) Eco-epidemiology: on the need to measure health effects from global change. Glob Change Human Health 3(1):58–66
Soskolne C et al (2007) Toward a global agenda for research in environmental epidemiology. Epidemiol Soc 18(1):162–166
Soskolne C et al (2009) Eco-epidemiological enquiry under global ecological change: an integrated assessment toolkit for beginners. Eur J Oncol
Sreenivasan G, Benatar S (2006) Challenges for global health in the 21st century: some upstream considerations. Theor Med Bioeth 27:3–11
Sterman J (2006) Learning from evidence in a complex world. Am J Public Health 96(3):505–514
Stern D et al (2011) Temperature and malaria trends in Highland East Africa. PLoS One 6, e24524
Susser M, Susser E (1996) Choosing a future for epidemiology: II. From black box to Chinese boxes and eco-epidemiology. Am J Public Health 86:674–677
Sutherst R (2004) Global change and human vulnerability to vector-borne diseases. Clin Microbiol Rev 17(1):136–173
Swart RJ, Raskin P, Robinson J (2004) The problem of the future: sustainability science and scenario analysis. Glob Environ Chang 14:137–146
Tesi M (2011) Global warming and health: the issue of malaria in Eastern Africa's highlands. Africa initiative discussion papers no. 2- Aug 2011. Africa Initiative and The Centre for International Governance Innovation
Trochim W et al (2006) Practical challenges of systems thinking and modeling in public health. Am J Public Health 96(3):538–546
UNEP (2002) Global environment outlook 3: past, present and future perspectives. Earthscan, London
UNEP (2007) IEA training manual. Module 6: scenario development and analysis. United Nations Environmental Programme, Nairobi
Van Asselt MBA, Rijkens-Klomp N (2002) A look in the mirror: reflection on participation in Integrated Assessment from a methodological perspective. Glob Environ Chang 12:167–184
WHO (2001) Macroeconomics and health: investing in health for economic development. World Health Organization, Commission on Macroeconomics and Health, Geneva
Wilcox B, Colwell R (2005) Emerging and reemerging infectious diseases: biocomplexity as an interdisciplinary paradigm. Ecohealth 2:244–257
Ye-Ebiyo Y, Pollcak R, Spielman A (2000) Enhanced development in nature of larval Anopheles arabiensis mosquitoes feeding on maize pollen. Am J Trop Med Hyg 63:90–93

Chapter 21
Mobility and Sustainability

Aaron Golub

Abstract Urban practices such as automobile dependence result from webs of institutions, from citizens and neighborhoods to city and state governments to federal policies. Effective action for achieving sustainability begins with understanding these institutions and how they respond to and resist change. In this chapter, we review those institutions involved with creating and preserving automobile use. This investigation illustrates that it is not enough to have a "right answer" be it a certain technology or a certain urban design proposal. The importance is in how these answers are implemented by citizens and governments – how visions are translated into interventions by real communities in various experiments and pilot projects which can help to illustrate pieces of those future states – today. In this chapter we review several cases of such proactive planning and policy which have been successful in enacting long-term visions for sustainable transportation. These include new urban planning paradigms based on transit-oriented design and accessibility, systems to facilitate sharing cars and to encourage cycling, and innovations in technology to improve the management of existing infrastructure.

Keywords Transportation • Mobility • Accessibility • Urban • Planning

1 Introduction

Some of the world's most pressing problems result from the manner in which urban systems operate. These systems consume huge amounts of energy and materials and create intense local "hotspots" for pollution, solid waste, congestion, safety problems, and other challenges to livability and sustainability. Urban mobility systems are often a leading cause of these challenges, and focusing on urban mobility is an effective approach to solving several key sustainability challenges (Black 2005; Golub 2012).

A. Golub (✉)
Toulan School of Urban Studies and Planning, Portland State University,
Portland, OR, USA
e-mail: agolub@pdx.edu

H. Heinrichs et al. (eds.), *Sustainability Science*, DOI 10.1007/978-94-017-7242-6_21

Urban *mobility*, in a broad sense, refers to the moving of people and goods between different destinations within the city, including residences, workplaces, shopping areas, warehouses, ports, and factories. Mobility is expensive, requiring resources and imposing various kinds of costs on society, including not only fees (e.g., tolls or parking) and fixed costs (e.g., costs of automobile ownership or infrastructures), but also time costs, and other costs, such as health or environmental damages. These costs of mobility, however, are often difficult for the average traveler to understand. As certain modes of travel are supported by investments in infrastructure (e.g., roads, trains) and institutional support (e.g., traffic engineering, zoning policies requiring minimum car parking supply), their costs to the average user may seem lower. Therefore, travel by a certain mode of transportation is not convenient in the absolute but is made convenient by a coordination of investments by a variety of social actors, from households to city governments, the national government, and private industries. For example, without significant public investments in traffic engineering, road construction, parking systems, and emergency systems, travel by automobile would be very expensive and inconvenient.

Related to mobility is the idea of *accessibility*, which considers more explicitly the objective of movement. Ultimately, the value of movement results from the value derived from the completed trip (unless the trip was made purely for leisure purposes). Accessibility is the attainment of that value from the trip – irrespective of how much travel that entails. Ultimately, accessibility is the aim of any mobility system. Thus, in urban areas where origins (say, residences) and destinations (say, workplaces) are far apart, accessibility results from being mobile. On the other hand, locating destinations close to origins, or placing them close to a coordinated public transit network, can improve access while reducing the need to travel.

Many urban mobility systems attempt to create ubiquitous, inexpensive mobility, typically based on the automobile (Cervero 1996). This mobility-focused approach creates significant external costs and unintended consequences. Furthermore, the size and extent of the roads and parking needed to support such an approach become a hindrance to the use of modes of transportation other than the automobile. From this excess need for travel stem many adverse effects, to be discussed below.

Efforts to enhance accessibility and transform urban mobility systems in order to control their detrimental effects focus on four core approaches: price signals, land-use changes, technology development, and communication. Pricing, which can include various types of taxes and fees, is used in mobility systems to manage demand, internalize externalities (e.g., environmental damages), fund infrastructure and operation of the systems, or subsidize other needs in society through general budgets (e.g., education, health). Changes to land uses generally transform urban places to include more activities in a given land area (increasing density) and allow for a greater "mixing" of uses (commercial, residential, light industrial) within a given area or even within a single development project (i.e., a "mixed use" project). Technological changes to mobility systems, such as pollution-control technology in automobiles, can reduce some environmental externalities per unit of travel (though total externalities may increase or decrease depending on the amount of total travel). Finally, an important but less commonly used approach to transformation includes attempts to affect the knowledge and attitudes of users or managers of mobility systems.

The state of the art is the understanding that these four approaches must be applied in combination to create net effects – no single approach will create significant transformations of existing mobility patterns. Also, because existing mobility systems are so resource intensive, there is significant inertia in continuing the existing socio-technological systems (Wachs 1995). Thus, even seemingly significant interventions may have little measurable effect on system-wide characteristics. A shift in practice toward more comprehensive "accessibility planning," to be introduced below, will require all four of these approaches at a variety of spatial and temporal scales to make long-term impacts on mobility systems.

- ***Task***: *Describe the different challenges to planning based on an accessibility paradigm compared to the mobility paradigm.*

2 Sustainability Problems Caused by Urban Mobility

2.1 Adverse Effects

Urban mobility is a significant direct and indirect cause of several detrimental effects in the city (Golub 2012).

Traffic Fatalities and Injuries In the United States, around 3000 people – roughly the same number that perished during the September 11, 2001 terrorist attacks – die every month on the nation's roadways from traffic accidents and have been dying at that rate for the past 700 months (ca. 60 years). For those that survive crashes, there are pain, suffering, and thousands of hours of lost work time and cost for physical rehabilitation, etc. Together, traffic fatalities and injuries impose costs on the US society, estimated to be between $46 and $161 billion per year (Delucchi and McCubbin 2010).

Social Inequality, Exclusion, and Isolation Planning a mobility system around the need to own and operate a personal vehicle means that, for those who are unable to do so, the system will be poorly configured. In most metropolitan areas in the United States, for example, around 25 % of the population is too old, too young, or not able to afford an automobile, and therefore, they can become isolated and excluded from the mainstream of society (Taylor and Ong 1995; Lucas 2012). For example, in many central cities where low-income populations lack access to automobiles, a lack of access to healthy food and grocery options results in what is known as a "food desert" (USDA 2009). Furthermore, transportation systems have been used to segregate or reinforce existing segregation in some cities (Golub et al. 2013).

Detrimental Health Impacts Studies have shown that mobility systems significantly impact peoples' activity levels, and in turn, their health. The lack of safe, walkable neighborhoods, or barriers in neighborhoods created by transportation infrastructure (such as busy streets or freeways), leads to low rates of cycling and walking. This lack of activity is linked to higher body–mass indexes (e.g., obesity), poorer health indicators (Frank et al. 2006; Keegan and O'Mahony 2003), and consequently, additional health costs for the society (Frumkin 2002).

Reduced Social Time Budgets and Productivity While in good traffic conditions, driving is normally the fastest way to travel in US cities; during rush hour, the average traveler can suffer from long delays which negatively affect personal life and social relations. At a value of $10 per hour, these delays are estimated to cost between $63 and $246 billion per year (Delucchi and McCubbin 2010).

Local Air Pollution In the United States, environmental legislation like the Clean Air Act, enacted in 1970, has reduced tailpipe pollution emissions by around 99 % for most pollutants. However, large increases in driving mean local air pollution remains a national problem. More than 120 million Americans live in counties which fail at least one of the National Ambient Air Quality Standards, imposing a health cost burden of around $60 billion per year (EPA 2010; Parry et al. 2007).

Greenhouse Gas Emissions Greenhouse gasses in the atmosphere manage the planet's greenhouse process, whereby the global climate is regulated. Most transportation systems other than bicycles burn fuel which creates greenhouse gas emissions such as carbon dioxide and methane. In the United States, transportation is responsible for about one-third of the national greenhouse gas emissions, imposing a total cost of around $9 billion per year (EPA 2011; Parry et al. 2007).

Over-Exploitation of Nonrenewable Resources Cars and lights trucks use a large amount of nonrenewable steel, glass, rubber, and other materials. Data from 2001 showed that automobile production in the United States consumed 14 % of the national consumption of steel, 32 % of its aluminum, 31 % of its iron, and 68 % of its rubber (McAlinden et al. 2003, pp. 21–23). Around ten million automobiles are retired and junked every year, with the majority of the built-in resources lost, worth around $3 billion.

Contamination of Habitats Negative environmental impacts occur throughout the petroleum supply chain – from spills and flares at the local sites of extraction to spills and toxic pollution emissions at ports and refineries to local service stations where fuels can cause groundwater contamination. Roughly ten million gallons are spilled into US waters every year (Etkin 2001). This does not include the large spills such as the Gulf (aka Deepwater Horizon) spill in 2010 of around 170 million gallons or the Exxon Valdez spill in 1989 of 11 million gallons. Worldwide, more than three billion gallons have been spilled into waters since 1970, with typical annual environmental damages costing around $3 billion (Parry et al. 2007).

Costs of Petroleum Dependence In the United States, around half of the country's petroleum needs are imported from other countries, resulting in significant costs, estimated to be between $7 and $30 billion per year (Delucchi and McCubbin 2010), from a lack of flexibility in the economy to respond to changes in price. The noncompetitive structure of the oil industry has resulted in artificially high prices, with costs estimated to exceed $8 trillion since 1970 (Davis et al. 2010). US military presence in locations of strategic importance to the oil industry amount to between $6 and $60 billion per year (Davis et al. 2010).

2.2 *Underlying Causes and Actors*

Urban mobility is driven by a complex set of practices, habits, norms and so forth driven by the transportation industry, planners, government, and consumers or some combinations of these, all of which give current planning paradigms great inertia (Geels et al. 2012). Here, we describe some of these processes in preparation for the next and final sections outlining the wide ranging decisions and behaviors of these key actor groups.

2.2.1 The Individual and the Household

The individual and households sit at the most micro level of activity yielding daily decisions about how to travel and less regular decisions about home location or vehicle purchases. Daily decisions are made, mostly on the rational maximization of perceived travel convenience. There are large constraints on these decisions, however, as significant costs sunk into automobile ownership compel people to drive, since they are paying for the vehicle (through depreciation, insurance, etc.) whether they use it or not. Home location decisions are rarely made to minimize travel, as many choose to locate themselves in particular school districts or in communities with particular demographics. Additionally, car ownership is a powerful tool of identity formation in the US society, where it's seen as a symbol of status and patriotism (Paterson 2007).

2.2.2 Planners and Developers

Early last century, most urban planners in the US felt that suburban-type development based on automobile transportation offered a better quality of life compared to the crowded and dirty industrial urban centers of the time (Foster 1981). Even today, most urban planning practices merely reproduce the suburban, automobile-oriented models. After all, planners are simply agents of the governments for which they work and rarely serve as forces for change.

Developers reproduce the suburban model, not out of a particular preference but mostly because it seems to be the least-risky investment (e.g., Levine 2005). Banks are more likely to lend construction loans to build traditional suburban developments, and developers find it easier to develop fresh "greenfield" sites on the edge of cities where they can avoid potential neighborhood rejection of their project and higher or unpredictable construction costs in urban infill sites. Furthermore, many developers feel local land-use zoning often prevents them from building more dense and walkable developments (Levine and Inam 2004).

2.2.3 The State and Federal Governments

The state governments have a special role in urban transportation systems in the United States, as they were tasked with overseeing the construction of the interstate highway system. Most states also collect their own gasoline taxes, mostly used for investment in roads, freeways, and bridges.

The federal government has an important role in supporting automobile use, as well as regulating it and supporting alternatives to the automobile. Together, the 1956 Interstate Highway Act, federal support for home mortgages, and a relative lack of investment in urban revitalization during the postwar era solidified Federal support for suburbanization and automobile-based mobility. Furthermore, the U S foreign and military policy is heavily tied to the stability of the oil supply, a key ingredient in mobility. Federal policies are also important for managing automobile use. These include regulations to control pollution from automobiles, fuel economy standards, and safety regulations. Federal funds also support public transportation systems and bicycle transportation, though in small amounts compared to the monies spent for roads.

2.2.4 Oil and Automobiles Industries

The oil and automobile industries are some of the most heavily concentrated in the entire US economy – a relatively small number of companies account for nearly all of their industry's production. This means that they can easily join together to coordinate their concerns, influence public policy, and shape consumer demands through organized action. Thus, we must see urban transportation systems' use and dependence on petroleum and automobiles as being tied directly into the needs of the oil- and automobile-related industrial pillars. In the United States, automobile manufacturers became the focus of the emerging mass-consumption economy during the interwar period (1920s-1930s), riding the wave of public investments in freeways and suburbia and overcoming competition from transportation alternatives such as streetcars in most cities in the country (Golub 2012).

- ***Task**: Describe the main factors that contribute most to the perpetuation of unsustainable mobility patterns. Provide an example from a specific city for each factor.*

3 Sustainable Solution Options for Urban Mobility from Around the World

Understanding the system driving urban mobility challenges is only a first step toward transforming urban mobility. A key next step is to create *visions* of sustainable mobility, highlighting the goals of safety, convenience for all travelers using all modes, acceptable external environmental and social costs (at many temporal and spatial scales), and efficiency in the use of public resources, among other things. The vision would also address the fairness of the manner in which mobility systems are planned and governed. Besides these broader issues, community-specific visions reflect the needs of specific urban neighborhoods while still complying with principles of sustainability (e.g., Machler et al. 2012), a process requiring deliberation and negotiation (Wiek and Iwaniec 2014).

A sound understanding of urban mobility challenges and a sustainable vision of urban mobility are critical ingredients, but they do not suffice. A third element is critical for transforming today's mobility system into one which can achieve the visions of sustainable mobility (Wiek et al. 2012; also see Chap. 3 in this book). Changes in trajectory result from *interventions*, which detail step by step how the current mobility system needs to be transformed. From our understanding of the status quo, we can determine effective intervention points and strategies at the multitude of scales at play in the system. For example, traffic engineering practice and norms are strong drivers of current mobility systems. Thus, experiments and pilot projects in traffic engineering may help transform the system.

As was mentioned at the start of this chapter, most solution options focus on a combination of four core areas: price signals, land-use changes, technology development, and communication. These domains of intervention are invoked at a variety of scales and by different actors in the urban development process. Experiments and pilot projects of various types create glimpses of future possibilities and allow the system to "learn" and transform (Geels et al. 2012). We have compiled some of the more promising solution options here.

3.1 Proactive Urban Planning Paradigms (Planners, Developers, and Governments)

Research shows that urban planning and its control of land-use and transportation systems can have profound effects on automobile dependence. Urban travel modes are more or less convenient, depending on the arrangement of land uses and the prices of using those modes, such as gasoline, parking, bus fares, tolls, etc. For example, one strategy is to join public transportation with land uses such as job and housing centers, often called transit-oriented development (TOD). TODs combine higher densities with the convenience of being colocated at a public transportation facility, such as a light-rail or bus rapid transit (BRT) station (GAO 2003). It has been shown that compact development approaches such as TOD reduce the need for driving by around 20–35 % in the United States (ULI 2010, p. 7). In fact, residents in one TOD area in Atlanta drive only one-third as much as the average Atlanta resident (ULI 2010, p. 7).

Implementing TOD while improving public transportation, reducing the rate of highway construction, and increasing fuel prices (whether by raising taxes or through the natural increase in petroleum prices) have been estimated to reduce total driving by about 38 % (ULI 2010, p. 12). Regions recognized worldwide for taking this combined approach include: Mexico City; Curitiba, Brazil; Bogotá, Colombia; Stockholm, Sweden; and Singapore (Cervero 1998). Government policies are often important for the success of these combined approaches by, for example, creating transit-oriented land-use zoning, funding public transportation investments, or regulating the use of streets.

Because of their role in regulating and funding transportation systems and regulating land uses, governments are in a particularly strategic position to affect advancements toward accessibility planning. City governments are increasingly attempting to leverage investments in public transit facilities such as light-rail by rezoning to encourage more intense urban development. They have also been leading the wave in investments in bike-sharing systems and often support for-profit and nonprofit car-sharing services.

3.2 Sharing Cars (For-profit and Nonprofit Businesses, Governments and Individuals, and Households)

At first glance, trading the convenience of one's private car for the occasional use of a shared car, located somewhere out in the public realm, seems countercultural in many places, especially the United States (Golub and Henderson 2011). It appears, however, that there are places all over the world where this idea makes sense and has increased in popularity. Car sharing is a system which allows members to use cars on a short-term rental basis – as short as 15 min in some systems. The cars are placed in public areas in cities, rather than in car rental agencies, and members can use them at any time of the day.

Car sharing dates back to the 1940s in Northern Europe (Shaheen et al. 2009). Though few car-sharing programs existed in North America before 1994, by mid-2009, following a decade of improvements in satellite communications technology, there were roughly 280,000 car-share members sharing about 5,800 vehicles in the United States (Shaheen et al. 2009), with these numbers growing roughly 20 % per year (Martin et al. 2010). Studies show that car sharing can reduce household car-ownership, user, and parking demand and increase demand for public transportation, cycling, and walking (Cervero et al. 2007). Even more vehicles were reduced because car-sharing households avoided the planned purchase of vehicles.

3.3 Fostering Bicycling (Government and Individuals and Households)

Representing only about 1 % of all trips, bicycling makes up a very small share of daily travel in the United States. But with increased gasoline prices and traffic congestion, and growing concern about climate change and health, bicycling has experienced a boom in many US cities (Golub and Henderson 2011). Chicago, New York, Portland, Seattle, and many smaller university cities have experienced significant increases in utilitarian bicycling. In San Francisco, it is estimated that 5 % of adults use bicycles as their main mode of transportation (up from 2 % in 2001) and 16 % ride a bike at least twice a week (SFMTA 2009, 22).

Bicycling is poised to be a substitute for many short-range automobile trips and has enormous potential to reduce automobile use. Nationally, roughly 72 % of all trips less than three miles in length are by car, a distance that an average cyclist can cover easily (USDOT 2010, 22). "Bicycle space," or an interconnected, coordinated, and multifaceted set of safe bicycle lanes, paths, and parking racks, and accompanying laws and regulations to protect and promote cycling, has been extremely difficult to implement in the United States (Henderson 2013). The lack of political will has been a major barrier; there are no nationally dedicated funding programs for bicycles, and advocacy for bicycling has been a largely local, grassroots, and fragmented effort.

Many cities in the United States, including Washington DC and New York City, are making large investments in bike lanes and bike-sharing systems. Places like Bogotá, Colombia, and Mexico City have implemented even more ambitious region-wide improvements in bike and pedestrian infrastructure with profound results in short timescales (ITDP 2013a, b).

3.4 Technology Innovations (For-profit and Nonprofit Businesses and Governments)

Governments and businesses have been pivotal in funding and deploying the research and development of technology, which have important effects on transportation sustainability. Technological improvements are already responsible for cutting the levels of local air pollution emissions per vehicle to a small fraction. They also show promise for reducing fuel use and thus carbon emissions. Prominent ongoing examples of technological developments include intelligent transportation systems (ITS), which use increased data processing capabilities from satellites, and wireless technologies to improve roadway and parking management and public transit services (ITSA 2013). ITS applications are now being applied to vehicles to make them communicate with the roadway and other vehicles, making traffic safer and smoother. Satellite communications were also pivotal in facilitating advancements such as London's congestion pricing scheme and most modern car-sharing systems.

Questions

1. What are the barriers to sustainable mobility solutions based on sharing (car sharing and bike sharing)? What kind of actions can be taken to overcome those barriers?
2. How might accessibility solutions vary from place to place? How do culture and history influence how accessibility planning needs to happen in a certain place?

4 Open Issues

Innovations and transformations away from automobile-based mobility systems face great challenges in making broad impacts. Still, important innovations are meeting with significant and rapid success in places like Bogotá, Colombia, and Mexico City, encouraging other cities to try similarly broad changes. Even in the United States, there is evidence that growth in automobile travel is finally stagnating and declining in some places (Millard-Ball and Schipper 2011). There are still open research questions, including: What policies have the largest effects on behavior?, For how long do changes endure?, How can policies balance social equity while altering travel behaviors?, and, Are there rebound effects or other unintended consequences? For example, TOD planning may lead to more congestion because of less road capacity and higher density, and a recent review of research about planning for bicyclists and pedestrians shows mixed results from approaches thought previously to be important (Forsyth and Krizek 2010). Furthermore, significant demographic changes are on the horizon in much of the developed world which may cause even greater changes in travel patterns, for better or worse (e.g., Nelson 2009).

5 Conclusions

There are several important lessons here for sustainability science and sustainable development. The larger lesson is that urban practices such as automobile dependence, water or energy use, pollution, etc., result from webs of institutions, from citizens and neighborhoods to city and state governments to federal policies. Effective action for achieving sustainability begins with understanding these institutions and how they respond to and resist change (Geels et al. 2012). Inertia in the maintenance of the status quo in the dependence on the automobile for urban mobility illustrates that it is not enough to have a "right answer," be it a certain technology or a certain urban density. The importance is in how these answers are implemented by citizens and governments – how visions are translated into interventions by real communities in various experiments and pilot projects which can help to illustrate pieces of those future states today. A turn toward sustainable mobility will be achieved when we join with others with similar visions and create the social change needed to challenge the dominant urban planning and practice of automobile dependence.

References[1]

Black WR (2005) Sustainable transport: definitions and responses. In: Transportation Research Board (ed) Integrating sustainability into the transportation planning process. Transportation Research Board, Washington, DC, pp 35–43

[1] The sources highlighted with an asterisk (*) are recommended as further readings.

*Black WR, Nijkamp P (eds) (2002) Social change and sustainable transport. Indiana University Press, Bloomington

*Bullard R, Johnson G, Torres A (eds) (2004) Highway robbery: transportation racism and new routes to equity. South End Press, Boston

Cervero R (1996) Paradigm shift: from automobility to accessibility planning. Working paper. IURD, Berkeley

Cervero R (1998) The transit metropolis: a global inquiry. Island press, Washington, DC

Cervero R, Golub A, Nee B (2007) City Carshare: longer-term travel demand and car ownership impacts. Trans Res Record J Trans Res Board 1992:70–80

Davis S, Diegel S, Boundy R (2010) Transportation energy data book, 29th edn, ORNL-6985. Oak Ridge National Laboratory, Oak Ridge. http://info.ornl.gov/sites/publications/files/pub24318.pdf

Delucchi M, McCubbin D (2010) External costs of transport in the U.S. In: de Palma A, Lindsey R, Quinet E, Vickerman R (eds) Handbook of transport economics. Edward Elgar Publishing Ltd, Cheltenham

EPA (2010) Our nation's air – status and trends through 2008. http://www.epa.gov/airtrends/2010/index.html

EPA (2011) Draft U.S. Greenhouse Gas Inventory report. http://www.epa.gov/climatechange/emissions/usinventoryreport.html

Etkin D (2001) Analysis of oil spill trends in the United States and Worldwide. Presentation to 2001 International Oil Spills Conference. http://www.environmental-research.com/publications/pdf/spill_statistics/paper4.pdf

*Featherstone M, Thrift N, Urry J (eds) (2005) Automobilities. Sage, London

Forsyth A, Krizek K (2010) Promoting walking and bicycling: assessing the evidence to assist planners. Built Environ 36:4

Foster M (1981) From streetcar to superhighway. Temple University Press, Philadelphia

Frank LD, Sallis JF, Conway TL, Chapman JE, Saelens BE, Bachman W (2006) Many pathways from land use to health: associations between neighborhood walkability and active transportation, body mass index, and air quality. J Am Plann Assoc 72(1):75–87

Frumkin H (2002) Urban sprawl and public health. Public Health Rep 117(3):201

*Geels F, Kemp R, Dudley G, Lyons G (2012) Automobility in transition? A socio-technical analysis of sustainable transport. Routledge, London

Golub A (2012) Sustainable transportation (Chap 7). In: Pijawka D, Gromulat M (eds) Understanding sustainable cities: concepts, cases, and solutions. Kendall-Hunt, Dubuque

Golub A, Henderson J (2011) The greening of mobility in San Francisco (Chap 6). In: Slavin M (ed) Sustainability in America's cities: creating the green metropolis. Island Press, Washington, DC

Golub A, Marcantonio R, Sanchez T (2013) Race, space and struggles for mobility: transportation impacts on African-Americans in Oakland and the East Bay. Urban Geogr 34(5):699–728

Government Accountability Office (GAO) (2003) Bus rapid transit offers communities a flexible mass transit option. United States General Accounting Office, GAO-03-729T, 24 June 2003

*Hanson S, Giuliano G (eds) (2004) The geography of urban transportation. Guilford Press, New York

Henderson JM (2013) Street fight: the politics of mobility in San Francisco. University of Massachusetts Press, Amherst

Institute for Transportation and Development Policy (ITDP) (2013a) Sustainable Transport Award – 2013: Mexico City, Mexico. http://www.itdp.org/get-involved/sustainable-transport-award/previous-award-recipients/user-nominee-submission32

Institute for Transportation and Development Policy (ITDP) (2013b) Sustainable Transport Award – 2005: Bogotá, Colombia. http://www.itdp.org/get-involved/sustainable-transport-award/previous-award-recipients/bogota-colombia

Intelligent Transportation Society of America (ITSA) (2013) ITS America Government Affairs Brochure. http://www.itsa.wikispaces.net/file/view/ITSA+Govt+Affairs+Brochure_1.5.pdf/419564912/ITSA%20Govt%20Affairs%20Brochure_1.5.pdf

Keegan O, O'Mahony M (2003) Modifying pedestrian behaviour. Trans Res Part A Policy Pract 37(10):889–901

Levine J (2005) Zoned out: regulation, markets, and choices in transportation and metropolitan land use. Resources for the Future Press, Washington, DC

Levine J, Inam A (2004) The market for transportation-land use integration: do developers want smarter growth than regulations allow? Transportation 31:409–427

Lucas K (2012) Transport and social exclusion: where are we now? Trans Policy 20:105–113

Machler L, Golub A, Wiek A (2012) Using a "sustainable solution space" approach to develop a vision of sustainable accessibility in a low-income community in Phoenix, Arizona. Int J Sustain Trans 6(5):298–319

Martin E, Shaheen S, Lidicker J (2010) Carsharing's impact on household vehicle holdings: results from a North American shared-use vehicle survey. Trans Res Record J Trans Res Board 2143:150–158

McAlinden S, Hill K, Swiecki B (2003) Economic contribution of the automotive industry to the U.S. economy – an update. Center for Automotive Research. http://www.cargroup.org/?module=Publications&event=View&pubID=57

Millard-Ball A, Schipper L (2011) Are we reaching peak travel? Trends in passenger transport in eight industrialized countries. Trans Rev 31(3):357–378

Nelson A (2009) The new urbanity: the rise of a New America. ANNALS, AAPSS 626:192–208

Parry IW, Walls M, Harrington W (2007) Automobile externalities and policies. J Econ Lit 45:373–399

*Paterson M (2007) Automobile politics - ecology and cultural political economy. Cambridge Press, Cambridge

San Francisco Municipal Transportation Agency (SFMTA) (2009) 2008 San Francisco State of cycling report. San Francisco Bicycle Program, Department of Parking and Traffic, Municipal Transportation Agency, San Francisco

Shaheen S, Cohen A, Chung M (2009) Carsharing in North America: a ten-year retrospective. Trans Res Record J Trans Res Board 2110:35–44

Taylor BD, Ong P (1995) Spatial mismatch or automobile mismatch? An examination of race, residence and commuting in US metropolitan areas. Urban Stud 32:1453–1473

United States Department of Agriculture (USDA) (2009) Access to affordable and nutritious food measuring and understanding food deserts and their consequences – report to congress. http://www.ers.usda.gov/media/242654/ap036_reportsummary_1_.pdf

United States Department of Transportation (USDOT) (2010) The national bicycling and walking study: 15-year status report. USDOT, Washington, DC

Urban Land Institute (ULI) (2010) Growing cooler: the evidence on urban development and climate change (Executive summary). http://www.smartgrowth.umd.edu/pdf/GrowingCooler-Ch1Overview.pdf

Wachs M (1995) The political context of transportation policy. In: Hanson S (ed) The geography of urban transportation, 2nd edn. Guilford Press, New York, pp 269–286

Wiek A, Iwaniec D (2014) Quality criteria for visions and visioning in sustainability science. Sustain Sci 9(4):497–512

Wiek A, Ness B, Brand FS, Schweizer-Ries P, Farioli F (2012) From complex systems analysis to transformational change: a comparative appraisal of sustainability science projects. Sustain Sci 7(Suppl 1):5–24

Chapter 22
International Development and Sustainability

Rimjhim M. Aggarwal

Abstract This chapter explores some global development challenges – such as that of extreme poverty, growing inequalities, and poor governance, from the perspective of sustainability. We begin by questioning what we mean by "development" and tracing the evolution of this concept from the monolithic vision of development as a linear process that characterized postcolonial era thinking on development policy to that of "sustainable development" and the current thinking in terms of development as a highly contested term. We then examine some of the major challenges at the interface of international development and sustainability, such as the need to delink resource-intensive growth from progress on human development indicators. This discussion then leads us on to exploring some of the innovative solution options that have been proposed by central planners as well as grassroots level searchers and the usefulness of different approaches, such as randomized control trials, to evaluate the effectiveness of these interventions. We conclude with a discussion of some open issues, such as the potential of human rights-based thinking about development and its implications for sustainability.

Keywords Sustainable development • Poverty • Human development • Governance • Human right

1 Development and Sustainability: Reflections on Key Themes and Trends

We generally understand "development" as a process of progressive change from "lower" to "higher" states. For biological organisms, this is easy to define as a linear process from childhood to adulthood. For societal evolution, what is meant by "higher," and what is meant by "lower"? Is this even a linear process? Who defines it?

R.M. Aggarwal (✉)
School of Sustainability, Arizona State University, PO Box 875502, Tempe, AZ 85287-5502, USA
e-mail: Rimjhim.aggarwal@asu.edu

H. Heinrichs et al. (eds.), *Sustainability Science*, DOI 10.1007/978-94-017-7242-6_22

From ancient times, philosophers, historians, and ordinary people have pondered over these questions. Ancient cultures embodied a diversity of values that shaped their visions about societal and human progress. What we understand by development today has been shaped largely by what Gunnar Myrdal, in his monumental work, *Asian Drama*, described as the "modernizations ideals" (Myrdal 1968). These ideals, rooted in Western Enlightenment, included the drive toward rationality in decision-making (seen as liberation from the hold of traditions and customs), application of scientific knowledge to increase material production, and control of nature in order to more efficiently service human needs. These ideas that originated in Europe shaped the process of industrialization in the Western world from the eighteenth century on.

After the Second World War, attention shifted to the former colonized nations in Asia, Latin America, and Africa, which were seen as "poor" and "uncivilized." The big question for development practitioners was how to get these nations to the same stage of "development" as the Western industrialized world. These efforts were based on the assumption that there was a universal, linear trajectory that each had to travel. Foreign aid and technology transfers were thought to be the elixirs. To date, trillions of dollars in foreign aid have been pumped into the world's poorest countries, yet around 1.1 billion people still live in extreme poverty and about one-sixth of the world's population is unable to meet their basic needs. Instead of convergence on a common path, we observe that differences among countries have widened. The income gap between the world's richest (20 %) and its poorest (20 %) increased from a ratio of 30:1 in 1960 to 60:1 in 1990 and widened further to 74:1 in 1997 (Pogge 2002: 265).

Given these trends, we have come to the realization that the universalist model of development based on a resource-intensive path of industrialization is unsustainable. We have to rediscover what "development" truly means and collectively envision our possible future states and how to navigate toward those that are socially desirable. The World Commission on Environment and Development famously put one such vision forward in 1987, in its pioneering manifesto, *Our Common Future*. The Commission coined the term "sustainable development" and defined it as meeting "the needs of the present without compromising the ability of future generations to meet their own needs" (WCED 1987:43). This definition brought issues of intergenerational equity to the forefront and underscored the idea that "development" is a highly contested term, and thus, negotiations and deliberations are critical.

More recently, sustainability science – with its emphasis on complexity, nonlinear dynamics, systems analysis, and futures (Kates et al. 2001; Wiek et al. 2011; Miller et al. 2014) – has offered new ways of thinking about core development problems, reimagining the future, and transformational change. Development studies and sustainability science originated as separate fields with different motivations, worldviews, and methodologies; yet it is obvious that sustainability without a vision of development has no meaning and development without being sustainable has no relevance. The field of development with its (a) focus on core human development values, poverty alleviation, justice, diversity of cultures, and institutions and (b) accumulated evidence on trajectories of socioeconomic development and a vast repertoire of field experiments has a lot to offer to advance sustainability science, as we explore in this chapter.

- **Task**: *What is your vision of a "developed" country? Develop a set of criteria for how to reliably distinguish a "developed" country from a "less developed" or "underdeveloped" country?*

2 Key Challenges at the Intersection of Development and Sustainability

2.1 Delinking Realization of Human Development Goals from Resource-Intensive Growth

The grand challenge of sustainability is often framed in terms of meeting the *needs* of a growing population – projected to reach 9.5 billion by 2050 – while maintaining the planet's life support systems and living resources (Kates and Parris 2003). An important underlying concept here is that of human needs, which, as Amartya Sen reminds us, gives a rather "meager view of humanity" (Sen 2004). He argues that "we are not only *patients*, whose needs demand attention, but also *agents*, whose freedom to decide what to value and how to pursue it can extend far beyond the fulfillment of our needs" (Sen 2004). Following from the work of Sen and others, a popular way to measure human development has been through the use of the Human Development Index (HDI), which is a summary measure of standards of living, education, and health (UNDP 1992).

Several researchers have attempted to relate the HDI to measures of sustainability in order to better understand the challenge of sustainable development. Neumayer (2012), for example, related HDI values for 1980–2006 to a measure of ecological footprint (EF) per capita. WWF (2008) estimates the globally available biocapacity to be 2.1 global hectares per person and categorizes countries with per capita EF greater than 2.1 ha as unsustainable. Using this measure, Neumayer found that *all* the countries with high and very high levels of HDI are not sustainable due to carbon dioxide emissions per capita that are far in excess of the natural absorptive capacity of the atmosphere. Thus, he argues "one of the biggest challenges of this century will be severing the link between high to very high levels of human development and strong unsustainability, particularly in the form of unsustainably high carbon dioxide emissions" (p. 576). This delinking will require an out-of-the-box rethinking of our developmental trajectory. Tragically, we do not have any good examples of countries that have been able to meet this challenge, although some have done better than others (Neumayer 2012).

2.2 The Devil is in the Distribution: The Challenge of Intra- and Intergenerational Equity

Many people would argue that the problem is not so much the overall limited availability of resources in relation to global population level, but more so about how these resources, and incomes, are distributed. The challenge is not just that of poverty, but that of poverty amidst growing islands of affluence, of deprivation with growing overconsumption. Besides raising several ethical issues, high inequality leads to social conflicts and lowers the incentive to invest in the future, specifically the future of the poor (Easterly 2001). This further exacerbates the cycle of growing poverty, inequality, and unsustainable resource use that needs to be addressed (Aggarwal 2006).

The ethical issues that are raised by these glaring rates of intragenerational inequities are also linked to intergenerational equity issues; sometimes, these complement each other, but they are often in conflict. Thus, for instance, an important question relates to whether redistribution to today's poor harms the future by enhancing current consumption spending and reducing investment for the future. Anand and Sen (2000) argue that this is not necessarily the case if assisting the poor helps them build up human capital, which will then also benefit the future. However, it is worth noting that not every policy will yield this double dividend for both intragenerational and intergenerational equity. For instance, there are growing concerns that increased spending on reducing greenhouse gas emissions will take financial resources away from assisting those who are currently poor (World Bank 2009). There are no easy answers here. An open and honest discussion of the complementarities and conflicts between these issues of inter-and intragenerational equity is urgently needed.

2.3 Unresponsive States and the Lack of Effective Participation: The Challenge of Governance

As we argued earlier, both development and sustainability demand deliberation and negotiations. However, in several developing countries, war, conflict, and failed states represent major threats to a sustainability transition (Kates and Parris 2003). At its peak in 1992, one-third of the countries of the world were ravaged by armed conflicts (Gurr et al. 2001). Armed conflicts threaten sustainability directly by destroying human lives, as well as infrastructure, and indirectly by encouraging exploitation of natural resources. Moreover, under conflict conditions, personal security issues dominate concerns for the common good and the thinking about the future (Kates and Parris 2003).

While armed conflicts get a lot of public attention, rampant corruption and rent seeking within large bureaucratic structures are other chronic problems that plague several developing countries and make the state largely unresponsive to the needs of its citizenry. Even in developing countries with representative democracies, there remains a large gap between formal legal rights in the civil and political arena and the actual capability to practice those rights effectively as citizens. There are two dimensions to this problem, as Heller (2009) observes: "On the one hand, there is the problem of *how* citizens engage the state. State-society relations in the developing democracies tend to be dominated by patronage and populism, with citizens having either no effective means of holding government accountable (other than periodic elections) or being reduced to dependent clients. On the other hand, there is the problem of *where* citizens engage the state. [...] Given that local government is often absent or just extraordinarily weak in much of the developing world, there are in fact very few points of contact with the state for ordinary citizens" (Heller 2009: 7).

Thus, it is not surprising, as Heller argues, that democracies in much of the developing world are characterized by both participatory failures (who participates and how they participate) and substantive failures (translation of popular inputs into concrete outputs). A poignant example here is the case of the metropolitan region of Mumbai, which has a population of over 20 million, but local bodies within the city are not accountable to the needs of local citizenry, such as protecting them against the vulnerabilities of monsoon flooding. The planning and management of basic services falls under various ministries that have statewide responsibilities and constituencies, as opposed to an elected body of local representatives who are responsive to local needs. Phatak and Patel (2005) examine how this lack of capacity and autonomy at the local level impacted the recovery effort during the 2004 floods in Mumbai and how the response may have been different under a more decentralized form of governance.

In the next section, we discuss examples of solutions that have evolved to address the challenges discussed above.

- **Task**: *Instead of convergence on a common path, as originally envisaged by economic theorists, we observe that differences among countries have widened since the start of industrialization. Why do you think this has happened? What implications might these differences have for global sustainability?*

3 Solution Options

To some, the challenges outlined above may paint a picture of gloom; to others, these represent possibilities. Paul Polak, a social entrepreneur who belongs to the latter group, eloquently describes the possibilities as follows: "Working to alleviate poverty is a lively, exciting field capable of generating new hope and inspiration [...]. Learning the truth about poverty generates disruptive innovations capable of enriching the lives of rich people even more than those of poor people" (Pollak 2009: ii).

The field of development has indeed been very lively and has inspired a wide range of solution seekers, who can be broadly classified into two categories: Planners and Searchers (Easterly 2007). Notable among the Planners is Jeffrey Sachs, who believes that the poor are caught in a trap and lack the minimum amount of capital necessary to get a foothold on the economic ladder, and so the richest countries need to make the investments necessary to give the poor countries "a boost up to the first rung" (Sachs 2006:244). The Planners thus advocate for the "big push" approach, which requires massive efforts in planning and coordination, as well as financing through foreign aid.

This paternalistic aid-giving approach, as traditionally practiced by Planners, has been critiqued by William Easterly (2007), who advocates instead for the Searchers' approach, which he contrasts with that of the Planners (see Box 22.1).

Planners as well as Searchers have offered several solutions. An interesting example, which somewhat reconciles the approaches of the two groups, is the "procedural" solution of public participation. For example, Brazil has made significant strides in using public participation successfully to address the complex challenge of poverty, corruption, and lack of accountability (see Box 22.2).

Box 22.1: In Search of Solutions: Planners Versus Searchers

"Planners announce good intentions but don't motivate anyone to carry them out; Searchers find things that work and get some reward. Planners raise expectations but take no responsibility for meeting them; Searchers accept responsibility for their actions. Planners determine what to supply; Searchers find out what is in demand. Planners apply global blueprints; Searchers adapt to local conditions. Planners at the top lack knowledge of the bottom; Searchers find out what the reality is at the bottom. Planners never hear whether the planned got what it needed; Searchers find out if the customer is satisfied" (Easterly 2007: 6).

Box 22.2: Participatory Reforms in Brazil

Significant participatory reforms in Brazil came in the form of the creation of various sectoral councils (e.g., in health, transport, education, environment) that were mandated by the constitution. The councils include representatives from sectoral interests, government, and civil society, thus creating "institutionalized spaces" for participatory action. The most significant of these local experiments has been *participatory budgeting*, which involves direct involvement of citizens at the neighborhood and city level in shaping the city's capital budget. Over 400 Brazilian cities have now adopted some form of participatory budgeting (Heller 2009).

Given that poverty alleviation is a complex process, with several interacting factors, it has not always been clear what works and what doesn't. One way of testing the effectiveness is through conducting a randomized control trial (RCT). Under the RCT method, the target population is split randomly into two parts: the treatment and control groups. The treatment group receives the treatment, while the control group receives a placebo. After enough time has elapsed for the treatment to work, results are compared between the control and treatment groups.

The RCT approach is now being widely adopted to test for alternative ways to reduce poverty. For instance, the microfinance agency BRAC, which has traditionally focused on giving small loans, decided to give assets, such as a few chickens, a cow, and a pair of goats, to the poor in the state of West Bengal in India. They also gave them training on how to take care of the animals and manage their finances. To test the results of the project, a team led by Esther Duflo compared the treated households with a random control group that did not get these assets.[1] The researchers found that, long after the treatment had ended, the treated groups ate 15 % more, earned 20 % more, and saved significantly more. These effects could not be explained by the direct effects of the treatment in terms of the extra earning from selling eggs, meat, and milk. The researchers argued that more than just the assets, the intervention gave the treated households "hope" for a better future. This may explain why the treated group worked harder – 28 % more – than the control group. The experiment helped clarify how lack of optimism may be an important reason why the poor are trapped in poverty and why small but carefully designed interventions, by offering help, can start a virtuous circle.

The RCT method has helped dispel several myths about the poor and the process of poverty alleviation. However, the approach also has several pitfalls. The fundamental problem is that it may not always be possible to create randomly selected control and treatment groups. An example here is the case of tourism programs which are selectively launched in specific sites with certain desirable characteristics. Finding reasonable alternative sites as controls may be difficult. In other cases, even if control and random sites can be identified, carrying out a selective intervention may not be politically feasible. Often, it may not be deemed ethical to deny project benefits to a section of the people. Finally, we need to keep in mind that, in field settings as opposed to laboratory settings, it may often be difficult to isolate the treatment and control groups. Social and economic interactions between groups may often be difficult to control, thus leading to spillovers (Taylor and Lybbert 2012). The main lesson here is that, just as there are multiple solutions, there are alternative approaches for evaluating impacts that need to be considered. Specifically in cases where society-wide effects of a complex nature are being evaluated, other statistical and sometimes qualitative approaches (such as narratives) may be helpful.

[1] "Hope springs a trap: An absence of optimism plays a large role in keeping people trapped in poverty," *The Economist*. May 12, 2012.

- **Task**: Paul Polak (2009) has argued that "learning the truth about poverty generates disruptive innovations capable of enriching the lives of rich people even more than those of poor people" (p. ii). Provide some examples of disruptive innovations that have transformed the lives of not only the poor but also the rich.

4 Open Issues

The solution options discussed above offer hope, but are these enough? The transition to sustainability cannot be complete without meeting at least the basic needs of the poor and, beyond needs, providing for their voices to be heard and upholding human dignity. Are we headed in the right direction toward this transition? Current trends suggest that the gap between have and have-nots is expanding and that struggles over natural resources are likely to intensify as we head precariously close to the planetary boundaries and face the threat of climate change. In such an environment, how can we guarantee that basic human development goals will be met and sustained? Whose responsibility is it?

In response to this challenge, several people have proposed that what is needed is an explicit normative approach that transforms the traditional thinking about poverty alleviation as "aid" or an "act of charity" to the framing of "freedom from poverty" as a "human right" that is guaranteed by law. As Irene Khan of Amnesty International vehemently argues, "human rights are claims that the weak advance to hold the powerful to account, and that is why poverty is first and foremost about rights" (Khan 2009: 21). Others opposed to this thinking have taken the view that rights can be effectively articulated only in combination with correlated duties and associated responsible parties; otherwise, the demands for human rights can be seen as just loose talk. Sen (2000: 203), on the other hand, argues that the "framework of rights-based thinking extends to ethical claims that transcend legal recognition. These rights can thus be seen as being prior (rather than posterior) to legal recognition. Indeed, social acknowledgement of these rights can be taken to be an invitation to the State to catch up with social ethics."

What does all this mean in terms of development action and practice? As Haglund and Aggarwal (2011) explain, rights seem to offer leverage that "development" alone has lacked in terms of providing "new discursive, normative, and morally compelling mechanisms that transcend framings of poverty in terms of neediness and charity and instead embrace the idea of firm obligations and the inalienability of rights." They show, through several cases, how rights-based thinking can powerfully shape behavior when backed by a range of accountability mechanisms. The process of rights-based policy formulation is not just an abstract idea; it is well underway in several countries. In a recent survey, Gauri (2004) found that, in a sample of 165 countries with written constitutions, 116 made reference to the right to education and 73 made reference to the right to health care. Rights to food, water, sanitation, and a clean environment have also been recently added to several constitutions.

Box 22.3: Social Guarantees for Fulfillment of Basic Needs

A "Social Guarantee" can be defined as a set of legal or administrative mechanisms that determines specific entitlements and obligations and ensures the fulfillment of those obligations on the part of the state. Social Guarantees have been instituted in a number of developing countries and cover a range of basic entitlements such as health (Chile, Peru), education (Peru, Guatemala, Uruguay), employment (India), housing (South Africa), and social protection (Uruguay). A system based on guarantees requires the following key elements (World Bank 2007):

- Normative (legal) framework (embodied in the constitution or specific policies) that clearly defines the rights.
- Financial mechanisms to secure the budget.
- Specific institutional arrangements to implement, monitor, and provide oversight.

One interesting example of how rights-based thinking has permeated into public policy is through the mechanism of "Social Guarantee" (World Bank 2007) (see Box 22.3).

Several studies have shown how the mechanism of Social Guarantee has been able to successfully bridge the gap between social rights norms and concrete public policies by (1) providing an innovative institutional design that emphasizes synergy and coordination among otherwise disparate agencies, (2) contributing to reducing gaps in opportunity among citizens by promoting universal access, and (3) strengthening democratic governance by engaging all citizens in collectively setting basic entitlement levels, monitoring that agreed-upon targets are met, and providing mechanisms of redressal (see World Bank 2007, for a review).

Finally, in examining such a rights-based approach, we have to ask how well such an approach compares with a goal-based approach, as, for example, is embodied in the Millennium Development Goals (MDGs) set by the UN. The MDG model sets national level targets and relies heavily on a top-down international transfer of resources, with very weak domestic and international accountability mechanisms. The Social Guarantee approach, on the other hand, delineates individual-level entitlements and relies on strong domestic accountability mechanisms but involves very limited engagement of foreign entities. This limited engagement could be a strength in cases in which this has led to development of domestic efforts at consensus building, mobilization, and accountability, but it could also be a weakness, particularly in the case of relatively poor countries, which could benefit from some foreign assistance. Thus, what may be needed to achieve human development goals is some kind of a hybrid approach, which builds on social learning – with engagement of all relevant stakeholders – about what works and under what contexts.

References

Aggarwal R (2006) Globalization, local ecosystems, and the rural poor. World Development Special Issue on "The Impact of Globalization on the World's Poor." 34(8):1405–1418

Anand S, Sen A (2000) Human development and economic sustainability. World Dev 28(12):2029–2049

Banerjee A, Duflo E (2011) Poor economics: a radical rethinking of the way to fight global poverty. Public Affairs, New York

Easterly W (2001) The elusive quest For growth: economists' adventures and misadventures in the tropics. MIT Press, Boston

Easterly W (2007) The white man's burden: why the west's efforts to aid the rest have done so much Ill and so little good. Penguin, New York

Gauri V (2004) Social rights and economics: claims to health care and education in developing countries. World Dev 32(3):465–477

Gurr TR, Marshall MG, Khosla D (2001) Peace and conflict 2001. Center for International Development and Conflict Management, College Park

Haglund LD, Aggarwal R (2011) Test of our progress: the translation of economic and social rights norms into practices. J Hum Rights 10:1–27

Heller P (2009) Democratic deepening in Brazil, India and South Africa: towards a comparative framework. Paper prepared for seminar on "Metropolis and Inequalities," São Paulo, Brazil. http://www.fflch.usp.br/centrodametropole/antigo/static/…/Patrick_Heller.pdf. Accessed 10 Oct 2013

Kates RW, Parris TM (2003) Long-term trends and a sustainability transition. Proc Natl Acad Sci U S A 100:8062–8067

Kates RW, Clark WC, Corell R, Hall JM, Jaeger CC et al (2001) Sustainability Science. Science 292(5517):641–642

Millennium Ecosystem Assessment (MEA) (2005) Ecosystems and human well-being: synthesis. Island Press, Washington, DC

Khan I (2009) The unheard truth: poverty and human rights. W. W. Norton & Company, New York

Miller TR, Wiek A, Sarewitz D, Robinson J, Olsson L, Kriebel D, Loorbach D (2014) The future of sustainability science: a solutions-oriented research agenda. Sustain Sci 9(2):239–246

Myrdal G (1968) Asian drama: an inquiry into the poverty of nations. Twentieth Century Fund, New York

Neumayer E (2012) Human development and sustainability. J Human Dev Capab 13(4):561–579

Phatak VK, Patel SB (2005) Would decentralisation have made a difference? Econ Polit Week India 3:3902–3904

Pogge T (2002) World poverty and human rights: cosmopolitan responsibilities and reforms. Polity Press, Cambridge

Pollak P (2009) Out of poverty: what works when traditional approaches fail. BK Currents, San Francisco

Sachs JD (2006) The end of poverty: economic possibilities for our time. Penguin, New York

Sen A (2000) Work and Rights. Int Labour Rev 139(2)

Sen A (2004) Why we should preserve the spotted owl. London Rev Books 26(3). http://www.lrb.co.uk/v26/n03/amartya-sen/why-we-should-preserve-the-spotted-owl. Accessed 10 Oct 2013

Taylor E, Lybbert T (2012) Essentials of development economics. Rebel Text, Berkeley

United Nations Development Programme (UNDP) (1992) Human development report: global dimensions of human development. Palgrave Macmillan, Basingstoke

Wiek A, Withycombe L, Redman CL (2011) Moving forward on competencies in sustainability. Environ Sci Policy Sustain Dev 53:3–13

World Bank (2007) Realizing rights through social guarantees: an analysis of new approaches to social policy in Latin America and South Africa, Social Development Department, report no. 40047. The World Bank Group, Washington, DC

World Bank (2009) World development report 2010: development and climate change. World Bank, Washington, DC

World Commission on Environment and Development (WCED) (1987) Our common future. Oxford University Press, Oxford

World Wildlife Fund (WWF) (2008) Living planet report 2008. World Wide Fund for Nature, Gland

Chapter 23
Tourism and Sustainability

David Manuel-Navarrete

> *One of the world's largest economic sectors, tourism is especially well-placed to promote environmental sustainability, green growth and our struggle against climate change through its relationship with energy. (Ban Ki-moon, UN Secretary-General, on World Tourism Day 2012)*

Abstract This chapter outlines specific sustainability challenges in tourism destinations and the sector's opportunities to contribute to global sustainability. The highly inequitable distribution of benefits among local actors, the energy-intensive character of most tourism activities, and the lack of systematic data on environmental and social impacts are identified as key challenges. Responses based on promoting "best practices" are useful and widely implemented by tourism corporations. Building on experiences from pioneering destinations, a case is made for sustainability solutions that go beyond the best practices approach and redefine tourism as a social activity that can actively promote broader sustainability transitions. This involves engaging local actors in the definition of "desirable or acceptable" tourism development objectives, as well as the identification of strategies that turn tourism into a social process that supports the emergence of new governance structures while questioning entrenched relations of power.

Keywords Sustainable tourism • Inequality • Tourism best practices • Sustainability solutions • Power

1 Sustainability Problems Associated with Tourism

Globally, the number of international tourist arrivals rose from 807 million in 2005 to 1,035 million in 2012, representing an average annual growth rate of 3.6 % (UNWTO 2013a). Emerging economies had a higher annual growth rate (4.8 %)

D. Manuel-Navarrete (✉)
School of Sustainability, Arizona State University,
800 S. Cady Mall, PO Box 875502, Tempe, AZ 85287-5502, USA
e-mail: davidmn@asu.edu

H. Heinrichs et al. (eds.), *Sustainability Science*, DOI 10.1007/978-94-017-7242-6_23

than advanced economies (2.6 %), and it is expected that, by 2015, for the first time in history, emerging economies will receive more international tourist arrivals than advanced economies (UNWTO 2011). Obviously, this dynamic contributes to global economic activity. For instance, in 2012, international tourism contributed about 9 % of world GDP and 1 in 11 jobs (direct, indirect, and induced). Also, it generated US$1.3 trillion in export and 6 % of the world's exports (UNWTO 2013b). Over the years, tourism has been one of the fastest-growing economic sectors in the world.

The data presented above indicates that tourism is, and will likely continue to be, a main driver of economic growth. However, this growth engenders sustainability-related challenges (UNEP and UNWTO 2012). The international tourism industry is often blamed for causing negative impacts on local cultures, people, and environments (Mowforth and Munt 1998). Furthermore, even culturally and ecologically oriented forms of tourism tend to capture and repatriate the majority of revenues, thus effecting only marginal improvements in terms of poverty alleviation. For instance, applications of value-chain analysis in 12 destinations in Asia and Africa estimated that the total income earned by the poor as a percentage of total tourism expenditure ranged between 5 and 30 % (Coles and Mitchell 2009).

Assessing tourism as a strategy that can promote sustainability requires considering its impacts on equity and other social criteria. For instance, the construction of new touristic complexes provides new jobs, but it can also limit access to certain areas (such as beaches or parks) and increase the prices of land and cost of living. Global environmental impact of tourism is another key aspect to consider when assessing sustainability. Even though tourism is not an intensively polluting sector, it is nevertheless an energy-intensive activity that contributes around 5 % to global carbon emissions (UNWTO et al. 2008).

The United Nations World Tourism Organization collects regular economic data on tourism across the world, including data on revenues and expenditures, overnights spent, and accommodation capacities. However, no systematic collection of data on environmental and social impacts of tourism activities is carried out at the global level. Even though there are numerous methodological proposals for developing indicators measuring the stress that tourism causes on socio-ecological systems (e.g., UNWTO 2004), the actual collection of data to populate these indicators is still scattered, centered on a few case studies, or largely anecdotal. Thus, there is a need for measuring sustainability indicators of tourism in systematic and reliable ways.

- **Task**: *Tourism growth has generally failed to contribute significantly to local sustainable development. Why is that? What indicators are needed in order to assess the contribution of tourism to local (un)sustainability? How is tourism progress currently measured? What new metrics of progress are needed?*

2 Solution Options Through Sustainable Tourism and Tourism Sustainability

Sustainable tourism seeks local incremental changes towards balancing the goals of economic growth, social equality and ecological integrity in the development and operations of tourism destinations (Hall 2011). Tourism is primarily conceived as an economic activity, the sustainability of which is contingent upon the reduction of associated social and environmental costs. The goal is to balance the well-being of host communities, the satisfaction of guests, and the profits of the industry while ensuring that the recreational services upon which the industry depends are maintained (Liu 2003). Such balancing has primarily been sought through the *definition and implementation of models of good practices*. Table 23.1 includes some "green ideas" already adopted by hotels according to the Green Hotels Association. In a more comprehensive effort, the Global Sustainable Tourism Council approved "Global Sustainable Tourism Criteria" in 2009, which were allegedly based on 4,500 industry best practices from around the world. At one level, "sustainable tourism" is a success story when judged by the concept's diffusion among industry, government, academics, and policy actors. Yet, it is simultaneously a policy failure given pervasive accumulation and growth of environmental degradation and social inequality (Hall 2011). In any case, balancing the multiple objectives of sustainability will require developing indicators that identify limits and opportunities and that help to design effective strategies toward minimizing the impacts of tourism on all scales.

There have been significant efforts toward drawing sustainable management guidelines, codes of practice, and corporate action programs that are potentially implementable by the private sector and eventually enforceable by public officials. This voluntary perspective has built on the concept of corporate social responsibility and the one-time adoption of some frameworks to evaluate the impacts of tourism on destinations, including Limits of Acceptable Change (LAC), Visitor Impact Management (VIM), or Visitor Experience and Resource Protection (McCool and Lime 2001). However, as stated above, no systematic collection of data on impacts

Table 23.1 "Green ideas" that hotels can implement

Idea	Already adopted by
Recycling stained tablecloths into napkins, chef's aprons, and neckties	Hotel in Toronto
Bought a mulcher to chop up garden clippings and create mulch	Hotel in Florida (mulcher paid for itself in 3 months)
Produces organically grown vegetables for its restaurant	Hotel in Pennsylvania
Solar energy is lighting signage and heating water	Hotels in tropical areas
Installed beautiful blue floor tile made from recycled automobile windshields	Wisconsin B&B
Bicycles are being loaned or rented to guests	Hotels across the globe

Source: Green Hotels Association. Available at: http://greenhotels.com/grnideas.php (accessed 07/14/13)

of tourism activities is carried out at the global level. Measuring sustainability indicators of tourism in systematic and reliable ways would not only help in monitoring the extent of sustainability challenges associated with tourism, but also account for the outcomes and performance of sustainable tourism efforts, be they voluntary or enforced.

The case of the Canary Islands in Spain is discussed here to explain the role of indicators in decision-making and management toward sustainability. The Canary Islands is one of the most popular and successful tourist destinations in Europe, but this success has provoked land use conflicts and problems related to the carrying capacity of infrastructure and public services (UNWTO 2004). These problems prompted the regional government to set up a working group to develop indicators applicable to tourism in general (see Sect. 1, above) and sustainable tourism in particular, based on UNWTO (1997). Proposed sets of indicators include, among others (UNWTO 2004):

- Tourist numbers (site pressure)
- Density of use and beds per km^2 (use intensity)
- Ratio of tourists to locals (social impact)
- Area covered by development plans (planning process)
- Percentage of tourists who agree they would like to return or recommend the destination to others (tourist's satisfaction)
- Percentage of population wanting more, same, or less tourists (resident's satisfaction)
- Value of tourist expenditures (contribution of tourism)

As a result of this assessment, the regional government initiated a tourism planning process to increase the management capacities of authorities and the industry. In this process, tourism growth began to be defined as wealth generated by tourism rather than just growth in the number of tourists. Furthermore, tourism growth was linked to the reconversion of existing infrastructures and the diversification and differentiation of the tourism offer, rather than to the mere construction of new accommodations (UNWTO 2004).

Tourism sustainability seeks transformational change by addressing the root causes of unsustainability. It involves the redefinition of tourism into a social activity that supports a global transition toward sustainability. Such redefinition requires, first of all, the *setting of a new policy paradigm veered away from traditional policy instruments and indicators* that are subordinated to a pro-growth model of governance. Instead, a polycentric and network-based model of governance for sustainability needs to be established. An example of an initial attempt to implement this type of governance is the "Natural Step" program, an international organization that uses a science-based, systems framework to help organizations and communities understand and move toward sustainability. The challenge is to institutionalize programs like Natural Step in order to foster the development of new destinations and the transformation of existing ones. A good example in this direction is the pioneering adoption since 2002 of the "Natural Step" framework by the Resort Municipality of Whistler in Canada (Gill et al. 2008). According to Gill and Williams (2011:

639): "The Whistler 2020 vision and principles emphasize that the traditional 'pillars of sustainability' are not distinct 'silos' for individual consideration but are interconnected components of the destination system. Framed in this manner, any action undertaken for a specific purpose […] must consider wider upstream and downstream implications."

Polycentric and network-based tourism governance systems may contribute to de-Westernizing tourism. That is, to stop assuming that Western culture is universal and modern, while non-Western cultural forms are provincial and backward. In turn, de-Westernized forms of tourism would contribute to de-Westernizing the world. The contemporary model of tourism governance is guided by values and philosophies cemented in a long history of colonization and expansion, which emphasizes individualism, hedonism, conquest, and exploitation (Higgins-Desbiolles 2010). Tourism sustainability scholars can increase the visibility of decolonizing alternatives both within Western culture and elsewhere. For instance, some forms of Islamic tourism are arguably geared toward spiritual growth and instilling solidarity within the community (Inayatullah 1995). Other examples include the values behind Polynesian hospitality, including generosity, reciprocity, and *Aroa* (a value full of complex meanings but possible to distill to love, kindness, and generosity), which can be seen as an alternative value system supporting alternative forms of engagement with tourism (Higgins-Desbiolles 2006). Within the USA, the Diné traditional living system *Sa'ah Naaghai Bik'eh Hozhoo* is grounded in Navajo cultural traditions and could contribute to place tourism development in harmony with the natural world and the universe. Unfortunately, there is little empirical evidence of successful experiences in this direction.

Tourism governance for sustainability requires a close academic scrutiny of the associations between (1) current power structures controlling tourism development, (2) the contribution of existing tourism policies toward the promotion of uneven and unequal development, and (3) the alliances of the tourism industry with governments and international institutions. Social sciences have developed a number of theories and analytical tools in recent decades, including actor-network, assemblages, non-representational, biopolitical, and entanglement of power approaches, oriented toward unraveling and understanding spatiotemporal patterns of power relations (Manuel-Navarrete 2012). Most of these theories have their origins in principles and frameworks put forth by or derived from the work of Karl Marx or Sigmund Freud. Yet, many of them, especially some of those often referred to as poststructuralist, have received strong influences from nonequilibrium thermodynamics and complexity theory (Manuel-Navarrete 2013). This link with complexity may facilitate a fluid dialogue between social science and sustainability scholars and practitioners that leads to combining a deeper knowledge of social power relations with structured visioning, constructive governance, and the promotion of a global transition.

At the very least, tourism development needs to be politicized. This is crucial in order to avoid sustainability being further hijacked by the prevailing model of development and continuing to fall into the service of the controllers of capital, the boards of directors of major transnational companies, and other organizations which manage the industry (Mowforth and Munt 1998). Research on social power and sustainability can work together to develop effective means of translating the critical politicization of tourism into actions and feasible policy options with impacts on the ground.

- **Task**: *Solutions to the problems of unsustainable tourism are under development. How can sustainable development and planning of touristic destinations be achieved? Provide some examples that illustrate the principles of sustainable tourism and tourism sustainability discussed above. How can polycentric and network-based models of governance be promoted in the context of tourism sustainability?*

3 Open Issues: Sustainability Challenges of Tourism

One of the main challenges of sustainable tourism consists of determining how much environmental and social stress is acceptable in any given tourism operation. This question cannot be meaningfully addressed without preestablishing the development objectives that are desirable for a destination. Thus, a key sustainability challenge that tourism faces consists of determining "desired or acceptable" development objectives through processes in which conflicting values, opposing interests, and controversial information are systematically and fairly considered.

A more serious challenge is to address the paradox that a very limited success in reducing environmental and social impacts of tourism has been achieved so far despite the fact that the industry, regulators, and other major tourism players have widely incorporated the sustainable tourism perspective into corporative discourse and adopted a number of green practices (as illustrated in Sect. 2). In recent years, particularly after the 2008 global financial crisis, tourism scholars have increasingly acknowledged that the attempts to balance the three dimensions of sustainability are not working. Economic growth is promoted in most destinations at the expense of the other two goals. This is often attributed to structural power asymmetries characteristic of neoliberal globalization that favor private sector needs over the demands of civil society and host communities. As a reaction, "limits to growth" arguments, articulated in the 1970s by the Club of Rome, are being dusted off, while sustainability is increasingly reformulated as a direct challenge to the hegemony of growth, to neoliberal ideologies, and to the "all that matters is profit" mentality that arguably still dominates the global tourism industry (Saarinen 2006; Matarrita-Cascante 2010; Gill and Williams 2011).

The idea of Tourism Sustainability brings about a new challenge: Can society use tourism as an instrument for promoting wider sustainability? Rather than trying to make tourism development and operations sustainable, tourism for sustainability is concerned with the role of tourism in promoting the broader transition toward a sustainable world. In this context, a number of tourism scholars are increasingly revisiting political economy approaches and placing power relations at the center of tourism sustainability analysis.

In a nutshell, the key tourism sustainability challenge for the twenty-first century consists of applying critical and relational theories of power in order to redirect societies away from uneven forms of development. Tourism has the unique potential to redistribute income toward regions that have been marginalized from the global economy until now. In order to realize this potential, it is crucial to promote a greater level of local involvement in the planning and development of destinations. This is basically a governance challenge that sustainability scholars are well equipped to address. A more meaningful involvement of local people in the planning of tourism destinations will require research that recognizes the complex and dynamic nature of destination development. Tourism resources are not static; they can be degraded, but also enhanced. Place-based and participatory approaches to tourism sustainability should be complemented with research on the trade-offs, synergies, and scenarios involved in the destination's process of development. In the case of tourism, it is crucial to establish the relationship between patterns of human activity and the social organization implications and types of environmental impact associated with these patterns. Thus, the destination process of development is not to be determined by the need to maximize economic growth. On the contrary, the creation of a destination becomes a social process and a democratic exercise in which tourism resources are consumed responsibly, as well as reproduced and enhanced through the social promotion and reinforcement of specific socio-ecological processes.

Some tourism scholars advocate the need to build on systems perspectives in order to improve our understanding of the characteristics and change patterns of tourism, as well as its dynamic interaction with the natural, technological, social, and economic environment (Liu 1994). This is a crucial step for the development of tourism sustainability as a field of study in which natural and social systems are studied as self-organizing, interdependent, and nonlinear systems that are in constant evolution. Farrell and Twining-Ward (2004) have emphasized the need to acknowledge the existence of "complex adaptive tourism systems." Research on tourism complexity may play a key role in envisioning and devising tourism systems that are more inclusive and socio-ecologically integrated. This type of research is key to situating tourism in a better position to act as a positive force in the transition toward sustainability (Farrell and Twining-Ward 2005). Yet, the task ahead goes beyond the analysis of the current situation and the visualization of future alternatives. In order to make the transition happen, tourism scholars can also support sustainability decisions, actions, and transformations, including strategies and tactics for transformation (Wiek et al. 2012).

The sustainability transformation of tourism encounters multiple forms of resistance due to the stability of dominant governance systems and policy paradigms. The current situation, not only in the case of tourism sustainability but also in the case of sustainability research in general, is one of impasse. There is an increasing acknowledgement that the dominant discourses and policies for promoting sustainable tourism are insufficient. Yet, the problem is that tourism sustainability requires the destabilization of current policy paradigms, away from the drive toward constant accumulation and away from persisting colonial forms of domination. The task ahead for tourism sustainability consists of unraveling social power relations that have persisted for many years, but doing so in a way in which the unraveling actually results in less exploitative and more sustainable forms of organization.

Sustainability scholars can devise strategies to facilitate the emergence of new governance structures, while carefully assessing the implications of changing existing patterns of power and dominance.

4 Conclusions

Sustainable tourism seeks to balance economic, social, and environmental goals in the local development and operations of tourism. Such balancing has primarily been sought through the definition and implementation of models of good practices. However, tourism sustainability cannot be meaningfully addressed without preestablishing the development objectives that are desirable for a destination. Thus, a key sustainability challenge that tourism is facing consists of determining "desired or acceptable" development objectives through processes in which conflicting values, opposing interests, and controversial information are revealed, discussed, and fairly considered. Another challenge concerns the ways in which tourism can become an activity that promotes the global transition toward a sustainable world. Addressing this second challenge involves redefining tourism into a social activity that supports positive social transformation. Such redefinition requires, first of all, the setting of a new policy paradigm veered away from traditional policy instruments and indicators that are subordinated to a pro-growth model of governance. Sustainability scholars can devise strategies for facilitating the emergence of new governance structures, while carefully assessing the implications of changing existing patterns of power and dominance. Yet, these strategies and assessments, as well as the actual harnessing of tourism sustainability, will inevitably be laden with power issues.

References[1]

Coles C, Mitchell J (2009) Pro-poor analysis of the business and conference value chain in Accra, report prepared for SNV. ODI, London

Farrell BH, Twining-Ward L (2004) Reconceptualizing tourism. Ann Tour Res 31(2):274–295

Farrell B, Twining-Ward L (2005) Seven steps towards sustainability: tourism in the context of new knowledge. J Sustain Tour 13(2):109–122

(*) Gill AM, Williams PW (2011) Rethinking resort growth: understanding evolving governance strategies in Whistler, British Columbia. J Sustain Tour 19(4–5):629–648

Gill AM, Williams PW, McCool SF, Moisey RN (2008) From "guiding fiction" to action: applying "The Natural Step" to sustainability planning in the resort of Whistler, British Columbia. In: McCool SF, Moissey RN (eds) Tourism, recreation and sustainability: linking culture and environment, 2nd edn. CABI, Wallingford, pp 121–130

Hall CM (2011) Policy learning and policy failure in sustainable tourism governance: from first- and second-order to third-order change? J Sustain Tour 19(4–5):649–671

[1]The sources highlighted with an asterisk (*) are recommended as further readings.

Higgins-Desbiolles F (2006) More than an "industry": the forgotten power of tourism as a social force. Tour Manage 27(6):1192–1208

(*) Higgins-Desbiolles F (2010) The elusiveness of sustainability in tourism: the culture–ideology of consumerism and its implications. Tour Hosp Res 10(2):116–129

Inayatullah S (1995) Rethinking tourism: unfamiliar histories and alternative futures. Tour Manage 16(6):411–415

Liu Z (1994) Tourism development – a systems analysis. In: Seaton AV et al (eds) Tourism: the state of the art. Wiley, Chichester, pp 20–30

Liu Z (2003) Sustainable tourism development: a critique. J Sustain Tour 11(6):459–475

Manuel-Navarrete D (2012) Entanglements of power and spatial inequalities in tourism in the Mexican Caribbean, desiguALdades.net working paper series 17. desiguALdades.net Research Network on Interdependent Inequalities in Latin America, Berlin

Manuel-Navarrete D (2013) Human-nature integration and global environmental change. In: O'Brien K, Wolf J, Sygna L (eds) A changing environment for human security: new agendas for research, policy, and action. Earthscan, London, pp 305–315

Matarrita-Cascante D (2010) Beyond growth: reaching tourism-led development. Ann Tour Res 37(4):1141–1163

McCool SF, Lime DW (2001) Tourism carrying capacity: tempting fantasy or useful reality? J Sustain Tour 9(5):372–388

(*) Mowforth M, Munt I (1998) Tourism and sustainability: new tourism in the third world. Routledge, London

(*) Saarinen J (2006) Traditions of sustainability in tourism studies. Ann Tour Res 33(4):1121–1140

UNEP, UNWTO (2012) Tourism in the green economy – background report. United Nations Environment Programme and World Tourism Organization, Madrid

UNWTO (1997) What tourism managers need to know: a practical guide to the development and use of indicators of sustainable tourism. World Tourism Organization, Madrid

UNWTO (2004) Indicators of sustainable development for tourism destinations. A guidebook. World Tourism Organization, Madrid

UNWTO (2011) Tourism towards 2030/global overview. World Tourism Organization, Madrid

UNWTO (2013a) UNWTO annual report 2012. World Tourism Organization, Madrid. Available at: http://dtxtq4w60xqpw.cloudfront.net/sites/all/files/pdf/annual_report_2012.pdf

UNWTO (2013b) Tourism highlights, 2013 edn. World Tourism Organization, Madrid. Available at http://mkt.unwto.org/en/highlights

UNWTO, UNEP, WMO (2008) Climate change and tourism: responding to global challenges. United Nations World Tourism Organization/United Nations Environment Program/World Meteorological Organization, Madrid/Paris/Geneva

Wiek A, Ness B, Brand FS, Schweizer-Ries P, Farioli F (2012) From complex systems analysis to transformational change: a comparative appraisal of sustainability science projects. Sustain Sci 7(Suppl 1):5–24

Chapter 24
Consumption and Sustainability

John Harlow, Michael J. Bernstein, Bastien Girod, and Arnim Wiek

Abstract Moving consumption toward sustainable patterns has been a key goal of sustainability science since the 1990s. However, a large knowledge gap remains between identified consumption problems that restrict social and ecological development and progress toward solutions. Unfortunately, "sustainable consumption" is generally discussed in a rational context and does not address how culture, pleasure, identity, and communication drive consumption. This exclusively rational framing limits innovation in problem-solving research methodologies based on the other drivers of consumption. Marketing, however, excels at capitalizing on all the drivers of consumption. Consumers are the targets of countless behavior change strategies, and this article offers perspective on how the nonrational drivers of consumption can be leveraged to instead pursue sustainable consumption. Bridging the behavior change knowledge gap can be relevant for many sustainability impasses, as many of them stem from human behavior, and boundary work drawing on behavioral science can effectively navigate norms and expectations at the interface of science and society. The better researchers understand the behaviors of relevant actors, the better their behavior change strategies will support sustainability transitions.

Keywords Behavior • Behavioral science • Consumption • Sustainable consumption • Sustainability

---

J. Harlow (✉) • M.J. Bernstein • A. Wiek
School of Sustainability, Arizona State University,
PO Box 875502, Tempe, AZ 85287-5502, USA
e-mail: john.harlow@asu.edu; michael.j.bernstein@asu.edu; arnim.wiek@asu.edu

B. Girod
Department of Management, Technology, and Economics, Swiss Federal Institute of Technology (ETH) Zurich, Weinbergstr. 56/58, CH-8092 Zürich, Switzerland
e-mail: bgirod@ethz.ch

H. Heinrichs et al. (eds.), *Sustainability Science*, DOI 10.1007/978-94-017-7242-6_24

1 Introduction

This chapter adopts Paul Stern's (1997, p. 20) definition of consumption: "Consumption consists of human and human-induced transformations of materials and energy. Consumption is environmentally important to the extent that it makes materials or energy less available for future use, moves a biophysical system toward a different state or, through its effects on those systems, threatens human health, welfare, or other things people value." Moving consumption toward sustainable patterns has been a key goal of sustainability science and practice since the 1990s (Kates 2000; Cohen et al. 2005). Herein, "sustainable consumption" will refer to the OECD's (2002, p. 9) definition: "The use of services and related products which respond to basic needs and bring a better quality of life while minimizing the use of natural resources and toxic materials as well as the emissions of waste and pollutants over the life-cycle of the service or product, so as not to jeopardise the needs of future generations."

There remains a large gap between knowledge of consumption-related problems and making progress on resolving them (Fischer et al. 2012; Thøgersen and Schrader 2012). For example, there are clear indications that consumption patterns, in particular, in industrialized and industrializing countries, have major impacts on greenhouse gas emissions and climate change (Harriss and Shui 2010). Because of the close link between consumption and production (i.e., consumption of produced goods), however, the majority of policy efforts so far have focused on regulating production (Aall and Hille 2010).

A series of recent studies call for a shift to addressing consumption instead. Harriss and Shui (2010) point out that global trade "offshores" the burden of greenhouse gas emissions from developed consumers to developing producers – but the consumers remain the primary driver of this scheme. Girod et al. (2010) emphasize the relevance of advanced consumption patterns, in which consumers initially save resources through technological advancements, but then re-spend the saved resources, and even more, due to more frequent use (rebound effects). Several studies propose methods for allocating greenhouse gas emissions to consumption categories (Spangenberg and Lorek 2002; Hertwich and Peters 2009), and a number of studies highlight the potential of influencing greenhouse gas emission profiles, and eventually climate change, by changing consumption behavior (Dietz et al. 2009; Girod and de Haan 2009).

With the intent to change consumption, special attention needs to be paid to the drivers of consumption. With focus on the drivers, Schaefer and Crane (2005) explore five views of "consumption," including rational, cultural, pleasurable, identity, and communication. They find that "sustainable consumption" is generally discussed within the rational view of consumption. Viewing consumption as simply rational is insufficient to address how culture, pleasure, identity, and communication drive consumption. To target behavior change strategies effectively, sustainability practitioners must synthesize views of consumption into a holistic problem map of consumption drivers. Problem maps display the root causes of problems and orient researchers to the most strategic intervention points for change (Wiek et al. 2012).

Consumers are the subject of countless behavioral change strategies (Michie et al. 2011). Most often they are the target group of marketing strategies, primarily designed to increase consumption. As an example, what goes where in supermarkets is the subject of many algorithms designed to increase consumption and retailer profits (Yang 2001; Hwang et al. 2005). Marketing is also effective at targeting views of consumption other than the "rational." Bertrand et al. (2010) found that a photo of an attractive woman was as effective in increasing demands for loans as a 25 % reduction in the interest rate.

In his seminal book *Influence: Science and Practice*, Cialdini (2009) lists "weapons of influence" with six principles and discusses "how their enormous force can be commissioned by a compliance professional who deftly incorporates them into requests for purchases" (p. xii). These tools of influence have been trained squarely on increasing consumption, which has manifested many sustainability challenges, including greenhouse gas emissions, waste, and inequalities (Jackson et al. 2004; Fischer et al. 2012).

There are numerous proposals on how best to change behavior in order to make progress toward sustainability (most recently in Cohen et al. 2013). In the following section, we provide examples of how to use insights from behavioral science (Whitley and Kite 2012), including the aforementioned tools of influence, to drive *sustainable* consumption.

- ***Task:*** *How does consumption contribute to sustainability challenges? What dimensions of sustainability are affected by over-consumption and how? As you reflect, consider the relationships between the resources we consume today and the resources that remain for future generations.*

2 Examples of Incentivizing Sustainable Consumption

2.1 Electricity Choices and Default Options

Pichert and Katsikopoulos (2008) examined home electricity consumption choices. They point out that 50–90 % of survey respondents in the USA, UK, and EU stated a preference for consuming renewable energy sources. Despite this, rates of adoption of renewable energy sources are abysmal, in the range of 0.4–1.0 %.

Conventional wisdom speculated that reluctance to adopt renewable energies is related to socioeconomic factors. However, Pichert and Katsikopoulos (2008) turned to psychology and behavioral science, hypothesizing that the electricity source chosen was a function of how electricity options were presented. The authors asked a simple question: what happens to "green electricity" when it is the default option? The results of two field and two laboratory studies show that simply changing the default source from "gray" to "green" results in a significantly higher percentage of customers buying green electricity.

The main behavioral science principle in this experiment is the "default option" made popular by Thaler and Sunstein's (2008) book *Nudge*. The default option is what individuals "choose" by doing nothing when presented with a decision. Changing defaults has been wildly successful in cases ranging from organ donation (Johnson and Goldstein 2003) to retirement savings (Madrian and Shea 2001). Making default consumption options more sustainable is a potent method for making consumption more sustainable.

2.2 Towels and Social Proof

There are over 4.5 million hotel rooms in the USA, and a hotel with 150 rooms can save about 6,000 gallons of water a month by reusing towels and linens (Cassingham 2006). The overall potential savings is 180 million gallons of water every month in the USA.

To nudge guests to reuse towels, Schultz et al. (2008) drew from research on the power of social proof to affect behavior (Cialdini 2009). Social proof tells us that the behavior of others is positive reinforcement for our behaviors. Schultz et al.'s study compared the effect of different messages on hotel guests. Messages that communicated both the value of conserving resources and the number of other guests who reused towels to help with conservation were the most effective in towel-use reduction (Schultz et al. 2008). In a similar study, such messages communicating that "the majority of guests in this room reuse their towels" reduced towel usage more than hotels' traditional appeals based on resource conservation (Goldstein et al. 2008). A recent study by Shang et al. (2010) concludes that responses to such social marketing programs are most positive when savings from the program are donated to charity. The more sustainable consumption becomes, the stronger the social proof feedback and support will become. Making sure that social proof feedback is built into sustainable consumption strategies can strengthen impacts and quicken sustainability transitions.

2.3 Plastic Bags and Loss Aversion

Because plastic bags are petroleum products with harmful environmental impacts (Derraik 2002), they have been banned in San Francisco and Los Angeles. In 2010, Washington D.C., began charging fees for disposable bags, using part of the funds to support cleanup efforts of the heavily polluted Anacostia River. Homonoff (2012) examined that program and the difference between a five-cent fee for plastic bags and a five-cent reward for reusable bags. Her analysis showed the fee to be effective. The reward reduced plastic bag use by only 2.4 %, whereas the tax reduced plastic bag use by 43.5 %.

What could be responsible for such a large difference with equal monetary incentives? First proposed by Kahneman and Tversky (1979), loss aversion postulates that we are more sensitive to the prospect of losing an amount of money than the prospect of gaining the same amount of money. The implications of loss aversion extend far beyond the decision of whether to use a plastic bag. Field (2009) showed that law students' career choices were dramatically affected by loss aversion. Lower tuition produced 36–45 % more students pursuing public interest law than debt remittance after graduation. Loss aversion is a powerful, well-evidenced behavioral tool with broad applications for making consumption more sustainable.

2.4 Cafeterias and "Shaping the Path"

Another common example of applying behavioral science to make consumption more sustainable comes from public health. Rozin et al. (2011) reduced overeating by demonstrating the power of the dining environment on food consumption. Subtly changing ease of access to certain foods or reducing the size of serving utensils reduced food intake by an average of 8–16 % (Rozin et al. 2011). The researchers go on to imply that altering the dining environment could also be leveraged to improve the quality of food eaten (think strategically placed fruits and vegetables). Heath and Heath (2010) call this type of environmental design "shaping the path."

There are plenty of examples of successfully shaping the path of dining environments to alter food consumption (Hanks et al. 2012; Schwartz et al. 2012; Kallbekken and Sælen 2013). Redesign of a decision environment (think of a cafeteria as a decision environment for what to eat) is a potent way of influencing consumption without wading into politically charged or otherwise controversial issues. Shaping the path can direct consumption toward products that are environmentally and socially responsible simply by changing the ways "easy options" are presented to consumers.

In summary, behavioral research provides insights into how tools of influence can be used to support sustainable consumption – by providing knowledge (towel reuse messages), incentives (plastic bag tax), or an environment conducive to sustainable consumption (cafeteria design). For an extensive list of behavioral science insights on "nudging," consult the Stirling Behavioural Science Blog/Nudge Database (Egan 2013).

- ***Task*** *: Reflect on the approaches to incentivizing sustainable consumption presented above – changing the default, using social proof, leveraging loss aversion, and shaping the path. Are your consumption decisions influenced by these principles, and if so, how? Try to think of examples of unsustainable consumption patterns. How might you use one or more of the approaches discussed to promote sustainable consumption in these cases?*

3 Open Issues and Future Research

The examples discussed above all repurpose tools of influence for sustainability objectives. Similar lines of research could be very fruitful for connecting proven behavioral tools with pressing sustainability problems. Of course, tools will have to be refined for new settings, which come with a host of challenges.

There is the challenge of transferability. Many behavioral science insights have been derived in specific cultural or experimental settings, leading to difficulties in reproducing results in new contexts (Henrich et al. 2010). Research on panaceas suggests, in general, exercising caution when transferring solution options from one specific context to another (Ostrom et al. 2007). Once a problem is chosen, tools are selected, and likely resistance is anticipated, testing assumptions in small scale or pilot interventions can communicate the effectiveness of behavioral tools within a specific problem's context.

A related challenge is selecting the appropriate suite of behavioral tools for each sustainability intervention. There are many more tools available than we have space to discuss here (Thaler and Sunstein 2008; Heath and Heath 2010; Kahneman 2011), and combinations will often be more effective than a single tool for a specific intervention. Future research can help determine the relative impacts of behavioral tools, find the most effective blends, and pair the right tools with the ripest problems.

Such intervention studies could support evidence-based policy-making for sustainability transitions. However, evidence alone is often insufficient to overcome social and political bias and the status quo (Bartels 2002; Baum and Gussin 2007; Nyhan and Reifler 2010). This is partially due to what psychology terms "motivated reasoning," which means that we use beliefs to filter evidence and choose which evidence to accept or reject (Kunda 1990; Redlawsk 2002). This indicates that interventions for sustainable consumption need to be embedded in a larger societal pursuit to overcome the widening imbalance between private and public interests (Reich 2008).

Evidence-based policy-making, using insights from behavioral research, faces additional challenges. Some behavioral interventions have been termed "libertarian paternalism," meaning that choice is offered, but the choice architecture aims at certain options (defaults, strategically placed food, etc.). Opponents of such behavioral interventions worry about the balance between freedom, democracy, and "libertarian paternalism." These arguments claim "nudges" are a slippery slope to social engineering-like "shoves" (Lott 2013). Supporters of "nudges" counter that the point of interventions is to balance existing influences and "use decision errors that ordinarily hurt people to instead help them" (Downs et al. 2009, p. 160). After all, it is called "libertarian" paternalism because the "nudged" are free to choose. "Nudges," in this sense, can be seen as recalibrating decision environments to account for biases and incentivizing preferred behaviors, such as sustainable consumption. Of course, preferred behaviors will depend on values and the "motivated reasoning" attached to those values. This is a reminder that tools of influence are *means* to an end. A broader societal discourse is necessary to converge on the end (goals) a society wants to pursue.

Behavioral intervention efforts also encounter problems familiar for intentional change toward sustainability (Beddoe et al. 2009). The status quo has culture, context, politics, and values that create winners and losers. Coal plants lose revenue when consumers choose "green" energy, and utility companies lose revenue when we wash fewer towels. Any sustainable behavior change will have to navigate the current setting strategically and anticipate likely points of resistance (companies whose food is no longer in the cafeteria hot spot). Choosing a scale where resistance is minimized, impact is maximized, and tools are available is key to successful interventions. This is especially important for making consumption more sustainable, as status quo market forces and capitalism will be extremely powerful at large scales (global trade and multinational corporations). Future research can establish which behavioral tools experience more or less backlash (and on what scale) and suggest which tools are sensitive to which political and cultural contexts.

- ***Task*** *: Pick an area of consumption (think of examples like automobiles, bottled water, computers, new homes, etc.) and think through (a) the negative impacts of this consumption activity, (b) who benefits from this consumption activity, (c) how the product is marketed to society, (d) what behavioral principles these marketing activities take advantage of, (e) what sustainable consumption alternatives exist, and (f) how you might use the behavioral principles discussed to promote these alternatives. What do you think of the idea of "nudging" people toward sustainable consumption? How would you make a case to someone who disagrees with your opinion?*

References[1]

Aall C, Hille J (2010) Consumption – a missing dimension in climate policy. In: Bhaskar R (ed) Interdisciplinarity and climate change – transforming knowledge and practice for our global future. Routledge, London, pp 85–100

Bartels LM (2002) Beyond the running tally: partisan bias in political perceptions. Polit Behav 24(2):117–150

Baum MA, Gussin P (2007) In the eye of the beholder: how information shortcuts shape individual perceptions of bias in the media. Q J Polit Sci 3(1):1–31

Beddoe R, Costanza R, Farley J, Garza E, Kent J, Kubiszewski I, Martinez L, McCowen T, Murphy K, Myers N, Ogden Z, Stapleton K, Woodward J (2009) Overcoming systemic roadblocks to sustainability: the evolutionary redesign of worldviews, institutions, and technologies. Proc Natl Acad Sci U S A 106(8):2483–2489

Bertrand M, Karlan D, Mullainathan S, Shafir E, Zinman J (2010) What's advertising content worth? Evidence from a consumer credit marketing field experiment. Q J Econ 125(1):263–306

Cassingham K (2006) Towel/sheet reuse program savings. Laudry Today, Sunday, 03 Dec 2006. Retrieved from: www.laundrytoday.com

[1] The sources highlighted with an asterisk (*) are recommended as further readings.

*Cialdini RB (2009) Influence: science and practice, 5th edn. Pearson, London
Cohen MJ, Comrov A, Hoffner B (2005) The new politics of consumption: promoting sustainability in the American marketplace. Sustain Sci Pract Policy 1(1):58–76
*Cohen MJ, Brown HS, Vergragt P (eds) (2013) Innovations in sustainable consumption –new economics, socio-technical transitions and social practices. Edward Elgar, London
Derraik JG (2002) The pollution of the marine environment by plastic debris: a review. Mar Pollut Bull 44(9):842–852
Dietz T, Gardner GT, Gilligan J, Stern PC, Vandenbergh MP (2009) Household actions can provide a behavioral wedge to rapidly reduce US carbon emissions. Proc Natl Acad Sci U S A 106:18452–18456
Downs JS, Loewenstein G, Wisdom J (2009) Strategies for promoting healthier food choices. Am Econ Rev 99(2):159–164
Egan M (2013) Nudge database. [Web log]. Retrieved from http://economicspsychologypolicy.blogspot.com/2013/03/nudge-database_3441.html
Field E (2009) Educational debt burden and career choice: evidence from a financial aid experiment at NYU law school. Am Econ Rev Appl 1(1):1–21
*Fischer J, Dyball R, Fazey I, Gross C, Dovers S, Ehrlich PR, Brulle RJ, Christensen C, Borden RJ (2012) Human behavior and sustainability. Front Ecol Environ 10(3):153–160
Girod B, de Haan P (2009) GHG reduction potential of changes in consumption patterns and higher quality levels: evidence from Swiss household consumption survey. Energy Policy 37:5650–5661
Girod B, Haan P, Scholz RW (2010) Consumption-as-usual instead of ceteris paribus assumption for demand. Int J Life Cycle Assess 16:3–11
Goldstein NJ, Cialdini RB, Griskevicius V (2008) A room with a viewpoint: using social norms to motivate environmental conservation in hotels. J Cons Res 35:472–482
Hanks AS, Just DR, Smith LE, Wansink B (2012) Healthy convenience: nudging students toward healthier choices in the lunchroom. J Public Health 34(3):370–376
Harriss R, Shui B (2010) Consumption, not CO_2 emissions: reframing perspectives on climate change and sustainability. Environ Sci Policy Sustain Dev 52(6):8–15
*Heath C, Heath D (2010) Switch: how to change when change is hard. Broadway Books, New York
Henrich J, Heine SJ, Norenzayan A (2010) The weirdest people in the world. Behav Brain Sci 33(2–3):61–83
Hertwich EG, Peters GP (2009) Carbon footprint of nations: a global, trade-linked analysis. Environ Sci Technol 43(16):6414–6420
Homonoff TA (2012) Can small incentives have large effects? The impact of taxes versus bonuses on disposable bag use. Job Market Paper
Hwang H, Choi B, Lee MJ (2005) A model for shelf space allocation and inventory control considering location and inventory level effects on demand. Int J Prod Econ 97(2):185–195
Jackson T, Jager W, Stagl S (2004) Beyond insatiability: needs theory, consumption and sustainability. In: Reisch L, Røpke I (eds) The ecological economics of consumption. Edward Elgar, Cheltenham, pp 79–110
Johnson E, Goldstein D (2003) Do defaults save lives? Science 302:1338–1339
Kahneman D, Tversky A (1979) Prospect theory: an analysis of decision under risk. Econometrica: J Econometric Soc 47:263–291
Kahneman D (2011) Thinking, fast and slow. Farrar, Straus and Giroux, New York
Kallbekken S, Sælen H (2013) 'Nudging' hotel guests to reduce food waste as a win-win environmental measure. Econ Lett 119(3):325–327
Kates R (2000) Sustainability science. Presentation at the world academies conference transition to sustainability in 21st century, Tokyo, Japan, 18 May 2000
Kunda Z (1990) The case for motivated reasoning. Psychol Bull 108(3):480
Lott M (2013) Gov't knows best? White house creates 'nudge squad' to shape behavior. FoxNews Politics. Retrieved from http://www.foxnews.com/politics/2013/07/30/govt-knows-best-white-house-creates-nudge-squad-to-shape-behavior/

Madrian BC, Shea DF (2001) The power of suggestion: inertia in 401 (k) participation and savings behavior. Q J Econ 116(4):1149–1187
Michie S, van Stralen MM, West R (2011) The behaviour change wheel: a new method for characterising and designing behaviour change interventions. Implement Sci 6(1):42
Nyhan B, Reifler J (2010) When corrections fail: the persistence of political misperceptions. Polit Behav 32(2):303–330
OECD (Organization for Economic Co-operation and Development) (2002) Policies to promote sustainable consumption: an overview, ENV/EPOC/WPNEP (2001)18/FINAL. OECD, Paris
Ostrom E, Janssen MA, Anderies JM (2007) Going beyond panaceas. Proc Natl Acad Sci U S A 104(39):15176–15178
Pichert D, Katsikopoulos KV (2008) Green defaults: information presentation and pro-environmental behaviour. J Environ Psychol 28(1):63–73
Redlawsk DP (2002) Hot cognition or cool consideration? Testing the effects of motivated reasoning on political decision making. J Polit 64(04):1021–1044
Reich RB (2008) Supercapitalism: the transformation of business, democracy, and everyday life. Random House Digital, New York
Rozin P, Scott S, Dingley M, Urbanek JK, Jiang H, Kaltenbach M (2011) Nudge to nobesity I: minor changes in accessibility decrease food intake. Judgm Decis Making 6(4):323–332
Schaefer A, Crane A (2005) Addressing sustainability and consumption. J Macromarket 25(1):76–92
Schultz WP, Khazian AM, Zaleski AC (2008) Using normative social influence to promote conservation among hotel guests. Soc Influence 3(1):4–23
Schwartz J, Riis J, Elbel B, Ariely D (2012) Inviting consumers to downsize fast-food portions significantly reduces calorie consumption. Health Aff 31(2):399–407
Shang J, Basil DZ, Wymer W (2010) Using social marketing to enhance hotel reuse programs. J Bus Res 63:166–172
Spangenberg JH, Lorek S (2002) Environmentally sustainable household consumption: from aggregate environmental pressures to priority fields of action. Ecol Econ 43(2):127–140
Stern PC (1997) Toward a working definition of consumption for environmental research and policy. In: Stern PC, Dietz T, Ruttan VR, Socolow RH, Sweeney JL (eds) Environmentally significant consumption: research directions. National Academy Press, Washington, DC, pp 12–35
Thaler RH, Sunstein CR (2008) Nudge: improving decisions about health, wealth, and happiness. Yale University Press, New Haven
*Thøgersen J, Schrader U (2012) From knowledge to action – new paths towards sustainable consumption. J Consum Policy 35(1):1–5
Whitley BE Jr, Kite ME (2012) Principles of research in behavioral science. Routledge, New York
Wiek A, Foley RW, Guston DH (2012) Nanotechnology for sustainability – what does nanotechnology offer to address complex sustainability problems? J Nanopart Res 14(9):1093
Yang MH (2001) An efficient algorithm to allocate shelf space. Eur J Oper Res 131(1):107–118

Chapter 25
Climate Change: Responding to a Major Challenge for Sustainable Development

Pim Martens, Darryn McEvoy, and Chiung Ting Chang

Abstract Climate change is emerging as one of the major challenges facing scientific and policy communities. The inherent complexity will ultimately require a much more integrated response scientifically to better understand multiple causes and impacts as well as at the scientific–policy interface where new forms of engagement between scientists, policymakers and wider stakeholder communities can make a valuable contribution to more informed climate policy and practice. The content of this chapter is considered particularly timely as scientific research and policy debate are shifting from one of problem-framing to new agendas that are much more concerned with implementation, the improvement of assessment methodologies from a multidisciplinary perspective and the reframing of current scientific understanding as regards mitigation, adaptation and vulnerability. A critical element of responding to the climate change challenge will be to ensure the translation of these new scientific insights into innovative policy and practice 'on the ground'.

Keywords Climate change • Adaptation • Mitigation • Policy

Based on: Martens P, McEvoy D, Chang C (2009). The climate change challenge: linking vulnerability, adaptation, and mitigation. Curr Opin Environ Sustain 1: 14–18

P. Martens (✉)
Maastricht University, Maastricht, The Netherlands
e-mail: p.martens@icis.unimaas.nl

D. McEvoy
Climate Change Adaptation Program, Global Cities Research Institute, RMIT University, GPO Box 2476, Melbourne, VIC 3001, Australia

C.T. Chang
Institute of Public Affairs Management, National Sun Yat-sen University, 70 Lienhai Rd, Kaohsiung 80424, Taiwan

H. Heinrichs et al. (eds.), *Sustainability Science*, DOI 10.1007/978-94-017-7242-6_25

1 Introduction

The consequences of rapid and substantial human-induced global climate change could be far-reaching, even leading senior commentators such as Sir David King to label it as one of the greatest threats facing future societies.[1] Until very recently, scientific and policy emphasis has focused on mitigation efforts, i.e. the reduction of anthropogenic greenhouse gas emissions. However, the success of global mitigation initiatives to date is questionable, and the impact of ever more stringent emission control programs could potentially have enormous social consequences. The efficiency of such action is also highly debatable. Whilst the characteristic of prompt costs and delayed benefits has resulted in early research which has concentrated for the most part on 'top-down' analyses of the cost-effectiveness of various greenhouse gas abatement strategies, little effort has been expended on the exploration of the interactions among the various elements of the climate problem, on a systematic evaluation of climate stabilisation benefits or on the costs of adapting to a changed climate, let alone an integration of different approaches. Crucially, these studies also do not assist decision-makers with the identification of climate change policy objectives; they only address the costs of meeting various abatement targets and the efficacy of different strategies.

2 Response Strategies

More mature climate strategies will require the integration of a wider range of mitigation, adaptation and vulnerability considerations, as well as responses more closely aligned with the objectives of other non-climate policy realms (McEvoy et al. 2006; Wilbanks and Sathaye 2007). In this regard, there is increasing recognition that, as policy evolves, new windows of opportunity may also emerge which allow for the articulation of integrated options for long-term policy on climate change mitigation and adaptation, as well as the building of local adaptive capacity and resilience in order to reduce vulnerability to climate change and variability (Neufeldt et al.2012). Ultimately, a coherent response will require consideration of all the facets of this issue (Fig. 25.1 shows a schematic representation of the climate change agenda).

Indeed, emerging international agendas are now reflective of a more holistic approach to responding to climate change. These are represented by funding agendas such as climate compatible development (development coupled with adaptation), green growth (development coupled with mitigation) and low carbon resilient development (an equal emphasis on the three agendas) (Fisher 2013).

The climate change issue has risen rapidly to the top of both research and policy agendas and is now the subject of widespread media coverage and increasing public

[1] http://news.bbc.co.uk/2/hi/science/nature/3381425.stm

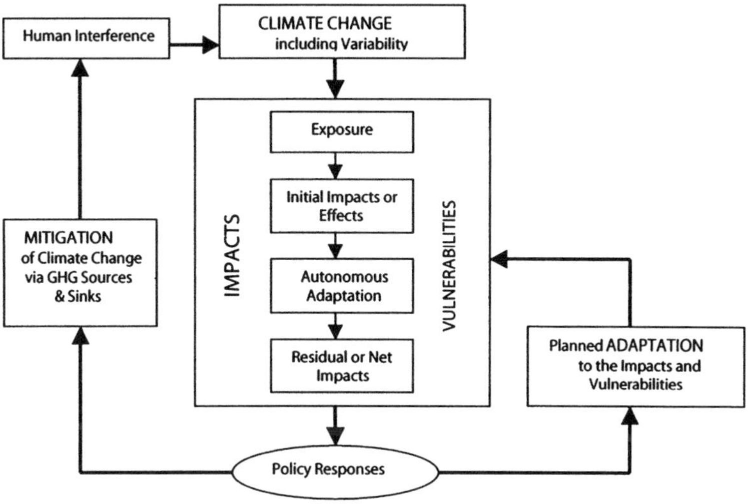

Fig. 25.1 The climate change agenda (Derived from Smit et al. 1999)

concern. To some extent, this is a consequence of important scientific advances in recent times. Several key developments are:

- *Increasing scientific consensus*: Concern about global warming, and the influence of human activity, can be traced back to the 1980s and is reflected in the creation of the Intergovernmental Panel on Climate Change (IPCC), an international attempt to consolidate the scientific community. Their most recent reports, published in 2007 and updated in AR5 in 2013, represent international state-of-the-art knowledge on climate change and its likely impacts. Through this forum, the collaborative efforts of scientists have concluded that climate change is happening and importantly that human activity is making a discernible contribution to this change.
- *From impacts to risk management*: Early scientific efforts concentrated on generating knowledge of the potential impacts of a changing climate and how to reduce anthropogenic greenhouse gas emissions. However, since the IPCC's Third Assessment Report in 2001, increasing emphasis has been placed on adaptation and the promotion of a risk management approach. This means going beyond mere consideration of climate-related hazards to more explicit consideration of issues surrounding the vulnerability and exposure of different elements at risk, as well as addressing conditions of uncertainty. This risk-based approach is embodied in state-of-the-art climate change strategies such as that recently adopted by London (Greater London Authority 2008). An important caveat also needs to be highlighted here. Whilst risk management has gained traction in the EU context, in many other regions of the world – particularly developing countries – other approaches continue to be valued, e.g. vulnerability

assessments which focus on current day adaptation deficits (Sattherthwaite and Dodman 2013).

- *Consideration of non-climate stressors*: When looking to the future, it also needs to be recognised that climate-related events will impact on societies that are likely to be very different than today's; hence, an integrated assessment of both climate and non-climate scenarios will ultimately be necessary to gain a better understanding of future risks (McEvoy et al. 2008).
- *Recognition of the need for greater interdisciplinary working*: Whilst mitigation has dominated policy and research agendas in recent years, there is an increasing recognition that actors also need to be preparing for change that is unavoidable. This has resulted in a greater consideration of vulnerability, adaptation and in many parts of the world already experiencing extreme events, disaster risk reduction. Drawing these different research domains closer together, with improved linkages between natural and social scientists, will be critical for effectively addressing the complexities of climate change.[2] New ways of working between scientists, policymakers and the wider stakeholder community will also be vital.

These advances have also filtered through to the public realm, as reflected in mainstream acceptance of human-induced climate change and the need for society to respond. Findings from actor-based research carried out for the EU-funded project ADAM (Adaptation and Mitigation: in support of European policy) suggest that, on the whole, the climate change issue is no longer questioned as it was in the past. Across all sectors, there is a common acknowledgement that climate change is happening and that we need to prepare for future change. Indeed, in the words of one interviewee, 'the world is changing fast and I no longer have to deal with sceptics as I have done in the past', whilst others have noted how 'rapidly the climate change issue has risen in profile over the past couple of years' and that there has been a 'rapid sea change in attitude following recent climate related events, such as the heat wave in 2003' (McEvoy et al. 2008). This shift in public perception has also been accompanied by a greater institutional impetus for change. Over the past couple of years, there has been an increasing recognition that climate change is not merely an environmental issue but one with important social and economic dimensions as well. This argument received much attention as a result of the Stern report in 2006 (Stern 2006). Although not the first economic report on this issue, it has become the most widely known document of its kind. In it, the author provides stark warning that climate change could result in a 'market failure on the greatest scale the world has seen'.

It is also important to recognise that these changes have taken place in a rapidly evolving policy context. Of the two mainstream agendas, mitigation is relatively 'mature' in comparison to adaptation, though new instruments continue to be developed and introduced. Perhaps the most high profile of these is carbon trading, a

[2] The Dutch NWO-funded VAM programme (Vulnerability, Adaptation, Mitigation and Adaptation-Mitigation) is a contemporary example of an attempt to encourage interdisciplinary working. See http://www.nwo.nl/nwohome.nsf/pages/NWOP_5XDGSK_Eng?OpenDocument

market-based mechanism given momentum by the signing of the Kyoto Treaty in 2005. Adaptation, on the other hand, can be considered an agenda still very much in its infancy. Politically, it is also recognised that there is a need for greater cohesion between climate change and sustainable development objectives. To date, climate change and sustainable development have tended to be treated as two distinct agendas; however, there is considerable added benefit to be gained by ensuring a more coherent approach. Not only will climate change have an adverse impact on progress towards a sustainable future, sustainable development activity can reinforce our response to climate change by both enhancing adaptive capacity and increasing resilience. As noted by the IPCC, however, few plans for promoting sustainability have explicitly included either adapting to climate change impacts or enhancing adaptive capacity.

3 Approaches to Climate Change Risks: The Way Forward

Moves towards considering both types of response as part of a more coherent policy programme represent an explicit acknowledgement by decision-makers that both mitigation and adaptation are important in reducing the risks associated with climate change, i.e. limiting the adverse effects of change and adapting to what is unavoidable. However, even though the preceding text has highlighted clear evidence of a desire to respond to the climate change issue, debate continues as to how best to approach this. Although linked in many ways, mitigation and adaptation have different problem structures with important implications for how political responses are framed (Klein et al. 2007). Therefore, whilst there is obvious interdependence between the two (they are both deliberate human responses aimed at reducing the risks associated with climate change), it is important to better understand some of the key synergies and conflicts between these two agendas. From McEvoy et al. (2006), some of these are:

- A common link between the two approaches is the capacity of a system to respond. For example, adaptive capacity can be simply defined as the ability of a system to adjust to climate change; this is thought to be determined by a range of factors, including technological options, economic resources, human and social capital and governance. Mitigation has similar determinants – in particular, the availability and penetration of new technology (although technological solutions have a role to play in both mitigation and adaptation, it should be recognised that 'soft engineering' has a particularly important role in adapting to climate change). The willingness and capacity of society to change is also critical (information and awareness-raising can be useful tools for stimulating individual and collective climate action);
- An integrated response is challenging as 'mitigation and adaptation are very different in what they mean and how they work'. Firstly, there is an obvious mismatch in terms of scale, both spatially and temporally. Mitigation efforts are

typically driven by national initiatives operating within the context of international obligations, whereas adaptation to climate change and variability tends to be much more local in nature, often in the realm of local/regional economies and land managers. As well as the spatial element, there are also differences in the timing of effects. As greenhouse gases have long residence periods in the atmosphere, the results of mitigation action will only be seen in the longer term. Adaptation, on the other hand, has a stronger element of immediacy.

- Disconnection in space and time can make it difficult for people to link the consequences of their activity with long-term environmental consequences. It also raises the question of environmental equity, i.e. who are the likely beneficiaries of the different types of response. Mitigation, being an action targeted at the longer term, attaches value to the interests of future generations and to some extent can be considered an altruistic response by society. Conversely, the impacts of climate change are felt more immediately by society, and adaptation is typically viewed as everyday 'self-interest'. As such, risk perception by individuals and organisations will be a critical influence on the acceptability and ultimate effectiveness of different responses.
- This inevitably leads to a consideration of trade-offs and distributional effects, in particular, who pays and who benefits, and whether there is a willingness to invest if the benefits of climate change response are perceived to be private. It is also important to note discrepancies in that those responsible for the majority of emissions (i.e. developed countries) also have the highest adaptive capacity, whilst the poorest countries, producing the lowest emissions, are most vulnerable to the impacts of a changing climate, and this has an influence on the urgency that is attached to any mitigation response. This also holds true within national territories, with uninsured, unaware and relatively immobile populations living in poorer quality accommodation often being hardest hit. In reality, those most vulnerable to climate change are those already at a socio-economic disadvantage in society.
- Another important difference between the approaches relates to those involved. Not only are decisions taken in different policy domains, but different stakeholder communities are also involved. Mitigation policy is primarily focused on decarbonisation and involves interaction with the large 'emitting' sectors such as energy, transport, etc., or else targets efficiency improvements according to specific end users – commercial, residential, etc. The limited number of key personnel and their experience of dealing with long-term investment decisions mean that the mitigation agenda can be considered more sharply defined. In contrast, multi-actors involved in the adaptation agenda come from a wide variety of sectors that are sensitive to the impacts of climate change. They also operate at a range of spatial scales. As a result, the implementation of adaptation measures is likely to encounter greater institutional complexity.
- It also needs to be recognised that adaptation agendas differ across regions of the world. In the European context, cities are well established with relatively stable populations, and therefore, responses to climate change are likely to focus on the medium to long term and involve retrofitting measures. In parts of the world already exposed to extreme events, adaptation inevitably overlaps with disaster

risk reduction, and in the developing world, adaptation not only needs to consider extreme events but also current-day deficits in infrastructure and adaptive capacity, as well as the pressure from rapidly growing populations.

Accounting for this complexity, combining the analyses of different dimensions of climate change, and highlighting the implications for policy and practice, will ultimately require a holistic and integrated approach (Van Zeijl-Rozema et al. 2008). The authors suggest that the science of integrated assessment (IA), with its combined application of modelling, scenario and participatory approaches, has considerable potential for both analysing the multiple causes and impacts of such a complex problem and informing the development of effective policy responses (Martens 2006). However, the generation of scientific knowledge alone will not suffice; information will also need to be translated into action 'on the ground'. Hence, as the policy debate moves from one of problem-framing to one more concerned with implementation, detailed political, ethical, social and normative analysis becomes increasingly important. A highly organised, multidisciplinary programme of research intended to add value to efforts to improve assessment methodologies, to contribute to the reframing of current scientific understanding and ultimately to provide new insights into innovative policy options will be required (McEvoy et al. 2013). Significant scientific and policy challenges remain ahead (Kemp and Martens 2007).

Questions:

1. Please identify interlinks among vulnerability, adaptation and mitigation in the context of climate change.
2. Give examples to describe the substitutability and complementarity between adaptation and mitigation measures.
3. Why is the framing of adaptation important?
4. All the agendas discussed have risk reduction and reducing vulnerabilities in common. What are the major differences between them?
5. Why does the adaptation agenda differ across different regions of the world?
6. Are climate scenarios necessary for adaptation planning? Why are they or why are they not?
7. Is it desirable to have a holistic climate action plan or should adaptation and mitigation be treated separately?

Further Reading

Dow K, Berkhout F, Preston BL, Klein RJT, Midgley G, Shaw MR (2013) Limits to adaptation. Nat Clim Change 3:305–307

Fuenfgeld H, McEvoy D (2011) Framing climate change adaptation in policy and practice, VCCCAR working paper. VCCCAR, Melbourne, Available at: http://www.vcccar.org.au/sites/default/files/publications/Framing_project_workingpaper1_240611_1.pdf

IPCC Fifth Assessment Report (AR5) (2014) Climate change 2014: impacts, adaptation and vulnerability. IPCC, Geneva

Weblinks

Green Climate Fund. http://gcfund.net/home.html
Intergovernmental Panel on Climate Change. http://www.ipcc.ch/
UNFCCC. http://unfccc.int/2860.php

References

CEC (2009) Adapting to climate change: towards a European framework for action COM 147 final. Commission of the European Communities, Brussels

Fisher S (2013) Low carbon resilient development in the least developed countries, IIED issue paper. IIED, London

Greater London Authority (2008) London climate change adaptation strategy – draft report. Greater London Authority, London. Available at: http://www.london.gov.uk/mayor/publications/2008/docs/climate-change-adapt-strat.pdf

Kemp R, Martens P (2007) Sustainable development: how to manage something that is subjective and never can be achieved? Sci Pract Policy 2:1–10

Klein RJT, Huq S, Denton F, Downing TE, Richels RG, Robinson JB, Toth FL (2007) Interrelationships between adaptation and mitigation. In: Parry ML, Canziani OF, Palutikof JP, van der Linden PJ, Hanson CE (eds) Climate change 2007: impacts, adaptation and vulnerability. Contribution of Working Group II to the Fourth Assessment Report of the Intergovernmental Panel on Climate Change. Cambridge University Press, Cambridge, pp 745–777

Martens P (2006) Sustainability: science or fiction? Sustain Sci Pract Policy 2:1–5

McEvoy D, Lindley S, Handley J (2006) Adaptation and mitigation in urban areas: synergies and conflicts. Proc Inst Civil Eng Munic Eng 159:185–191

McEvoy D, Lonsdale K, Matczak P (2008) Adaptation and mainstreaming of EU climate change policy CEPS policy briefing note for the European Commission. European Commission, Brussels. Available at: http://shop.ceps.eu/BookDetail.php?item_id=1593

McEvoy D, Bosomworth K, Fünfgeld H (2013) Resilience and climate change: the importance of framing. Plann Pract Res 28(3):280–293, http://dx.doi.org/10.1080/02697459.2013.787710

Neufeldt H, Jochem E, Hinkel J, Huitema D, Massey E, Watkiss P, McEvoy D, Rayner T, Hof A, Lonsdale K (2012) Climate policy and inter-linkages between adaptation and mitigation (Chapter 1). In: Hulme M, Neufeldt H (eds) Making climate change work for us: European perspectives on adaptation and mitigation strategies. Cambridge University Press, Cambridge, pp 3–30

Sattherthwaite D, Dodman D (2013) Towards resilience and transformation for cities within a finite planet Environment & Urbanization Brief – 28

Smit B, Burton I, Klein RJT, Street R (1999) The science of adaptation: a framework for assessment. Mitig Adapt Strat Glob Chang 4:199–213

Stern N (2006) The economics of climate change (The Stern Review) report for the Cabinet Office– HM Treasury. University Press Cambridge, Cambridge

Van Zeijl-Rozema A, Cörvers R, Kemp R, Martens P (2008) Governance for sustainable development: a framework. Sustain Dev 16:410–421

Wilbanks TJ, Sathaye J (2007) Integrating mitigation and adaptation as responses to climate change: a synthesis. Mitig Adapt Strat Glob Chang 12:957–962

Chapter 26
Art and Sustainability

Heather Sealy Lineberry and Arnim Wiek

> *The role of artworks is no longer to form imaginary and utopian realities, but to actually be ways of living and models of action within the existing real, whatever the scale chosen by the artist.*
>
> (Bourriaud, *Relational Aesthetics*, 2002, p. 13)

> *A more functional relationship between art and the everyday is urgently needed, through which artists can act as interlocutors…intervening in the debate itself and mediating new forms of acting and living.*
>
> (Teddy Cruz in Thompson, *Living as Form*, 2012, p. 58)

> *Artists cannot change the world…alone. But when they make a concerted effort, they collaborate with life itself, working with and between other disciplines and audiences, and given the chance to be seriously considered outside the rather narrow world of art, they can offer visual jolts and subtle nudges to conventional knowledge.*
>
> (Lippard, *Weather Report*, 2007, p. 6)

Abstract Over the past four decades, approaches to persistent and complex sustainability challenges have relied on solutions developed through scientific problem analysis and subsequent decision-making. Recently, this assumption has been exposed to various criticisms pointing out flaws and a lack of success. Art occupies a different intellectual, creative, and social space that can allow for surprising and promising perspectives and outcomes, offering innovative approaches to address sustainability problems. Since the 1990s, there has been a surge in interest among artists, curators, and theorists in collaborative art practice. Engaging directly with specific audiences and with pressing issues, the artists produce works that range in

H.S. Lineberry (✉)
Arizona State University Art Museum, PO Box 872911, Tempe, AZ 85287-2911, USA
e-mail: heather.lineberry@asu.edu

A. Wiek
School of Sustainability, Arizona State University,
PO Box 875502, Tempe, AZ 85287-5502, USA
e-mail: arnim.wiek@asu.edu

H. Heinrichs et al. (eds.), *Sustainability Science*, DOI 10.1007/978-94-017-7242-6_26

their intent from encouraging reflection, conversation, and learning to developing concrete solutions. This chapter focuses on the confluence of our heightened sustainability challenges with an increasing willingness among artists to address them and socially engaged practice as a particularly conducive art form. We focus on visual art and artists, although the most successful projects span disciplines and engage constituencies to challenge existing assumptions and propose new models.

Keywords Relational aesthetics • Social practice • Socially engaged art • Art and sustainability • Art and environment

1 Introduction

Today, we are confronted with complex, urgent, and persistent sustainability challenges that threaten the viability and integrity of societies across the world. Over the past four decades, approaches to these challenges have relied on a one-dimensional knowledge-to-action assumption, which suggests that solutions are best developed through scientific problem analysis and subsequent decision-making. Recently, this assumption has been exposed to various criticisms pointing out flaws and a lack of success, as well as suggesting alternatives and often more effective ways of developing robust solutions to sustainability challenges, including, among others, experimentation with alternative practices and rapid trial and error procedures (Sarewitz et al. 2012; Lang et al. 2012; Wiek et al. 2012). Among these, we are now recognizing that art offers innovative approaches for addressing sustainability problems and facilitates collective deliberation, learning, and transformation (Benessia et al. 2012; Lineberry et al. 2010; Kagan and Kirchberg 2008; Smith 2005).

Art occupies a different intellectual and creative space—more open-ended, somewhat outside of existing behavioral patterns, and often subversive—that can allow for surprising and promising perspectives and outcomes. It has the ability to engage the mind and the body, the imaginative and cognitive, the individual and the community with complex ideas, vivid representations, and experiences. Making us more conscious of accepted systems, it can facilitate deep collaboration across disciplines and social groups to deconstruct existing power structures and propose new paradigms.

Since the 1990s, there has been a surge in interest among artists, curators, and theorists in collaborative art practice, called by a variety of names, including "social practice," "new genre public art," "relational aesthetics," "participatory art," and "dialogic art." Engaging directly with specific audiences and with pressing issues, the artist or artist collectives produce works that range in their intent from encouraging reflection, conversation, and learning to developing concrete solutions by means of new objects, services, and practices. Consequently, their artworks can take the form of social events, gathering spaces, marketing campaigns, publications, workshops, websites, meetings, and performances.

This chapter focuses on the confluence of our heightened sustainability challenges with an increasing willingness among artists to address them and social practice as a particularly conducive art form. We will focus on visual art and artists, although socially engaged practice can be found in theater, dance, music, design, and architecture. The most successful artists draw upon methods and theories from a range of fields—performance, sociology, linguistics, urban planning, collaborative dynamics, and community organizing—and upon their experiences and resulting social intelligences. The best work spans disciplines and engages constituencies to, as the Danish art collective Superflex states, challenge existing models and propose new ones.

2 Precedents in Art, Social Practice, and Sustainability

Social practice is a complex and diverse field with no single definition and no linear trajectory, but rather a web of art influences and precedents paralleling broader social and cultural shifts. Since the 1960s, contemporary international art has operated in what influential critic Rosalind Krauss describes as the expanded field, "beyond the modernist demand for the purity and separateness of the various mediums" (Krauss in October, p. 42). Artists began exploring and combining new media and processes, leaving behind the discrete object and working instead in video, performance, language, and environmental installations. They often worked outside of the rarified space of art institutions (museums, galleries, etc.) and strove to avoid the influence of the art market. Land artists sculpted the very earth in remote locations, and conceptual artists prioritized idea over materiality. The international movement Fluxus and the Tropicalia artists in Brazil emphasized interactivity and spectator involvement in performances and performative spaces. Joseph Beuys considered public discourse and teaching to be at the center of his art practice and advocated for a radically expanded notion of art. Art activists championed racial equality and feminism, lobbied for gay rights, and raised awareness of the AIDS pandemic. New genre public art engaged diverse, urban communities around pressing social, economic, and political concerns with traditional and nontraditional media. In general, there has been a drive to have greater agency and impact, which was limited within the existing traditions and systems. Together, these artists and movements, and others, laid the groundwork for emphasis on the idea, the public realm, and the social—and the desire to merge art and life.

Art focusing on the environment and ecological systems also surged in the late 1960s and 1970s. Paralleling activism of the time, artists such as Agnes Denes, Helen Mayer Harrison and Newton Harrison, Hans Haacke, and Alan Sonfist tended to present nature as a separate sphere needing conservation and protection from the human impact of pollution and industrial destruction. Sonfist's stated goal with his work *Time Landscape* was "to elevate disappearing native landscapes to the status of historical monuments…." Although they brought much needed attention to neglected problems and started to change the conversation, the work of these eco-art pioneers often presented utopian myths of the natural, objectified nature or offered a closed response (see Demos in *Radical Nature*, 2009). Other and more recent artists, such as Mel Chin and Mierle Laderman Ukeles, move beyond objectifying nature and address the web of ecological, social, political, and economic issues. Ukeles and Chin are bridging figures, exemplars of successful collaborations across disciplines and communities, and offer interventions into stalled or dysfunctional processes.

Over a period of 11 months, Ukeles shook hands with and personally thanked all 8500 New York City sanitation workers, saying, "Thank you for keeping New York City alive." This text, image, video, and durational performance piece, called *Touch Sanitation* (1977–1980), strove to recognize the stigmatized and anonymous service workers who make our cities habitable. The first artist-in-residence at the New York City Department of Sanitation, with an office in their headquarters, Ukeles has created a powerful series demanding a complete shift toward viewing waste management as the primary maintenance system of our cities. *Flow City* (1985) established a visitor center at the 59th Street Marine Transfer Station, providing an on-site look at the process of treating urban waste water and recyclable materials. Ukeles' work has been called "exercises in outreach" (Thompson, p. 233) and makes visible the range of social, ecological, and economic forces in urban waste management.

Hyperaccumulators are plants that leach heavy metals from contaminated soil, and artist Mel Chin considers both the plants and toxic earth as his material to "sculpt a site's ecology." His installation *Revival Field* (1991–ongoing) was created in collaboration with US Department of Agriculture scientist Dr. Rufus Chaney. The process behind the piece included building his own knowledge of the science, building trust with Dr. Chaney, working with the sponsoring art museum, and negotiating with various government agencies for funding and site approval on the Pig's Eye Landfill in St. Paul, Minnesota. The work consisted of a 60-square-foot enclosure planted with six types of plants. After the second year of planting, the test results indicated enough success to inspire an international work group at the US Department of Energy. Dr. Chaney has said that it took an artist and an artwork to further the research on hyperaccumulators, which had been stalled due to politics and the resulting lack of funding (Finkelpearl, pp. 385–417).

Chin has gone on to address soil contamination in the ambitious and expansive project *Operation Paydirt/Fundred Dollar Bill Project* (2006–ongoing). Working with teams of scientists, volunteers, activists, teachers, and school children, Chin seeks to support a solution to lead-contaminated soil in post-Katrina New Orleans and help end this form of childhood lead poisoning. With a media campaign, scientific studies, and a nationwide participatory art project—drawing Fundred dollar bills to present to Congress to "pay" for treating the soil—Chin brings attention to the politics behind the refusal to act in low income and racially diverse areas and the social and economic impact on societies of lead contamination on young minds and bodies.

Touch Sanitation, *Revival Field*, and *Operation Paydirt* address the interrelatedness of social, economic, political, and ecological processes. The artists take on local and global topics of immediate and future concern and examine them critically and ethically. As such, they meet theorist Sacha Kagan's indicators for sustainability in the arts (Kagan and Kirchberg, pp. 17–18). We propose that they are also examples of artists exploring new collaborative methods, across disciplines and communities, to bring about changes in values and behaviors.

3 Today and Tomorrow, Here and There

In the boxes below, we present five recent projects by artists utilizing a variety of social practice methodologies to address sustainability challenges (for a comprehensive listing, see Thomson 2012). The projects range widely in scale, content, implementation, and intent, from the largely symbolic to the practical. They have been called micro-utopias or hands-on utopias, connecting forcefully to our challenges and offering new and surprising perspectives. Cuban artist Tania Bruguera calls for *arte útil* or "useful art" in her *Immigration Movement International Project* in Queens, New York. The art collective Superflex refers to their projects as "tools" for exploring new socioeconomic models.

Social practice projects can be durational, embedded in specific communities and tied to locations, or brief interventions that may not continue beyond an artist's temporary residency. They can be open to unlimited discussion and participant input or directed by the artist and key agents. Often commissioned and supported by art institutions or cultural agencies, they begin with a central question or problem which morphs through the participatory process and the impact of the location and context. They are social experiments that strive to build connections and dialogue and open up new, previously unforeseen pathways for societal development.

- ***Tasks*** *: Within the art field, there is much debate on social practice, and the challenges are only amplified when viewed from sustainability fields. Reflect on the following questions:*

  1. *When does the project become social service, political activism, or scientific experimentation as opposed to art (for a delineation of this debate, see Bishop* 2006*)?*
  2. *Is it more effective or appropriate for art to visualize and occupy problems or to propose practical solutions?*
  3. *Is it possible to accomplish change through short-term art projects or intense engagement with small groups?*
  4. *How legitimate is the change proposed in social practice projects?*
  5. *How can we measure the success of these projects and based on what criteria (aesthetics, awareness, social change)?*
  6. *How sustainable are the solutions proposed?*
  7. *Finally, artists are often viewed by the sciences as the communicators, illustrating complex ideas for a broad public, rather than bringing new knowledge and strategies to the research process. What is necessary for true collaboration between artists and sustainability scientists?*

4 Conclusions

Social practice has become a powerful approach for addressing sustainability challenges societies face around the world. Unlike conventional forms of problem-solving, social practice engages stakeholders, ranging from government and businesses to nonprofit organizations and civil society, in exploring and experimenting with alternative practices that pursue sustaining the viability and integrity of our societies and natural environments. These projects create often uncomfortable, subversive, and potent spaces for novel experiences that challenge conventions, habitual practices, and the preference to sustain the status quo. There is a significant overlap in intentions between social practice and transformational sustainability science efforts; synergies could emerge from combining and integrating both approaches. They could fulfill different roles while pursuing the same objectives of sustainability. Transformational sustainability research is uniquely positioned to partner in social practice projects with evidence-based proposals, project components, or actor networks; through monitoring and evaluating impacts of social practice projects, including distribution of benefits, as well as unintended consequences; and, finally, through support in improving social practice projects, based on evidence and best practices derived from comparative evaluative studies. Both fields have a track record in interdisciplinary cooperation, which could serve as a solid fundament for such synergistic efforts.

Future challenges will pertain less to definitions and demarcations concerned with questions such as: is social practice still art, or is transformational sustainability research a basic science endeavor? In fact, both fields often substitute such academic debates with a pragmatic attitude that focuses on: whatever is needed, whatever works, and whatever art and science can contribute to sustainability transformations. The key challenge, however, will remain the degree of real progress toward sustainability, in all its forms and facets, that can emerge from novel alliances between art and sustainability science.

Edible Park by Nils Norman
2010–ongoing
The Hague, Netherlands
Commissioning agent/collaborator: Stroom den Haag, Foodprint project

Norman considers his practice to be critical public art, investigating the possibility of creating change through disruptive and experimental models and methodologies in urban environments. For *Edible Park*, the British artist collaborated with a range of art, architecture, gardening and food activist organizations, and community volunteers to experiment with socio-ecological permaculture ideas and gardening methods in a public green space. The project is conceived as an experiment with a sustainable alternative to the current

(continued)

way of treating such urban spaces and partially as a response to a grandiose development plan by architect Rem Koolhaas for a high-impact "spontaneous city" that would have included an amusement park, beach, skyscrapers, and a Formula 1 race track. *Edible Park* functions as a gathering space for the neighborhood residents, visitors, and schools, with a pavilion, playground, educational demonstrations, and shared work space managed by a local gardening association. Under their auspices, the garden has begun to grow into surrounding areas.

Find more information in: Nils Norman, Peter de Rooden, Taco de Neef (eds), Nils Norman: Edible Park (Valiz and Idea Books: Amsterdam, 2012)

Superkilen by Superflex
2012–ongoing
Copenhagen, Denmark
Commissioning agent: City of Copenhagen and Realdania

The work of the Danish art collective Superflex examines underlying structures, reconfigures spaces and expectations, and stands back to see what happens. The *Superkilen* urban park has a series of 100 objects and pieces of furniture proposed by people representing more than 50 nationalities in the surrounding neighborhoods.[1] The artists called for submissions, selected five proposals, and then traveled to Palestine, Spain, Thailand, Texas, and Jamaica to acquire the specific objects or plan their design. Commissioned by the City of Copenhagen, in collaboration with architectural firms, the artists proposed and implemented a new system of community input in urban planning and completely reinvented the approved furniture for city parks.

Find more information at www.superflex.net.

[1] The Danish word "kilen" means "wedge" in English.

Superflex
Superkilen/Swing from Baghdad, Iraq, 2012
Nørrebro, Copenhagen, Denmark
Photo: Torben Eskerod

Superflex
Superkilen/Octopus from Tokyo, Japan, 2012
Nørrebro, Copenhagen, Denmark
Photo: Iwan Baan

Guaraná Power by Superflex
2004–ongoing
Maués, Brazil

Guaraná Power was a collaboration initiated by Superflex with the guaraná farmers' cooperative in the Brazilian Amazon. Responding to a cartel monopoly that controlled the use of guaraná berries in energy drinks and drove prices for the crop down by 80 %, Superflex worked with the farmers to organize and develop their own alternate product. Guaraná Power included package design, marketing campaigns, and distribution venues. The product is currently sold in convenience stores in Denmark and exhibited in art museums.

Find more information at www.superflex.net.

Superflex
Guaraná Power, 2003
Production/Bar at the Venice Biennale
Bottling and sales of Guaraná Power
Photo: Superflex

The Land by Rirkrit Tiravanija and Land Foundation
1998–ongoing
Chiang Mai, Thailand

Imagine visiting a contemporary art gallery in New York or London and being invited to sit down to a meal of Pad Thai prepared and served in the middle of the gallery by an artist or curator. Rirkrit Tiravanija is considered a pioneer in social practice for staging such gatherings in high art venues and encouraging visitor engagement and participation to an unusual degree. In the late 1990s, he expanded his experiments in building social connectivity by purchasing a faltering rice farm in Chang Mai with artist Kamin Lertchaiprasert, The area had been plagued by floods and high water and threatened with development. The project is conceived as an ongoing lab for a self-sustaining community centered around discussion and experimentation. In collaboration with University of Chiang Mai students, they have experimented with a year-round rice crop. The harvest is distributed among participants, as well as to families impacted by HIV/AIDS. Other projects led by international artists, like Superflex, Tobias Rehberger, and Arthur Meyer, have worked with community members, schools, and students to experiment with biogas and solar systems.

Find more information at thelandfoundation.org.

It's Not Just Black and White by Gregory Sale
2011
Tempe, Arizona
Commissioning agent: Arizona State University Art Museum

In the USA, more than seven million citizens are incarcerated or under the jurisdiction of the justice system, and the country spends $80 billion to keep them there. Gregory Sale's project, *It's not just black and white*, built a charged but safe and welcoming gathering place where inmates, crime victims, their families, corrections and law enforcement officers, activists, academics, and the general public could gather to examine the underlying cycles of poverty, racism, and politics in incarceration. He staged his project in an art museum in Arizona, home to the internationally known Sheriff Joe Arpaio and boot camp-style prisons with black and white striped uniforms, tent housing, and chain gangs. Through a long process of negotiation and trust building, the artist received approval for jail inmates to visit the museum, with their armed guards, to work with him in the gallery to create a powerful, black and white striped setting for 52 meetings, workshops, and talks on this pressing issue. Participants ranged from Sheriff Joe to activist and former Black Panther member Angela Davis. *It's not just black and white* is an example of a social practice project, the public form of which is short—3 months—but which develops out of extensive work with diverse communities and

(continued)

stakeholders. The artist has gone on to do additional work in Arizona and beyond that addresses the broader forces in incarceration.

Find more information at http://asuartmuseum.org/socialstudies-projects/5/Gregory-Sale-It-s-not-just-black-and-white.html.

Gregory Sale
It's not just black and white, 2011, Arizona State University Art Museum, Tempe, AZ
Artist-in-residence and social art project with 18 individual and 37 institutional collaborators including Arizona State University Humanities Project, AZ Common Ground, Gina's Team, Maricopa County Sheriff's Office, University of Arizona Poetry Center, Elizabeth Johnson and others (Curator: John Spiak, ASU Art Museum)
Photo: Stephen Gittins

Immigrant Movement International (IM) by Tania Bruguera
2011–ongoing
Corona, Queens, New York
Commissioning agents: Queens Museum of Art and Creative Time

Cuban artist Tania Bruguera conceived of Immigrant Movement International as an art project in the form of an artist-initiated sociopolitical movement. The IM headquarters is a flexible community space in a multinational neighborhood where 167 languages are spoken. Working with residents, social service organizations, elected officials, and artists, the project focuses on issues of immigration reform, as well as the daily needs of immigrants. "As migration

(continued)

becomes a more central element of contemporary existence, the status and identity of those who live outside their place of origin increasingly become defined not by sharing a common language, class, culture, or race, but instead by their condition as immigrants," states the project's mission, citing the need to recognize and redefine the immigrant as the new global citizen. Activities range from a transdisciplinary convening that generated the Migrant Manifesto to legal services, slogan writing workshops, youth music lessons, and a women's health program. After the first year, the initial funding came to an end, yet the community rallied to keep the center open. In 2012, a sister organization was opened in Mexico City, el Partido del Pueblo Migrante (PPM), and the artist hopes to open other locations around the world. Bruguera is well known for her concept of *arte útil*, or "useful art", which takes on social and political challenges through direct engagement in people's lives.

Find more information at http://immigrant-movement.us/wordpress or http://www.taniabruguera.com.

Tania Bruguera
Immigrant Movement International, 2011
A class from the Paper Orchestra
http://immigrant-movement.us/
Photo: Tania Bruguera

Tania Bruguera
Immigrant Movement International, 2010–2015
Corona, Queens, New York, USA
http://immigrant-movement.us/
Photo: Tania Bruguera

References and Additional Readings[2]

Andrews M (ed) (2006) Land, art: a cultural ecology handbook, 1st edn. RSA/Arts Council, London

Benessia A, Funtowicz S, Bradshaw G, Ferri F, Medina CP, Raez-Luna EF (2012) Hybridizing sustainability: towards a new praxis for the present human predicament. Sustain Sci 7(1):75–89

Bishop C (2006) The social turn – collaboration and its discontents. Artforum 2006:178–183

Bishop C (2012) Artificial hells. Verso, London/Brooklyn, 390 p

Bourriaud N (2002) Relational aesthetics. Les Presses Du Reel, France

*Demos TJ (2009) The politics of sustainability: art and ecology. In: Manacorda F, Yedgar A (eds) Radical nature: art and architecture for a changing planet 1969–2009. Koenig Books/Barbican Art Gallery, London, pp 17–30

Demos TJ (ed) (2013) Contemporary art and the politics of ecology. Third Text 27(1):1–9

Finkelpearl T (2001) Dialogues in public art. The MIT Press, Cambridge, MA

Finkelpearl T (2013) What we made: conversations on art and social cooperation. Duke University Press, Durham/London

[2] The sources highlighted with an asterisk (*) are recommended as further readings.

Frieling R, Atkins R, Groys B, Manovich L (2009) The art of participation: 1950 to now. San Francisco Museum of Modern Art/Thames & Hudson, San Francisco/New York
Giannachi G (2012) Representing, performing and mitigating climate change in contemporary art practice. Leonardo 45(2):125–131. MIT Press
*Helguera P (2011) Education for socially engaged art: a materials and techniques handbook. Jorge Pinto Books, New York
Jackson S (2011) Social works: performing art, supporting publics. Routledge, New York
Kagan S, Kirchberg V (2008) Sustainability: a new frontier for the arts and cultures. VAS, Frankfurt am Main
Kastner J (ed) (2010) Land and environmental art. Phaidon Press, New York
Kester G (2008) Conversation pieces: community and communication in modern art. The University of California Press, Berkeley
Kester G (2011) The one and the many: contemporary collaborative art in a global context. Duke University Press, Durham/London
Kwon M (2004) One place after another. The MIT Press, Cambridge, MA
Lacy S (ed)(1995) Mapping the terrain: new genre public art. Bay Press, Seattle
*Lang DJ, Wiek A, Bergmann M, Stauffacher M, Martens P, Moll P, Swilling M, Thomas C (2012) Transdisciplinary research in sustainability science – practice, principles and challenges. Sustain Sci 7(Suppl 1):25–43
*Lineberry HS (ed) (2010) Defining sustainability. ASU Art Museum, Tempe
Lippard LR (2007) Weather report: art and climate change. Boulder Museum of Contemporary Art with EcoArts, Boulder
Sarewitz D, Clapp R, Crumbley C, Kriebel D, Tickner J (2012) The sustainability solutions agenda. New Solut 22(2):139–151
*Smith S (2005) Beyond green: toward a sustainable art. Smart Museum of Art, the University of Chicago, Chicago
*Thompson N (2012) Living as form: socially engaged art from 1991–2011. The MIT Press, Cambridge
*Wiek A, Ness B, Brand FS, Schweizer-Ries P, Farioli F (2012) From complex systems analysis to transformational change: a comparative appraisal of sustainability science projects. Sustain Sci 7(Suppl 1):5–24

Chapter 27
Teaching and Learning in Sustainability Science

Matthias Barth

Abstract The concept of sustainability does not present *the* pathway or distinctive solution which needs to be followed and is defined differently by different actors around the globe. Thus, the transition towards sustainability relies on constant negotiation and societal learning processes. To achieve this, education and learning must be seen as key processes. It is the area of education for sustainable development that is concerned with aspects of learning that enhance the transition towards sustainability – an area that can best be described as a vision of education that seeks to balance human and economic well-being with cultural traditions and respect for the Earth's natural resources. This chapter elaborates upon how education for sustainable development translates research outcomes of sustainability science into educational practices and guides the selection of learning objectives, relevant content and appropriate forms of teaching and learning.

Keywords Education for sustainable development • Competence development • Self-directed learning • Collaborative learning • Problem-based learning

1 Introduction

In sustainability science, a consensus exists that we are living in a time of transformation, in which a global range of social, economic, cultural and ecological changes occur on levels rarely seen before, threatening a number of 'planetary boundaries' in the long term (Rockström et al. 2009). Over the last few decades, a growing awareness of these unsustainable changes has emerged within the general public, as well as in politics, and the concept of sustainable development was introduced into the public discourse. This concept offers an orientation towards what a transformation within 'safe and just boundaries' (Raworth 2012) could look like. At the same time, sustainability does not present *the* pathway or distinctive solution which needs

M. Barth (✉)
Institute for Integrative Studies, Leuphana University of Lüneburg,
Scharnhorststr. 1, 21335 Lüneburg, Germany
e-mail: matthias.barth@leuphana.de

H. Heinrichs et al. (eds.), *Sustainability Science*, DOI 10.1007/978-94-017-7242-6_27

to be followed. As an 'ill-defined concept' (Laws et al. 2004) and a 'moving target' (Hjorth and Bagheri 2006) that is defined differently by different actors around the globe, the transition towards sustainability relies on constant negotiation and *societal learning processes*.

Both at the individual and global levels, what we need to learn is how to improve our capacity to adapt to inevitable changes and to mitigate the future consequences of today's actions. This necessitates a shift in mindsets that goes beyond 'doing things better' or 'doing things differently' towards a paradigm change of learning to alter the way we look at things completely. To achieve this, education and learning must be seen as key. Not surprisingly, education features prominently in various declarations, visions and missions around the world as a soft measure for bringing about change.

Important Facts
Education is repeatedly referred to as a *soft* measure for achieving sustainability. This is based on a distinction between 'hard' instrumental measures and 'soft' persuasive measures seeking to bring change. Hard measures include legislative, regulatory and juridical, as well as financial and market instruments – many of which are discussed in earlier chapters of this book. Besides education, it is, e.g. social marketing and media campaigning that comprises persuasive approaches of soft measures.

It is the area of education for sustainable development that is concerned with aspects of learning that enhance the transition towards sustainability. Education for sustainable development can best be described as a vision of education that seeks to balance human and economic well-being with cultural traditions and respect for the Earth's natural resources. It translates research outcomes of sustainability science into educational practices and guides the selection of learning objectives, relevant content and appropriate forms of teaching and learning.

2 What Are We Learning For?

- **Task**: *Before we elaborate on the role of education and learning objectives in education for sustainable development, take a moment to reflect for yourself: what is the role you think education can or should play for a more sustainable future? What are the learning objectives you would hope to be met in education for sustainable development? Make notes of what you think of as most important in this regard, then read this chapter and come back to your notes. Would you reconsider your first impression?*

As the answer to this question will ultimately influence the way content and learning and teaching methods are chosen and designed, we need to think carefully

Box 27.1: The Opposing Poles of Instrumental Versus Emancipatory Approaches

Instrumental approaches, which are most commonly found in policy papers and among politicians, focus on the achievement of sustainable development. Here, it is argued that sustainability is an important societal objective and education thus must contribute to achieving this objective. Consequently, education is interpreted as a means to achieving an end – that of sustainability.

On the other side in that debate, *emancipatory approaches* argue that what must be considered is the free will of the autonomous learner. Education in that sense is not about giving directives, but about offering learning opportunities in which the individual can develop. Sustainability thus is not the ultimate goal of education, but a learning context to support broader educational goals.

about what we try to reach with education for sustainable development. In the academic world, a debate arose about the question of how education should relate to the concept of sustainable development and what outcomes education should be aiming at. This question reconsiders the role education in general can and should play for a more sustainable future. Two opposing positions, deeply critical of each other, inform this debate, namely, the instrumental and the emancipatory (see Box 27.1).

The two positions mark fundamentally different approaches and can be found on opposite poles. However, in reality, it is not so much of an either/or situation, as there is a variety of approaches that lie in between, some of them of a more instrumental and others of a more emancipatory nature (Wals et al. 2008). And indeed, both sides have significant arguments in their favour, as sustainability, on the one hand, will not take place if fundamental action is not taken, though on the other hand, we neither can nor should prescribe specific activities for the individual, bearing in mind the complexity and uncertainty of future developments.

- **Task**: *Discuss in small groups with your peers: what are the arguments for and against instrumental or emancipatory approaches? Where would you position yourself and why?*

In an attempt to reconcile means and ends in education for sustainable development, Vare and Scott (2007) distinguish between 'ESD-1', which promotes certain behaviours and ways of thinking, and 'ESD-2', which focuses on 'building capacity to think critically about [and beyond] what experts say and to test sustainable development ideas', as well as 'exploring the contradictions inherent in sustainable living' (Vare and Scott 2007). They argue that while ESD-1 is a necessary form of learning to achieve sustainable development, it is ESD-2 which complements the learning process, as it supports the learner's capability to analyse, question alternatives and negotiate decisions.

It is the concept of key competencies that has recently gained ground in the debate on intended learning outcomes and brings together both forms of learning. But what exactly do we mean by key competencies? The answer to this question is far from easy. From a rather broad perspective, competencies can be understood as 'a roughly specialised system of abilities, proficiencies or skills that are necessary or sufficient to reach a specific goal' (Weinert 2001: 45). Competencies are developed as a response to complex demands that necessitate the interplay of cognitive, emotional and motivational dispositions (Klieme et al. 2007). The term '*key* competencies' highlights the significance of certain competencies. Key competencies are relevant across different spheres of life and for all individuals (Rychen and Salganik 2003). They do not replace but rather comprise domain-specific competencies, which are necessary for successful action in certain situations and contexts.

If we now go even one step further and ask about key competencies to be able to contribute to a more sustainable future, the literature on education for sustainable development offers a number of frameworks that define such learning objectives. These approaches use a number of different abstract concepts, such as skills, literacy, competencies or capabilities. What they have in common, though, is the goal of enabling people not just to acquire and generate knowledge, but also to reflect on further effects and the complexity of behaviour and decisions in a future-oriented, global perspective of responsibility. They share a broad consensus on the 'key ingredients', they focus on the aspects that are important for future change agents and key actors in different and sustainability-related contexts, and they intend to reconcile instrumental and emancipatory approaches.

It is the work of Wiek et al. (2011) that provides us with insights into what such a set of sustainability-related key competencies might look like, enabling students, especially those of sustainability science, to analyse and solve sustainability problems and thus to create opportunities for sustainability. In a systematic approach, they derived a set of key competencies, specified as sustainability research and problem-solving competence (see Table 27.1).

This set of competencies is based on the insight that sustainability problems have specific characteristics, and therefore, analysing and solving sustainability problems require a particular set of interlinked and interdependent key competencies. As students in sustainability science should be enabled to plan, conduct and engage in sustainability research and problem-solving, it is precisely the interplay of systems thinking, anticipatory, normative, strategic and interpersonal competencies upon which higher education for sustainable development needs to focus.

3 How Can the Development of Competencies Be Supported?

Having defined sustainability-related key competencies as intended learning outcomes in education for sustainable development, the question remains as to how to facilitate the development of such competencies. There are two areas in which

Table 27.1 Sustainability research and problem-solving competencies (Wiek et al. 2011)

1. Systems thinking competence:
The ability to collectively analyse complex systems across different domains and across different scales, thereby considering cascading effects, inertia, feedback loops and other systemic features related to sustainability issues and sustainability problem-solving frameworks
2. Anticipatory competence:
The ability to collectively analyse, evaluate and craft rich 'pictures' of the future related to sustainability issues and sustainability problem-solving frameworks
3. Normative competence:
The ability to collectively map, specify, apply, reconcile and negotiate sustainability values, principles, goals and targets
4. Strategic competence:
The ability to collectively design and implement interventions, transitions and transformative governance strategies towards sustainability
5. Interpersonal competence:
The ability to motivate, enable and facilitate collaborative and participatory sustainability research and problem-solving

competence development takes place that can be distinguished, namely, formal and informal learning.

3.1 Competence Development in Formal Learning Settings

When we focus on competence development in formal education, it is necessary to consider new ways of teaching and learning as key competencies can 'be learnt but hardly be taught' (Weinert 2001). Such an orientation challenges traditional views of the relationship of learning outcomes, topics and teaching and learning methods and comes with various shifts: from teacher to learner-centred pedagogies, from input to output orientation and from a focus on content and topics to a focus on problem-solving and processes. This is based on an understanding of learning as situated and as an active construction, in which the emphasis is not exclusively on knowledge creation, but takes in various forms of experience-oriented and problem-based learning.

There are three key principles by which learning processes for supporting competence development can be characterised (see Fig. 27.1). The first principle is *self-directed learning*, which is based on a view of learning not directly linked to teaching and which emphasises the active development of knowledge rather than its mere transfer. It is an approach 'where learners are motivated to assume personal responsibility and collaborative control of the cognitive (self-monitoring) and contextual (self-management) processes in constructing and confirming meaningful and worthwhile learning outcomes' (Garrison 1997: 18). The central role of the learner is explicitly acknowledged, which also calls for a new role for the teacher, who needs to focus on coaching and moderating the learning processes of the students

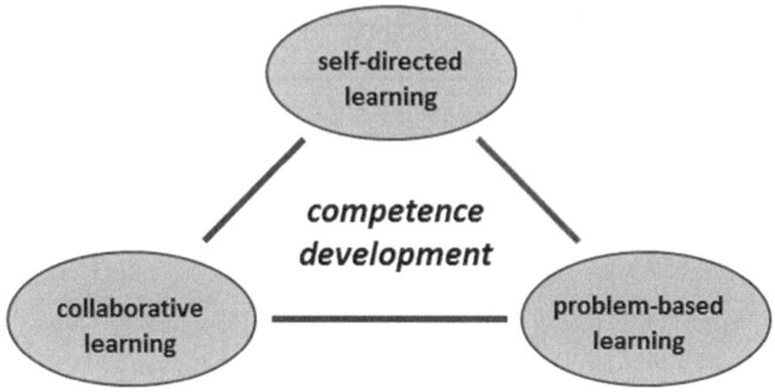

Fig. 27.1 Key principles of learning and competence development

who take ownership of their learning. The aim is then to stimulate learning processes in which students construct their knowledge independently.

Collaborative learning is the second important principle, as the acquisition of competencies is both an individual and a social activity. In collaboration processes, learners not only have to deal with different perspectives but are forced to elaborate and defend their own perspective, which increases their social and discursive abilities. This is of utmost importance for competence development, as it addresses both cognitive and social-affective aspects of learning. Learning is based on shared experiences and jointly accepted learning objectives and happens individually and in the group on the basis of collaborative experiences. Thus, knowledge is seen as the result of shared group processes, and different opinions and approaches are not only tolerated but appreciated and even encouraged.

The third principle, that of *problem-oriented learning*, focuses on complex real-world situations and the development of creative solutions to trigger competence development (Brundiers and Wiek 2013). While traditional learning processes often encounter problems because of their exclusive focus on factual knowledge, which cannot be used for action in specific situations, a problem-oriented approach is especially suited to supporting action-relevant procedural knowledge and skills. Problem-oriented learning provides a motivating context, as students experience authentic situations in which they do not just learn 'dry' theory but tackle the respective issues on their own. It is facilitated by complex 'real-world' problems and different approaches and perspectives. Thus, the first two principles of self-directed learning and collaboration can be seen as preconditions for a problem-oriented approach. See chap. 29 in this book.

While these three key principles provide a strong base for different approaches, it goes without saying that there is no simple formula that fits all contexts and situations. Instead, the successful support of competence development in formal education relies heavily on pedagogical creativity to create learning environments that are supportive, motivating and challenging for students.

- **Task**: *Think about your own experiences as a student in higher education. Where have you experienced learning processes, which relied on the principles outlined above? If this was not (always) the case, how would you picture a really* *supportive learning environment? What change would be needed to offer the best opportunities for competence development?*

3.2 Informal Learning and Competence Development

From an educational perspective, formal learning within schools and universities is of primary interest. But if you think of the manifold opportunities in which learning takes place in different environments, formal learning is no longer the only perspective. Students' learning also takes place as informal learning, and it is of great importance that the institutional context is experienced as a learning environment that offers opportunities to engage with sustainability-related issues.

In many case studies, it has been analysed as to how the institution itself can be used as just such an informal learning space. Informal learning that is always self-directed and often appears as incidental and experiential learning contributes to developing competencies, as it is linked to active involvement of the learner and always happens in contexts that are meaningful for that learner. To support such a form of learning, it is important to provide time and space for the form, to examine the environment with regard to learning opportunities and to create an atmosphere of cooperation and confidence in which the learning process can be reflected (Barth et al. 2007).

In the higher education context, it is the metaphor of the 'living lab' that expresses well how sustainability can be dealt with in the campus community and in the wider community of an institution (Brundiers et al. 2010). It refers to the practice of students having to deal with challenges directly connected to their own lifeworld during their learning process, in which solutions are developed and projects are started. Such learning is especially powerful, if links between formal and informal learning are utilised. For students, it is then highly motivating to have manifold opportunities for self-directed learning while support and guidance from lecturers are available.

4 What Can Educational Science Contribute to Sustainability Science?

So far, we have looked at how competence development can be supported in formal and informal learning and thus how education is responding to the challenges that arise from striving for sustainability. But this is only one side of the complex relationship between education and sustainability science. Educational science, in turn, also has much to offer to the transdisciplinary arena of sustainability science. For the ongoing discourse within sustainability science, an educational perspective offers new insights into the understanding of both individual and social transformation processes and broadens the variety of disciplinary contributions. This is even more important when we remember that it is increasingly acknowledged that the transformation towards a more sustainable future will call on all our resources to learn and adapt. When we look at the strive for sustainability as a shared learning process, insights into how best to support learning processes can and must be added from an educational perspective.

Further Reading

Barth M (2014) Implementing sustainability in higher education: learning in an age of transformation, Routledge studies in sustainable development series. Taylor & Francis, London

Much of what could be only touched upon in this overview is elaborated on in my latest monograph, so I hope you forgive me a bit of self-promoting here

Orr DW (2004) Earth in mind: on education, environment, and the human prospect. Island Press, Washington, DC

David Orr's wonderful book is one of the more elaborate contributions to the debate about what education can or should be. He refers specifically to environmental education, but this, of course, is even more important in education for sustainable development

Rychen DS, Salganik LH (eds) (2001) Defining and selecting key competencies. Hogrefe & Huber Publishers, Seattle

A thorough understanding of key competencies and how to define and select them is provided in this book from Dominique Rychen and Laura Salganik

Wiek A, Withycombe L, Redman CL (2011) Key competencies in sustainability: a reference framework for academic program development. Sustain Sci 6(2):203–218

Arnim Wiek and his colleagues provide us with an overview of how the concept of competencies is used in education for sustainable development and offer a concise concept of sustainability related competencies

References

Barth M, Godemann J, Rieckmann M, Stoltenberg U (2007) Developing key competencies for sustainable development in higher education. Int J Sustain High Educ 8(4):416–430

Brundiers K, Wiek A (2013) Do we teach what we preach? An international comparison of problem- and project-based learning courses in sustainability. Sustainability 5:1725–1746

Brundiers K, Wiek A, Redman CL (2010) Real-world learning opportunities in sustainability: from classroom into the real world. Int J Sustain High Educ 11(4):308–324

Garrison DR (1997) Self-directed learning: toward a comprehensive model. Adult Educ Q 48(1):18–33

Hjorth P, Bagheri A (2006) Navigating towards sustainable development: a system dynamics approach. Futures 38(1):74–92

Klieme E, Hartig J, Rauch D (2007) The concept of competence in educational contexts. In: Hartig J, Klieme E, Leutner D (eds) Assessment of competencies in educational contexts. Hogrefe & Huber Publishers, Cambridge, MA, pp 3–22

Laws D, Scholz RW, Shiroyama H, Susskind L, Suzuki T, Weber O (2004) Expert views on sustainability and technology implementation. Int J Sustain Dev World Ecol 11(3):247–261

Raworth K (2012) A safe and just space for humanity. Can we live within the doughnut? Oxfam discussion papers. Oxfam International, Oxford

Rockström J, Steffen W, Noone K, Persson A, Chapin FS, Lambin EF, Lenton TM, Scheffer M, Folke C, Schellnhuber HJ, Nykvist B, de Wit CA, Hughes T, van der Leeuw S, Rodhe H, Sorlin S, Snyder PK, Costanza R, Svedin U, Falkenmark M, Karlberg L, Corell RW, Fabry VJ, Hansen J, Walker B, Liverman D, Richardson K, Crutzen P, Foley JA (2009) A safe operating space for humanity. Nature 461(7263):472–475

Rychen DS, Salganik LH (2003) A holistic model of competence. In: Rychen DS, Salganik LH (eds) Key competencies for a successful life and well-functioning society. Hogrefe & Huber Publishers, Cambridge, pp 41–62

Vare P, Scott WAH (2007) Learning for a change: exploring the relationship between education and sustainable development. J Educ Sustain Dev 1(2):191–198

Wals AEJ, Geerling-Eijff F, Hubeek F, van der Kroon S, Vader J (2008) All mixed up? Instrumental and emancipatory learning toward a more sustainable world: considerations for EE policymakers. Appl Environ Educ Commun 7(3):55–65

Weinert FE (2001) Concept of competence. In: Rychen DS, Salganik LH (eds) Defining and selecting key competencies. Hogrefe & Huber Publishers, Seattle, pp 45–66

Wiek A, Withycombe L, Redman CL (2011) Key competencies in sustainability: a reference framework for academic program development. Sustain Sci 6(2):203–218

Chapter 28
Education for Sustainable Development

Niko Roorda and Han van Son

Abstract Education is to play an essential role for sustainable development (SD). The chapter dedicated to "Education for Sustainable Development" (ESD) explains why and how this can be done.

First of all, a distinction is made between several levels of organizational change within a university, varying between minor changes and all-out transformation processes, leading to a state of "System Integration of Sustainable Development" (SISD). In such a process, not only the main aspects and activities of a university are transformed but even the very identity of the institution. In order to achieve this, a university has to act as a "learning organization," as is described using the concrete example of a Dutch university for applied sciences.

The key role of a university toward sustainable development is its education, as is argued. In order to describe which elements the transformation of the education consists of, the so-called Tree Model is used.

The "roots" of this tree represent the educational goals, i.e., a description of the type of professional the university wishes to deliver to the society and the professional fields. For this purpose, a tool is available called "RESFIA+D," which offers a method to develop or improve the professional competences of study programs, making use of a description of a number of levels of competence.

Another set of tools is described to develop the "trunk of the tree," i.e., a general introduction to sustainable development for all academic disciplines, consisting of a textbook and a website offering accessories, e.g., exercises, serious games, video clips, etc.

Other elements of a tree are used to describe more aspects of ESD, such as the branches, which represent the disciplinary integration of SD within modules and topics throughout the curriculum.

Finally, the chapter describes a way to raise the expertise of the teaching staff, which is quintessential to achieve the desired ESD transformation. Together with an integration of this development with the quality management, making use of ESD assessment tools such as STARS (Sustainability Tracking, Assessment & Rating

N. Roorda, Ph.D. (✉)
Roorda Sustainability, Tilburg, Netherlands
e-mail: nroorda@planet.nl

H. van Son, MA
School of International Studies, Avans University, Breda, Netherlands
e-mail: jwa.vanson@avans.nl

H. Heinrichs et al. (eds.), *Sustainability Science*, DOI 10.1007/978-94-017-7242-6_28

System) or AISHE (Assessment Instrument for Sustainability in Higher Education), the goal of SISD can be reached.

Keywords Education • System integration • Transformation • Competences • Assessment

1 Levels of Change: From Minor Additions to System Integration of SD

Education is an essential contributor to sustainable development (SD). This is expressed in many sources, e.g.:

> Education is critical for promoting sustainable development and improving the capacity of the people to address environment and development issues. [...] Education [is] indispensable to changing people's attitudes so that they have the capacity to assess and address their sustainable development concerns. It is also critical for achieving environmental and ethical awareness, values and attitudes, skills and behaviour consistent with sustainable development and for effective public participation in decision-making. (UNCED 1992: Agenda 21, §36.3)

In order to contribute to SD, education will have to change drastically. The science that investigates this change process is "Education for Sustainable Development" (ESD).

When educational institutions start ESD activities, this usually initially leads to minor changes, in which SD elements are added, "bolted-on," to the education without deeply changing the existing. In an ESD assessment instrument, AISHE (Roorda 2001; Roorda et al. 2009), this development stage is described as "stage 1: activity oriented." As the development proceeds, the institution may enter stage 2, "process oriented," in which SD becomes more and more integrated into the curriculum and in the institutional vision, policy, and operations.

A crucial next stage is "system oriented." If a university or school reaches this stage on a wide range of criteria, it realizes a state of "SISD": *System Integration of Sustainable Development*. This stage is described as:

> SISD not only means a systematic integration of sustainable development into an educational organization (or a functional unit within it, e.g. a faculty, a school, or a study program), but also, and even primarily, at integration at a systems level. The latter implies that sustainable development has become a part of the fundamental characteristics of the organization, of its very identity. If this is the case, it will be observed that sustainability has become a part of all or most activities, or at least of the thoughts and philosophies behind those activities. (Roorda 2010, p. 138)

Question

Consider your own university, or a university that you are familiar with. If you were to express the present state of this university as a percentage of a full SISD, what percentage would you choose? Do you think that everybody would roughly agree with your estimate?

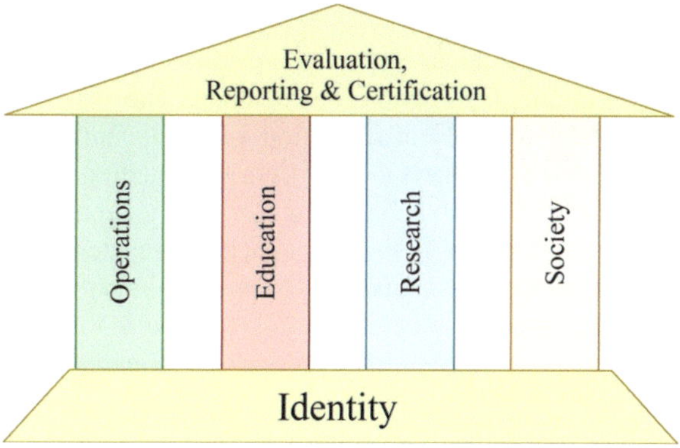

Fig. 28.1 The main functions of an educational institution

The development stages of AISHE show resemblance to the levels of change defined by Sterling (2004). A comparison can be found in Roorda (2010, p. 139).

Thorough ESD integration will have consequences for all four main functions of a university (Velazquez et al. 2006), as Fig. 28.1 shows, i.e., on its operations (Clugston and Calder 2000), its education, its research, and its community outreach (Megerle and Megerle 2000). More fundamentally, realizing SISD has consequences for the identity of a university and on its quality management and public reporting.

Some aspects of SISD are (Roorda 2001; Roorda et al. 2009):

> The organization visions itself as a key player for sustainable development [...]. Staff and students are actively involved in the continuous development and improvement of the vision and policy on sustainable development. The organization can be characterized as a *learning organization*. (AISHE 2.0, criterion I1: *Vision and policy*, stage 3)

> SD is implemented systematically in the entire curriculum, in accordance with the graduate profile. (AISHE 1.0, criterion 4.1: *Curriculum*, stage 3)

> All environmentally related topics are part of an integrated environmental management system (EMS). This EMS is fully functional within all parts of the organization. The environmental reporting is an integrated part of the annual reporting of the organization. (AISHE 2.0, criterion O4: *Ecological sustainability*: stage 3)

Details about the application of AISHE are described below.

Avans University hosts 27,000 students, over 100 study programs, and 2200 employees. Its ambitions regarding sustainability can be summarized in two statements:

Avans post-graduates contribute actively to sustainable development by combining entrepreneurial spirit with sustainable awareness and engagement.

Supported by its knowledge of and engagement with society, Avans participates in solving major societal issues.

(continued)

In order to realize this, the university board has formulated its vision, mission, and goals regarding ESD in a vision document. The main targets are also agreed with the Dutch Ministry of Education as a legal contract.

In 2012 Avans took off to meet its ambitions by installing several multidisciplinary groups of lecturers to enhance and sustain the process.

One group received targets to develop educational materials for SD to be used in all study programs: basic materials to be used as an introduction to SD in the first year and building blocks covering specific SD issues, e.g., C2C, circular economy, sustainable finance, bio-based energy, and scenario thinking.

Another group was trained in using assessment instruments (described below), such as RESFIA+D for the educational goals, the C-scan for the curriculum contents, and AISHE for the overall ESD strategy. In these assessments all stakeholders of Avans are represented: students, lecturers, management, and the professional field.

In 2013, the various initiatives are integrated into an all-encompassing program to implement SD, both in the curricula and the organization. The aim is to evoke awareness and build commitment for SD, with clear goals and quality indicators for competences of staff, students, and organization.

If all ambitions prove to be successful, around 2018 Avans will have realized SISD in all aspects: its education, operations, research, community outreach, and – last but not least – its identity.

2 The Learning Organization

The organization development of a university toward SISD can hardly be described as a project or a program, as the final goals cannot be described and planned conclusively right from the beginning. A better qualification is an explorative journey, or an adventure, in which the goals and the strategy are redefined continuously, in an iterative process. Therefore, in order to realize SISD, an institution has to behave like a *learning organization* in which the SISD development process is performed as action research, with the institution itself as its object of study. This implies

> … a participatory, democratic process concerned with developing practical knowing in the pursuit of worthwhile human purposes, grounded in a participatory worldview which we believe is emerging at this historical moment. It seeks to bring together action and reflection, theory and practice, in participation with others, in the pursuit of practical solutions to issues of pressing concern to people, and more generally the flourishing of individual persons and their communities. (Reason and Bradbury 2001)

The complexity of the SD concept and of an educational institution, as well as the fact that ESD is not value-free but has the ambition to achieve societal improvements, requires a new scientific paradigm for this kind of action research. One such paradigm is “transdisciplinary science” (Roorda and Rachelson 2016).

> The birth of science is based on a strict dissociation of scientific knowledge from the various aspects of practical knowledge. The ideal of scientific knowledge as it was shaped in antiquity is still influential today, although the conception of science and the relationship between science and the life-world has undergone major changes. (…) Transdisciplinary research is challenged by the following requirements:
>
> - To grasp the complexity of the problems,
> - to take into account the diversity of scientific and societal views of the problems,
> - to link abstract and case specific knowledge, and
> - to constitute knowledge with a focus on problem-solving for what is perceived to be the common good.
> - (Hirsch Hadorn et al. 2008)

Other related scientific approaches are "mode-2 science" (Gibbons et al. 1994), case-based research (Yin 2009), and post-normal science (Funtowicz and Ravetz 1993; Martens 2006). Together, they form a "cluster concept" for a new, society-oriented scientific paradigm (Roorda 2010, pp. 22–23).

3 Focus on Education: The Tree Model

Of the four roles of a university toward ESD, shown in Fig. 28.1, education is by far the most important. This is due to the fact that, through its education, a university creates a strong multiplier effect: if, over the years, hundreds of thousands of its graduates possess the knowledge, insights, skills, and attitudes to involve sustainability aspects into their profession, the effect on society is immense.

Question
Consider your own university, or a university that you are familiar with. Which of the four roles gets the highest attention there? Why do you think this was decided? Is this all right, in your opinion, or would you shift focuses if you could?

The main educational aspects of ESD can be understood with the "Tree Model"; see Fig. 28.2. An overview of the elements of this model is shown in Table 28.1.

3.1 The Roots: The Sustainably Competent Professional

The roots of the tree symbolize the "roots" of a study program, i.e., the educational goals, answering the question: "What kind of professionals do we want to deliver to society?" In other words, what exactly defines a "sustainably competent professional"?

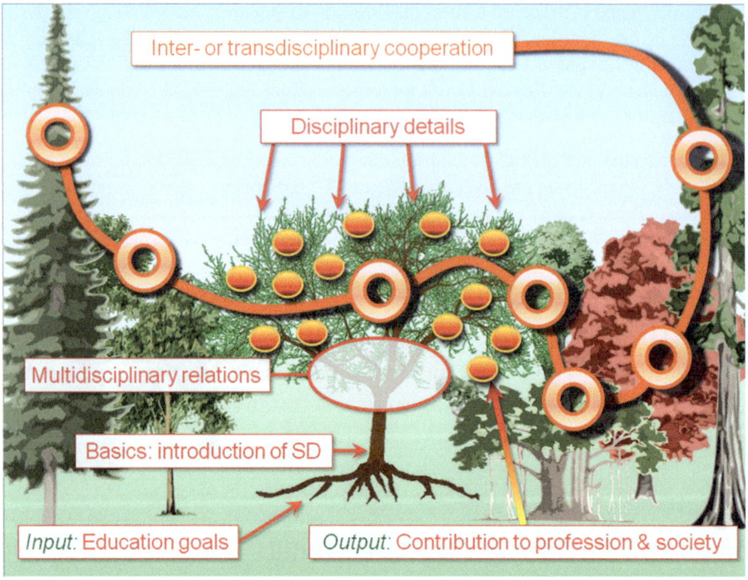

Fig. 28.2 The tree model

Table 28.1 The tree model: defining the sustainability strategy

Tree element	Topic
The *genotype*	The university mission and identity
The *roots*	The graduate profile, i.e., the education goals, e.g., the competence profile
The *trunk*	The basics: what *every* student should learn
The *branches*	The disciplinary details of SD in the curriculum
The *biochemistry*	Didactics: methodologies for the learning process
The *ecosystem*	Inter- and transdisciplinary cooperation
Sprouting and growing	ESD strategy and assessment
Reaching maturity	System Integration of Sustainable Development (SISD)
The *fruits*	Sustainably competent professionals

For this purpose, an assessment tool was developed called RESFIA+D (Roorda 2012). The model consists of seven SD competences, shown in Table 28.2: six generic competences – each subdivided into three more detailed competences – plus an unlimited set of more specific, discipline-dependent competences. The RESFIA+D model is a synthesis of a wide range of earlier sets of competences for sustainable development as described in Roorda (2016) An ample series of case studies (in the Netherlands and Belgium) further defining and illustrating the RESFIA+D model is available in Roorda (2015) another series of such case studies (in the USA and Canada) is in Roorda and Rachelson (2016).

To each of these competences, a set of four competence levels was added: varying from "apply" (implying that a student or a professional is able to do what he/she

Table 28.2 RESFIA+D: Professional competences for sustainable development

Competence R: responsibility	*Competence E: emotional intelligence*
A sustainably competent professional bears responsibility for his or her own work	A sustainably competent professional empathizes with the values and emotions of others
That is, the sustainable professional can…	*That is, the sustainable professional can…*
R1. Create a stakeholder analysis on the basis of the consequence scope and the consequence period	E1. Recognize and respect his or her own values and those of other people and cultures
R2. Take personal responsibility	E2. Distinguish between facts, assumptions, and opinions
R3. Be held personally accountable with respect to society (transparency)	E3. Cooperate on an interdisciplinary and transdisciplinary basis
Competence S: system orientation	*Competence F: future orientation*
A sustainably competent professional thinks and acts from a systemic perspective	A sustainably competent professional works and thinks on the basis of a perspective of the future
That is, the sustainable professional can…	*That is, the sustainable professional can…*
S1. Think from systems: flexibly zoom in and out on issues, i.e., thinking analytically and holistically in turn	F1. Think on different time scales – flexibly zoom in and out on short- and long-term approaches
S2. Recognize flaws in the fabric and sources of vigor in systems; have the ability to use the sources of vigor	F2. Recognize and utilize nonlinear processes
S3. Think integrally and chain oriented	F3. Think innovatively, creatively, and out of the box
Competence I: personal involvement	*Competence A: action skills*
A sustainably competent professional has a personal involvement in sustainable development	A sustainably competent professional is decisive and capable of acting
That is, the sustainable professional can…	*That is, the sustainable professional can…*
I1. Consistently involve sustainable development in the own work as a professional (sustainable attitude)	A1. Weigh up the unweighable and make decisions
I2. Passionately work toward dreams and ideals	A2. Deal with uncertainties
I3. Employ his or her conscience as the ultimate yardstick	A3. Act when the time is right, and not go against the current: "action without action"
Plus: *Disciplinary competences* for sustainable development (differing for each course, discipline, or profession)	

has learned – nothing less, nothing more) to "innovate." In RESFIA+D, a set of "competences cards" was created, of which an example is shown in Table 28.3.

When RESFIA+D is used as an assessment tool for an individual study program, a heterogeneous group is composed, in which the professional field, the education management, the teaching staff, and the students are represented. During a group meeting, for each of the 6 × 3 competences, they answer three questions:

Table 28.3 An example of a competence card

Competence F: future orientation. *a sustainably competent professional works and thinks on the basis of a perspective of the future*			
F1. Think on different time scales – flexibly zoom in and out on short- and long-term approaches			
Apply	*Integrate*	*Improve*	*Innovate*
In concrete working situations, you recognize and describe operational methods for the performance and improvement of your work	In the case of concrete work-related problems, you recognize and describe the differences between short-term methods aiming at reducing the symptoms and long-term methods aiming at eliminating causes	In the case of work-related problems, you contribute to the design of a solution strategy based on a carefully selected combination of short- and long-term methods	You contribute to the (re)definition and the application of the mission and of the strategic policy of the organization you belong to
You contribute to the application of these methods, and thus contribute to short-term improvements	You contribute to the application of symptom reducing methods based on the operational policy of the organization or team you belong to	You contribute to the design of symptom reducing methods based on the tactical policy of the organization or team you belong to	You involve present and expected future trends in your working field and in society

1. What – in your opinion – should the *minimum competence level* be for each student at the moment of graduation?
2. Which level is presently described or demanded in the *current competence profile* of the study program?
3. Which level is actually reached in the *current curriculum*?

In all cases in which the model was used, significant differences were found between the answers to these three questions. As the conclusions are based on consensus, the support for the resulting desire for improvement is strong, setting a development task for the next 1-2 years.

Question

RESFIA+D can also be used to define the competence level of individual professionals and students. Please, download the RESFIA+D documents (see the end of this chapter). Score your own competence levels. Based on the outcomes, define a personal development plan to reach higher levels according to your own ambitions.

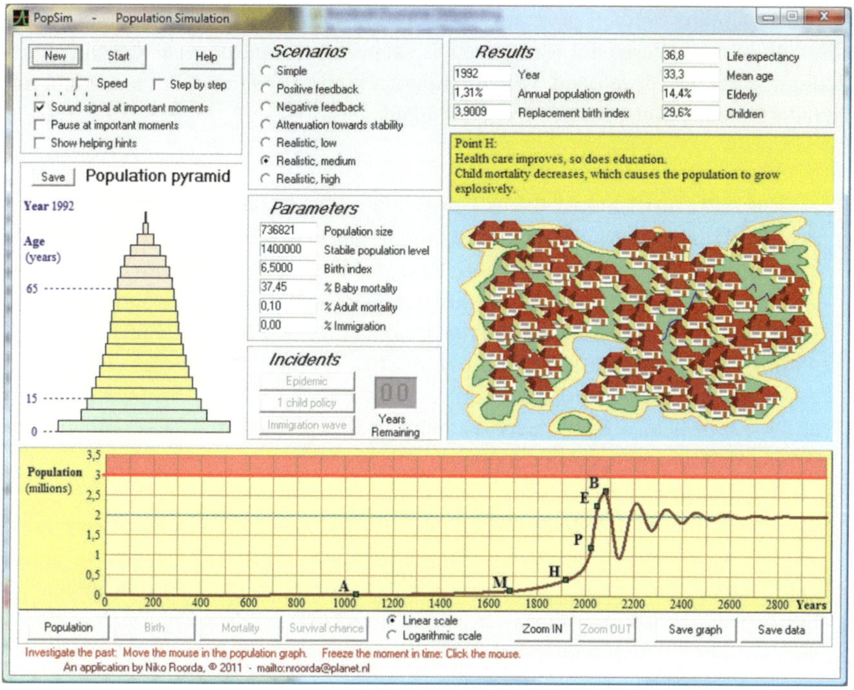

Fig. 28.3 A serious game: a simulation of population growth

3.2 *The Trunk: Fundamentals of Sustainable Development*

If a university wishes to educate all of its students in SD, it is essential that both the lecturers and the students speak the same language. So a general introduction to SD is needed, preferably in the first year of all study programs.

Several such introductions exist. Examples are Rogers et al. (2008) and Blewitt (2008). They offer many details and are mainly appropriate for students who want to become real SD experts.

For a more general introduction, to be used in every academic discipline at an undergraduate level, Roorda (2012) is suitable. The book comes with a website containing a lot of extra materials, e.g., hundreds of exercises, video clips, informational spreadsheets, serious games (Fig. 28.3), overviews of learning goals per chapter, etc.

3.3 *The Branches: SD in the Curriculum*

Apart from an SD introduction early in the curriculum, SD should ideally not be treated in separate education modules newly inserted throughout the curriculum. Rather, it should be integrated as a range of aspects and topics into existing modules, in such a way that the complexity and the multidisciplinarity increase in the course of the study program.

Most, perhaps even all programs of every university, do contain such elements, although in many cases not recognized as such. For this purpose, a "C-scan" (curriculum scan) was developed, which renders a kind of "SD map" of a curriculum, including focus points, connections, and blank spots.

The study programs of Avans University all have started the implementation of SD into their curricula, based on the outcomes of assessments. Some program teams have decided to start "bottom-up" with the C-scan, which gave them an overview of the present SD elements within the curriculum and a series of recommendations for improvement. Other program teams started with the RESFIA+D assessment, because they wanted first to have clarification of the educational goals. Yet other teams started top-down, using AISHE, an assessment tool which enables them to define an overall ESD strategy before making any operational plans.

Which approach is the better? This appears to depend on the specific structure and culture of the university department. Whenever the approach is selected that – according to the team – suits best in their organization, the approach is effective, whether it starts with the curriculum details, the educational goals, or the overall strategy. In other words: *the best approach is the one that those who are going to do the work believe in.*

Question
Please think of a university program that you are familiar with. Which approach would you recommend? Why?

4 Raising the Expertise

A crucial factor in any ESD strategy is the required level of expertise of the teaching staff, who are not only to instruct and guide the students but also to develop the education.

Avans University has decided to use a staff development plan on SD consisting of three circles (see Fig. 28.4). The first or "inner" circle will in some years reach the level of genuine experts on ESD (broad but not deep expertise); they will be the task force that "teaches the teachers." The second circle consists of those who possess or acquire expertise in various specific SD topics (deep but not necessarily broad expertise). The remainder will, in some years, have at least basic knowledge about SD and about its relations to their discipline and study program.

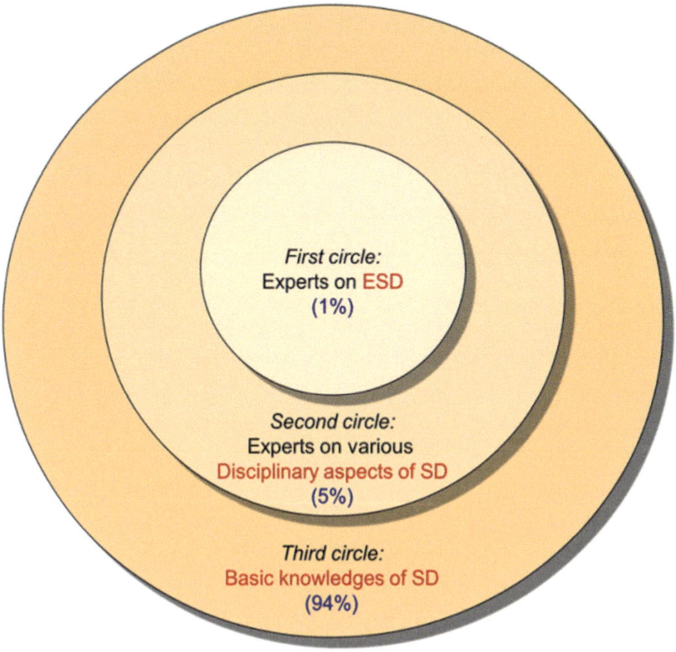

Fig. 28.4 The three circles of SD expertise

5 System Integration: Assessment and Quality Management

In order to assess the results of ESD projects in universities, several instruments have been developed, such as the STARS system in the USA, the LiFE system in the UK, the Plan Vert (Green Plan) in France, and AISHE (2nd version) by an international European group. AISHE, as already mentioned, makes use of a series of levels of organizational change, e.g., "activity oriented" and "system oriented." The latter level defines SISD.

Question
Please think of a university program that you are familiar with –not necessarily your own, rather one that does not focus specifically on SD. Download the AISHE 2.0 document (see the end of this chapter). Imagine that AISHE was to be applied to this program. What effects do you expect it would have?

Most of these assessment instruments are derived from or inspired by well-known models for quality management, such as ISO or EFQM (European Foundation for Quality Management, see: EFQM 2009). The application of such instruments enables a university to integrate ESD into the general quality management and thus to enter a cycle of continuous improvement: plan–do–check–act, the famous "Deming Cycle" (Deming 1986). This is an essential step, without which SISD will never be fully realized.

Question

Could you imagine the concept of "SISD" to be applied to organizations that are very different from universities, e.g., industrial companies, banks, and governmental departments? Do you know any such organization that is near or even has arrived at SISD?

Recommendations and References

Recommended Books

Corcoran PB, Wals AEJ (eds) (2004) Higher education and the challenge of sustainability. Kluwer, Dordrecht

de Kraker J, Lansu A, van Dam-Mieras R (2007) Crossing boundaries. Innovative learning for sustainable development in higher education. Verlag für Akademische Schriften, Frankfurt am Main

Roorda N (2010) Sailing on the winds of change. The Odyssey to sustainability of the universities of applied sciences in the Netherlands. PhD dissertation, Maastricht University. To be retrieved from https://www.box.net/shared/nz75typdk5

Roorda N (2012) Fundamentals of sustainable development. Routledge, London/New York. Online accessories: to be retrieved from http://www.routledge.com/cw/roorda-9781849713863

Journals (Fully or Partly) Dedicated to ESD

International Journal of Sustainability in Higher Education, ISSN 1467–6370: www.emeraldinsight.com/products/journals/journals.htm?id=ijshe

Journal of Cleaner Production, ISSN 0959–6526: www.journals.elsevier.com/journal-of-cleaner-production

Journal of Education for Sustainable Development, online ISSN 0973–4074; print ISSN 0973–4082: http://jsd.sagepub.com

More Weblinks

AISHE 2.0 manual: to be retrieved from https://www.box.net/s/0dglhugzyyzta4kkfb83

Avans University: www.avans.nl/international
RESFIA+D model: to be retrieved from https://www.box.com/s/04xy2xss5mpz5i0vk13u (the model) and https://www.box.com/s/c3amfguepxh0t0l0oav7 (the tables)
The Platform for Sustainability Performance in Education: www.eauc.org.uk/theplatform. ESD assessment tools: www.eauc.org.uk/theplatform/the_tools. This page contains not just the internationally available tools, such as AISHE, LiFE and STARS, but also links to (inter) national ESD organizations, such as AASHE and EAUC

Other References

Blewitt J (2008) Understanding sustainable development. Earthscan, London
Clugston RM, Calder W (2000) Chapter 2: critical dimensions of sustainability in higher education. In: Leal Filho W (ed) Sustainability and university life. Peter Lang, Frankfurt am Main
Deming WE (1986) Out of the crisis. MIT Press, Cambridge
EFQM (2009) EFQM excellence model 2010. European Foundation for Quality Management, Brussels
Funtowicz S, Ravetz J (1993) Science for the post-normal age. Futures 25(7):739–755
Gibbons M, Limoges C, Nowotny H, Schwartzman S, Scott P, Trow M (1994) The new production of knowledge: the dynamics of science and research in contemporary societies. Sage, Newbury Park
Hirsch Hadorn G, Biber-Klemm S, Grossenbacher-Mansuy W, Hoffmann-Riem H, Joye D, Pohl C, Wiesmann U, Zemp E (2008) The emergence of transdisciplinarity as a form of research. In: Hirsch Hadorn G, Hoffmann-Riem H, Biber-Klemm S, Grossenbacher-Mansuy W, Joye D, Pohl C, Wiesmann U, Zemp E (eds) Handbook of transdisciplinary research. Springer, Dordrecht
Martens P (2006) Sustainability: science or fiction? Sustain Sci Pract Policy 2(1, Spring):36–41
Megerle A, Megerle H (2000) University support to local and regional agenda initiatives for sustainable development. In: Leal Filho W (ed) Sustainability and university life. Peter Lang, Frankfurt
Reason P, Bradbury H (eds) (2001) Handbook of action research: participative inquiry and practice. Sage, London
Rogers PP, Jalal KF, Boyd JA (2008) An introduction to sustainable development. Earthscan, London
Roorda N (2001) AISHE – Assessment Instrument for Sustainability in Higher Education. Stichting Duurzaam Hoger Onderwijs (DHO), Amsterdam
Roorda N (2015) De zeven competencies van de duurzame professional. Garant Uitgeverji, Antwerpen
Roorda N (2016) The seven competencies of sustainable professional. The RESFIA-D model for HRM, education and training. In: Machado C, Devim P (eds) Management for sustainable development. River Publisher, Aarlborg
Roorda N, Rachelson A (2016) The seven competencies of sustainable professional. Wageningen Academic Publishers, Wageningen
Roorda N, Rammel C, Waara S, Fra Paleo U (2009) AISHE 2.0 manual: Assessment Instrument for Sustainability in Higher Education, Edition 2.0
Scholz RW, Lang DJ, Wiek A, Walter AI, Stauffacher M (2006) Transdisciplinary case studies as a means of sustainability learning. Int J Sustain High Educ 7(3):226–251
Sterling S (2004) Higher education, sustainability and the role of systemic learning. In: Corcoran PB, Wals AEJ (eds) Higher education and the challenge of sustainability. Kluwer, Dordrecht, pp 47–70
UNCED (1992) Agenda 21. United Nations Conference on Environment and Development, Brazil
Velazquez L, Munguia N, Platt A, Taddei J (2006) Sustainable university: what can be the matter? J Clean Prod 14:810–819

Chapter 29
Problem-Based and Project-Based Learning for Sustainable Development

Ron Cörvers, Arnim Wiek, Joop de Kraker, Daniel J. Lang, and Pim Martens

Abstract Universities hold a crucial responsibility and role to contribute to sustainable development, also in their education task. The concept of competencies for sustainable development and the idea of using real-world sustainability issues in education are promising approaches to transform sustainability programmes at universities into student-centred learning environments. Especially the educational formats of problem-based learning and project-based learning foster such a process of educational innovation towards student-centred learning. Moreover, hybrid forms of problem-based and project-based learning offer added value, but challenges for PPBL courses in sustainability remain salient.

Keywords Competencies • Real-world problems • Student-centred learning • Problem-based and project-based learning

---

R. Cörvers (✉)
International Centre for Integrated assessment and Sustainable development, Maastricht University, Kapoenstraat 2, P.O. Box 616, 6200 MD Maastricht, The Netherlands
e-mail: r.corvers@maastrichtuniversity.nl

A. Wiek
School of Sustainability, Arizona State University,
PO Box 875502, Tempe, AZ 85287-5502, USA
e-mail: arnim.wiek@asu.edu

J. de Kraker
International Centre for Integrated assessment and Sustainable development, Maastricht University, Kapoenstraat 2, P.O. Box 616, 6200 MD Maastricht, The Netherlands

Management, Science and Technology, Open Universiteit,
Valkenburgerweg 177, 2960, 6401 DL Heerlen, The Netherlands
e-mail: j.dekraker@maastrichtuniversity.nl

D.J. Lang
Institute for Ethics and Transdisciplinary Sustainability Research, Leuphana Universität Lüneburg, Scharnhorststrasse 1, 21335 Lüneburg, Germany
e-mail: daniel.lang@leuphana.de

P. Martens
Maastricht University, Maastricht, The Netherlands
e-mail: p.martens@icis.unimaas.nl

H. Heinrichs et al. (eds.), *Sustainability Science*, DOI 10.1007/978-94-017-7242-6_29

1 Introduction

Given the educational and research capacity, the external partnerships, and the position of higher education as an influential voice in society, there is ample opportunity for universities to help shift societal norms, practices, products, and services towards sustainability (Cortese 2003; Rowe 2007). Universities therefore hold a crucial responsibility and role to contribute to sustainable development within their scope of Community engagement, Operations, Research, and Education; the so-called CORE activities (Jenssen 2012). However, in practice only few universities excel in meeting this challenge as most universities lack a systematic approach to implement sustainability in their CORE activities (Ferrer-Balas et al. 2008; Jenssen 2012; Lang and Wiek 2012). Right now, sustainability is treated by many as an add-on, as another item on an already full plate.

In this chapter we focus on the *education* task of universities. In particular, we discuss the concept of competencies for sustainable development and the idea of using real-world sustainability issues to transform sustainability study programmes at universities into student-centred learning environments. The educational formats of problem-based learning and project-based learning foster such a process of educational innovation by viewing learning as a student-centred, experiential, interactive, situated, and social process, instead of a unilateral process in which the teacher imparts knowledge on the students.

2 Competencies in Sustainability

Given the challenges of sustainable development and the need for policy and behavioural change, Rowe (2007) and others argue that universities need to reconsider the competencies students are expected to acquire. The emphasis needs to shift from descriptive-analytical knowledge, logical reasoning, and critical thinking alone to the inclusion of normative competence and effective change-agent skills. This implies a re-emphasis on the role of universities in not only educating academics and professionals, but in educating responsible citizens and decision makers. And even the profile of academics needs to get revisited and revised. For moving forward to a sustainable world, Martens and Rotmans (2012) argue that it is time for many (more) scientists to become 'scientivists' (scientists-activists): individuals and groups that are engaged in systematic knowledge acquisition and generation (the scientist part), as well as in promoting and directing societal change (the activist part). See Box 29.1 on the roles of sustainability experts (academics and professionals).

- *Task*: Give pros and cons for the statement that sustainability scientists should be 'scientivists'. In your answer you should also reflect on the variety of roles sustainability experts can fulfil in modern society amidst other actors such as representatives from the government, business community, non-governmental organisations, and citizens.

Box 29.1: Roles of Sustainability Experts
Sustainability experts (academics and professionals) work at the interface between science, policy, and society when helping to solve problems of sustainable development. However, the roles they perform differ depending on the nature of the problem and the type of knowledge needed to solve it, as well as on the institutional setting of their work and their personal and professional values. The latter implies that, to some extent, it is up to the individual professional to choose a role. Despite the variety of roles, certain patterns can be distinguished. Here we present five different roles of sustainability experts:

1. The pure scientist, who restricts himself to explaining the state of affairs. The pure scientist delivers the facts, but does not interpret them in the light of a policy question.
2. The arbiter, who tries to provide the best available knowledge to answer policy questions, with the aim to inform, not to advise. Arbiters typically communicate their expertise in the form of 'if-then' statements.
3. The advocate, who uses knowledge to argue for a specific course of action. The advocate goes a step further than the arbiter and adopts a normative position.
4. The broker, who attempts to open up decision-making processes by exploring multiple perspectives and alternatives and by integrating stakeholders' concerns with available knowledge. However, the broker takes the primacy of scientific knowledge over other types of knowledge for granted and maintains a distinction between knowledge and action.
5. The participatory expert, who engages in participatory knowledge production in which the distinction between knowledge and action and between scientific and lay knowledge is blurred.

In principle, each of these roles can contribute effectively to addressing sustainability issues, depending on the factors mentioned above.
Source: Broekhans and Turnhout (2012) and Pielke (2007)

The majority of graduates will not work as academics, but will be employed in government, business, or civil society and will have to deal with diverse challenges in pursuit of sustainable development. If universities want to prepare students to contribute to sustainable development in their working life, the key question is: What competencies are critical to acquire for the upcoming sustainability professionals?

There is a growing body of literature on *competencies* for sustainable development, but most proposals are lists of isolated competencies, instead of integrated frameworks (Wiek et al. 2011a). To overcome this gap, Wiek et al. (2011a) reviewed the literature and synthesised a framework of key competences in sustainability (Fig. 29.1). Competence is defined as 'a functionally linked complex of knowledge,

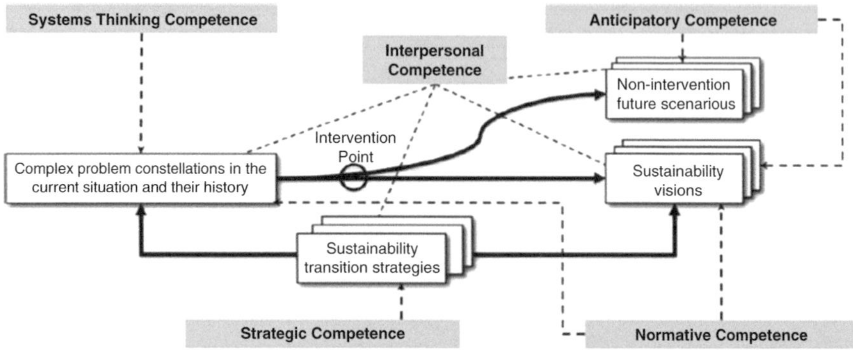

Fig. 29.1 The interplay of five key competencies for sustainability research and problem-solving (Wiek et al. 2011a)

skills, and attitudes that enable successful task performance and problem solving' (p. 204). The overarching sustainability competence is being able *to mitigate and solve sustainability problems* (*through transformational research or professional practices*). Five key competencies are required to get integrated:

- *Systems-thinking competence*: being able to understand structure and dynamics of complex system
- *Anticipatory competence*: being able to anticipate possible and/or sustainable futures
- *Normative competence*: being able to differentiate, justify, and apply values and goals for sustainability
- *Strategic competence*: being able to create transition and intervention strategies to enact change
- *Interpersonal competence*: having communicative and collaborative skills

The interplay of the five key competencies in sustainability enables graduates and professionals to mitigate and solve sustainability problems (in research and professional practices).

Other studies largely align or support the concept proposed by Wiek et al. (2011a). The International Society of Sustainability Professionals (ISSP) surveyed about 400 sustainability professionals (mainly sustainability consultants and managers working for or in corporations) to find out what they believe are important competencies for their jobs (Willard et al. 2010). The results indicate that sustainability professionals mainly promote the value of sustainability concepts and deal with climate change and energy issues. Top skills were considered to have good skills for communicating with internal and external stakeholders and to be able to inspire and motivate others (interpersonal competence). Additional skills considered important are strategic planning, systems thinking, and project management. A study by the Institute of Environmental Management & Assessment (IEMA) suggests that *leadership (for change)* is a key competence for working in environmental or sustainability professions (IEMA 2011).

In summary, we see emerging agreement on key competencies of sustainability that bridges the often existing gap between sustainability education and professional practice. A key factor is that mitigating and solving complex sustainability problems becomes the common goal across the community of sustainability academics, students, and professionals.

- *Task*: In the discussion on competencies for sustainable development, interpersonal competence plays an important role (e.g. leadership, good communication, strong teamwork, effective networking). To what extent do you think that this competence can be acquired through formal training, and how much is 'natural talent'? Justify and illustrate your answer.

3 Problems and Projects: Learning through Real-World Sustainability Issues

As indicated in the previous section, the complexity of sustainability challenges asks for experts (academics and professionals) who are able to 'cross borders': to work with colleagues from different backgrounds, to judge the value of different types of knowledge, to participate in multi-stakeholder processes, to analyse a problem from a systemic perspective, to envision sustainable future states, and to generate evidence-based solution options (De Kraker et al. 2007; Wiek et al. 2011b). The development of such competencies requires learning environments that combine actual practice ('learning by doing') and explicit reflection on what and how to learn from that practice ('learning by reflection').

To meet these challenges De Kraker et al. (2007), Rowe (2007), Brundiers et al. (2010), and others propose to address *real-world sustainability issues* in different educational settings. In this way, it might be possible to develop and renew sustainability education at universities in a fundamental way, 'from the outside in'.

Recent literature suggests that a *student-centred learning* strategy is a powerful setting to building students' competencies and that it should be based on four learning principles (Dolmans and Schmidt 2010):

- *Constructive learning*: students should learn constructing their knowledge base by connecting new information with existing knowledge (e.g. through discussion).
- *Collaborative learning*: students should learn collaborating with each other in order to maximise learning effects through peer-to-peer teaching (sharing knowledge, challenging, negotiating, etc. in small-group work).
- *Contextual learning*: students should learn to consider relevant context of cases and problems in order to be able to transfer and apply insights and knowledge to different cases (e.g. through real-world problems).

– *Self-directed learning*: students should learn to regulate their learning by playing an active role in planning, monitoring, and evaluating their learning process (e.g. through elaborations on subject matters of own interest).

The educational models of *problem-based* and *project-based learning* incorporate these four learning principles and have been further developed in direct response to the calls for innovation and transformation in sustainability education. Accordingly, sustainability programmes around the world have begun to offer problem-based and project-based learning courses (Lang and Wiek 2012; Brundiers and Wiek 2011; Brundiers et al. 2013; Wiek et al. 2014). In these settings, student learning shifts from passive (course instructors deliver, students receive) to active (students deliver, course instructors provide feedback), and students work on real-world problems by engaging in small-group work (ideally in interdisciplinary teams) and often collaborating with stakeholders on developing solution options to the identified problems (Brundiers and Wiek 2013). Both educational models have many features in common, but display also some differences (Fig. 29.2).

- *Task*: What are the key differences between the educational model of problem-based learning and project-based learning? Justify and illustrate your answer.

Brundiers and Wiek (2013) argue that in education for sustainability hybrid forms of problem-based and project-based learning – *PPBL courses* – offer added value. First, a hybrid PPBL course adopts the problem inquiry as in problem-based learning and, in order to develop solution options, the product orientation from project-based learning. Second, a hybrid PPBL course expands the engagement structure of problem-based learning, wherein stakeholders are mostly not actively

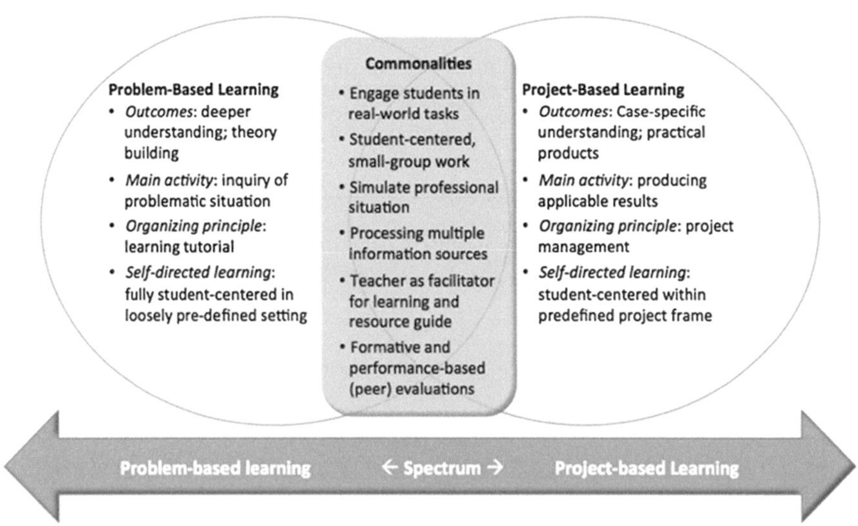

Fig. 29.2 Commonalities and differences between problem-based learning and project-based learning (Brundiers and Wiek 2013)

involved, to a collaborative process of knowledge generation and critical reflection with stakeholders, which is much more common in project-based learning (Brundiers and Wiek 2013). See Box 29.2 for two examples of problem-based learning and project-based learning courses.

Box 29.2

European Virtual Seminar on Sustainable Development

In this course, students from 12 universities in Europe work together in international, multidisciplinary teams. The learning objectives are (1) that students gain an understanding of the concept of sustainable development and apply it to a case study in a European context and (2) that students learn to collaborate with students from other disciplines, countries, and cultures, using internet technology. The topics of the case studies range from sustainable urban waste management to sustainable regional tourism and sustainability communication. The ultimate goal of the case study is to come up with evidence-based recommendations on the topic. The challenge is to fully utilise the international and multidisciplinary diversity in the group in terms of different perspectives and expertise among the group members. The case study groups are responsible for their learning process: keeping the learning and research process going and delivering integrated group products on time. Each group is coached by a tutor, who oversees the process, and an expert, with in-depth knowledge of the case study. Students work in 'virtual teams' – a recent trend in global sustainability education (Wiek et al. 2013). The majority of team members do not meet each other in person. Yet, supported by Web-based communication tools, such as discussion forums, chat, and tools for joint document writing, most case study groups develop into well-functioning teams. Such learning environments offer a diversity of interacting perspectives that is often difficult to achieve in face-to-face settings (Cörvers and De Kraker 2009).

Sustainability Assessment Project at Maastricht University

Problem-based learning has been at the core of all study programmes at Maastricht University since it was founded in 1974. Also the Master of Science programme in Sustainability Science and Policy (which started in 2011) fully embraces this educational model, while its Sustainability Assessment Project is a hybrid form of problem-based and project-based learning. In this course students work together in small groups on real-world sustainability problems (case studies) commissioned by different stakeholders such as local and regional governments, small and medium-sized enterprises, multinational corporations, research institutes, and non-governmental organisations. The course builds on previous courses in the Master of Science programme and acquired skills are applied to the case. Research is carried out by student project teams that are coached and assessed by faculty members, the external client, and, if appropriate, other stakeholders. The project results,

(continued)

Box 29.2 (continued)
including policy recommendations, are presented live and in project reports. Examples of sustainability problems examined in 2013 are sustainable car use and parking in Maastricht (commissioned by Q-Park, an international parking organisation), a strategy for an energy-neutral Maastricht in 2030 (commissioned by the city of Maastricht), waste management at Maastricht University (commissioned by Green Office, a student-driven department of the university), conflict mineral-free products at Philips (commissioned by the sustainability department of Philips), and sustainable management of the Rijn river (commissioned by Deltares, an independent research institute).

From a comparative study of six PPBL courses in sustainability programmes at universities in Europe, North America, and Australia, Brundiers and Wiek (2013) conclude that three challenges for PPBL courses in sustainability remain salient:

- Define learning objectives that directly and comprehensively aim at the acquisition of sustainability competencies.
- Design strong transacademic settings that allow meaningful and continuous collaboration between students and stakeholders.
- Fully account for key principles such as self-directed learning, advanced team working, and solution-oriented research with real-world impacts (Brundiers and Wiek 2013).

Problem-based and project-based learning both foster student-centred learning, and hybrid PPBL courses can combine the best of both worlds for acquiring competencies in sustainability. The idea of using real-world sustainability issues in university programmes is innovative and powerful for building competencies that are applicable in professional settings; yet, in practice, it is not always easy to comply with all guiding principles of problem-based and project-based learning.

- *Task*: Why is it difficult to design and implement problem-based and project-based learning that fully embrace all guiding principles discussed? Think about critical factors such as the prevailing educational model at the university, the diversity in prior knowledge of students, a potential tension between theory and practice in the offered content, and the role of course instructors, students, and stakeholders in the learning process. Justify and illustrate your answer.

4 Conclusions

We summarised competencies in sustainability and offered a brief insight into real-world learning settings in sustainability programmes at universities in Europe, North America, and Australia. The educational models of problem-based and project-based learning, and their hybrid forms, can support universities in changing towards student-centred learning and realising a strong sustainability mandate to educate the next generation of sustainability experts (academics and professionals).

Three issues require further attention:

- An integrated framework of key competencies in sustainability finds more and more acceptance in academia and professional circles; it can act as a focal point for education and professional development. There is still more effort required to fully disseminate and incorporate this framework in academic programmes and professional settings (Wiek et al. 2011b).
- Sustainability professionals should be competent in the key competencies outlined above. There is still more structured and shared experience to be gained in order to fully understand how these competencies (individually and collectively) can be acquired through formal training and learning-on-the-job and for designing sustainability education accordingly.
- There is wide agreement that problem-based and project-based learning are powerful settings for sustainability education. Yet, there are still various institutional inertia and obstacles to fully incorporate these settings in sustainability programmes around the world; innovative coping strategies have been developed, but more efforts are needed to fully utilise this potential.

Further Reading

Roorda N (2010) Sailing on the winds of change. The Odyssey to sustainability of the universities of applied sciences in the Netherlands. PhD thesis Maastricht University, Universitaire Pers Maastricht

Wiek A, Withycombe L, Redman CL (2011) Key competencies in sustainability: a reference framework for academic program development. Sustain Sci 6:203–218

Weblinks

Centre for PBL & Sustainability. http://www.pbl-sus.plan.aau.dk/

European virtual seminar on sustainable development. http://www.openuniversiteit.nl/eCache/DEF/90/085.html

MSc Sustainability Science and Policy, Maastricht University. http://www.maastrichtuniversity.nl/masterssp

UNESCO programme Teaching and Learning for a Sustainable Future. http://www.unesco.org/education/tlsf/index.html

References

Broekhans B, Turnhout E (2012) Roles of environmental experts in practice. In: Tuinstra W et al (eds) Environmental problems: crossing boundaries between science, policy and society. Open University, Heerlen, pp 89–105

Brundiers K, Wiek A (2011) Educating students in real-world sustainability research – vision and implementation. Innov High Educ 36:107–124

Brundiers K, Wiek A (2013) Do we teach what we preach? An international comparison of problem- and project-based learning courses in sustainability. Sustainability 5:1725–1746

Brundiers K, Wiek A, Redman CL (2010) Real-world learning opportunities in sustainability – from classroom into the real world. Int J Sustain High Educ 11:308–324

Cortese AD (2003) The critical role of higher education in creating a sustainable future. Plan High Educ 31(3):15–22

Cörvers RJM, De Kraker J (2009) Virtual campus development on the basis of subsidiarity: the EVS approach. In: Stansfield M, Connoloy T (eds) Institutional transformation through best practices in virtual campus development: advancing e-learning policies, Information science reference. IGI Global, Hershey/New York, pp 179–197

De Kraker J, Lansu A, Van Dam-Mieras MC (2007) Competences and competence-based learning for sustainable development. In: de Kraker J, Lansu A, van Dam-Mieras MC (eds) Crossing boundaries. Innovative learning for sustainable development in higher education. Verlag für Akademische Schriften, Frankfurt am Main, pp 103–114

Dolmans D, Schmidt H (2010) The problem-based learning process. In: Van Berkel H, Scherpbier A, Hillen H, van der Vleuten C (eds) Lessons from problem-based learning. Oxford University Press, Oxford

Ferrer-Balas D, Adachi J, Banas S, Davidson CJ, Hoshikoshi A, Mishra A, Motodoa Y, Onga M, Ostwald M (2008) An international comparative analysis of sustainability transformation across seven universities. Int J Sustain Higher Educ 9(3):295–316

Institute of Environmental Management & Assessment (IEMA) (2011) Environmental skills map. IEMA, Lincoln. http://www.iema.net/skills

Jenssen S (2012) Sustainability at universities. An explorative research on assessment methods and tools for sustainability implementation at universities. Master thesis, Maastricht University

Lang DJ, Wiek A (2012) The role of universities in fostering urban and regional sustainability. In: Mieg HA, Töpfer K (eds) Institutional and social innovation for sustainable urban development. Earthscan, London, pp 393–411

Martens P, Rotmans J (2012) Scientivists urgently needed. Blog http://ideas4sustainability.wordpress.com/2012/09/20/scientivists-urgently-needed/

Pielke RA (2007) The honest broker. Making sense of science in policy and politics. Cambridge University Press, Cambridge

Rowe D (2007) Education for a sustainable future. Science 317(5836):323–324. doi:10.1126/science.1143552

Wiek A, Withycombe L, Redman CL (2011a) Key competencies in sustainability: a reference framework for academic program development. Sustain Sci 6:203–218

Wiek A, Withycombe L, Redman CL, Banas Mills S (2011b) Moving forward on competence in sustainability research and problem solving. Environ Sci Policy Sustain Dev 53(2):3–12

Wiek A, Bernstein M, Laubichler M, Caniglia G, Minteer B, Lang DJ (2013) A global classroom for international sustainability education. Creat Educ 4(4A):19–28

Wiek A, Xiong A, Brundiers K, van der Leeuw S (2014) Integrating problem- and project-based learning into sustainability programs – a case study on the School of Sustainability at Arizona State University. Int J Sustain High Educ 15(4):431–449

Willard M, Wiedmeyer C, Flint RW, Weedon JS, Woodward R, Feldman I, Edwards M (2010) The sustainability professional: 2010 competency survey report. Environ Qual Manage 20(1):49–73

Chapter 30
Science for Sustainability – A Societal and Political Perspective

Günther Bachmann

Abstract A changing world calls for advanced sustainability thinking. Recently, the notion of sustainability gains ever more momentum in the German entrepreneurial and political context. Science and the humanities can and should increase delivery against the Sustainable Development Goals in the post-2015 development agenda. But still, society needs broader and multiplied hubs for advanced sustainability thinking. Therefore, transformational research schemes must be part of the top agenda. Transformation must be made part of any institution's performance. Thus, twofold approach suggests fostering both "science for sustainability" and "sustainability in science." The German Sustainability Code and compatible schemes might be used as reference. More evidence based input into the ways and means societies use for choice editing, e.g., in consumption and production, but also in education and visionary thinking may prove as a major leverage to overcome mental path dependencies.

Keywords Carlowitz • Resource crisis • Sustainable Development Goals • Transformation

When Carl von Carlowitz first talked about the need for sustainable forest management 300 years ago, he followed an evidence-based approach. The mining business relied heavily on the limitless availability of wood for smelter facilities and other mining practices. Carlowitz observed a dramatic dwindling of forest resources. It was evident that resource depletion was driving societal prosperity and well-being to a brink. The environment set limits to growth. He came to the conclusion that, for the Saxon economy and society, the resilience and vulnerability of timber made it necessary to change the way of sourcing timber, and in general, of handling natural resources. The same happened in various places throughout Europe.

Then, instead of turning into a sustainable economy, history went another way. For hundreds of years, coal and oil, and finally also nuclear energy, made people

G. Bachmann (✉)
German Council for Sustainable Development, Potsdamer Platz 10, 10785 Berlin, Germany
e-mail: info@nachhaltigkeitsrat.de

H. Heinrichs et al. (eds.), *Sustainability Science*, DOI 10.1007/978-94-017-7242-6_30

forget about finite resources. That turned those preindustrial limits to growth into new frontiers for growth and what appeared (and still appears) to be an unlimited development. Politically, that marginalized the notion of sustainability and basically laid it to rest for a long time.

In our days, however, we are again experiencing crises, but this time on both a larger scale and systemic scope, globally and regionally. It is all about the quality and quantity of natural resources, and in some cases, the limits set by quality are outscoring those set by quantity. For example, the emissions of carbon dioxide are likely to be more restrictive than are the limits exposed by the remaining volume of known fossil resources. It is good to have the historical reference of Carlowitz. It shows to us that change in the direction of sustainability concepts is nothing out of touch, and reasonable thoughts have been presented on this matter in the past. Although, the materiality of today's challenges makes more advanced concepts mandatory. Increasingly, the case of sustainability is relevant for the agenda of science and research. Measuring and managing natural resources requires scientific input. Assessing impacts on social, economic, and ecological goods is getting prime importance and requires new methods. Furthermore, research is required to deliver solution options and even guide the way onto sustainability trajectories. Also, when it comes to action the institutions of science are actors in their own case. The careful use of energy and resources, the switch to renewable energy supplies, the dealing with ecosystem services, and the social dimension of sustainability are challenges the scientific institutions must be facing like any other organization or company. Thus, sustainability science is necessary in order to keep pace with societal (including economic) and political demands and to renew and strengthen the credibility and political acceptance. What the private sector calls the "license to operate" is increasingly relevant also for the social and political perspective of science.

This article expresses a practitioners' view on the societal and political perspective of sustainability science. That does not mean to underestimate the growing discourse on theory and methodological implications of transformational science. The importance of sustainability and the natural, social, and human sciences and engineering is currently underscored by the appointment of a UN Secretary-General's Scientific Advisory Board. The board is tasked with strengthening the connection between science and policy by giving advice to the United Nations on science, technology, and innovation for sustainable development.

Global policies require substantial scientific input. Knowledge is required in order to inform the deliberations on Sustainable Development Goals and the post-2015 development agenda. Stakeholders from all societal fields including the academic community are currently invited to input into the intergovernmental process on Sustainable Development Goals[1]. It can be assumed that sustainability science acquires more traction. At the same time the pressure to deliver increases. In Germany, however, the national Council for Sustainable Development, on request

[1] German Council for Sustainable Development (2015) Germany's sustainability architecture and the SDGs. Statement to Federal Minister Peter Altmaier, dated 2015, May 26. http://www.nachhaltigkeitsrat.de/uploads/media/20150526_German_RNE_recommendation_on_national_SDG_implementation.pdf

of the Government, has already presented a pre-SDG analyses on how the German national architecture on sustainability will cope with SDGs.

The term "science for sustainability" (or sustainability science) is new, as is the theory-driven thinking around the issues of sustainability and knowledge transfer. As the notion of sustainability addresses natural science and engineering as well as social sciences and the humanities, the English term "science" is used in an all-inclusive way (in the sense of the German "Wissenschaften"). The term sustainability science emerges in the context of politics and in the academic context, both interlinked and mutually enforcing. The political framing goes back to the report of the UN Commission lead by Gro Harlem Brundtland that prompted the "Rio Process" and the Earth Summit in 1992. A number of multilateral environmental agreements and a series of global conferences followed. Enforcing the notion of sustainability as a global, regional, and local way of tackling the challenges of growth, resource depletion, and climate (to name only a few) has been placed on the agenda. Although much has been achieved, more leaves to be done. There is progress in terms of awareness raising and capacity building. Referring to some specialized aspects, the international community even agreed on meaningful measures. But the overall picture is threatening. The changing Earth, the extent to which humankind interferes with the geosphere (expressed so nicely by the notion of the Anthropocene), and the global food disaster are more challenging than ever before. And the progression of sustainability concepts is slow. It confronts the actors with unprecedented challenges.

The academic framing evolved from major scientific programs on global change and its human dimension. Domestically in Germany, long-standing work profiles of environmental sciences and in particular on the connectedness of social–ecological research successfully laid out the ground for developing first principles and practices of sustainability science. This is work in progress.

1 The Normative Implication

Sustainability science is shorthand for the way science, social sciences, and the humanities address the issues that are placed right in front of us by the unsustainable way the development of humankind is currently headed. Those issues may be addressed by using all kinds of already established methods and procedures and to redesign and elaborate new ones. What proved to be successful in the past must not be dismissed or sidelined. The complex dynamics of the mentioned problems, however, demand an in-depth rethinking of programming and procedures. What delivered in the past does not automatically deliver in the future. This is the normative imperative to sustainability science.

It is safe to assume that any one-size-fits-all approach to sustainability science will most probably fail. It will most likely add to the problems instead of being part of the solution. Sustainability science will rather be sketchy and bound to trial and error. Already now, we can build on practical experience and initiatives at universities and research facilities. Advanced universities and academic institutions create some sustainability science institutes or centers in order to provide a gateway to research, education, and activities in sustainability. Public research institutions begin to report back to the public about their ecological and social footprint. They access sustainability reporting schemes that are well known in the private sector. Frontrunner companies benefit from sustainability reporting since several years. First movers gain competitive advantages by applying the German Sustainability Code or compatible schemes.

All best practices granted, the overall state of the art is currently nowhere near to sufficient or satisfactory. It now even lives up to the possibilities. Students and researchers should feel encouraged to get themselves more engaged in shaping a meaningful relationship between research and sustainability. There is the chance that more emphasis on sustainability and research may also add value to career tracks.

2 Freedom of Research

In general, we find the term sustainability used fairly often in the public and private sector, as well as in academic discourses. This is both an achievement and a challenge. In any case it must be viewed critically. There is always a chance that the term is used for window-dressing purposes or that it is used in a shallow-brained way to catch up with the talk of the town. Used more seriously the term gives access to the benefits of out-of-the-box thinking. Building on disciplinary excellence, the term's rationality reaches out beyond the limits of disciplinary excellence. The first advanced sustainability institute in Germany, the Institute for Advanced Sustainability Studies, IASS, shows the essence of sustainability science: Transformation is not only described and analyzed scientifically, but transformation is made part of the Institutes' performance.

The notion of freedom of research might be seen as controversial to sustainability science. Some advocates of freedom of research may ward off the idea of science for sustainability because, to them, this would spoil pure (and even applied) science. A diversion of the scientific agenda is being feared when special interests use the normative imperative as a backdoor to the agenda setting and to tapping financial grants.

This raises important points, but nevertheless on the basis of a misconception. Freedom of research is a historic icon with lasting merits. It successfully liberated science and the humanities from religious doctrines and lobbyist influence and still does so. Whether the freedom of science was ever applied full scale may be questioned, though. However, the concept clearly has its merits. It should be defended. But it must not be held against the notion of sustainability science. I rather suggest a bridging link. Freedom is a moving target. Freedom is a social category that emerges and changes with the contingency of social development. In Gutenberg times the freedom (of information, of press) was different from the freedom axiom in times of Web-based big-data information. The same is true for science and the science–society interface.

Path dependency and captured science are keywords, as is the reflexivity in choice editing. Life cycle analysis and other complexities, e.g., trans-sectoral cooperation, add to the variety of what allegedly are state-of-the-art results. In modern times, the process of sorting out scientific "truth" is not exclusively an academic business. Significant aspects are being relocated from the inner field of the academic community to the midfield of the science–society interface. Of course, this is not a smooth process. Controversies and challenging issues are most likely to surface as the example of the climate change science demonstrates.

I suggest differentiating between two anchoring concepts for linking science and sustainability. "Science *for* sustainability" relates to the output scientific research delivers in terms of results, insights, long-term forecast, assessment of basic processes and impacts, and designing solutions. "Sustainability *in* science" relates to the input side. It addresses the working modality and the experiment design, e.g., the energy intensity of server farms and modeling, and the resource efficiency of experiments including livestock and animal welfare. It also includes a look at the social side, e.g., on working conditions in research facilities. Thus, the latter aspect addresses the housekeeping, with virtually no difference to the corporate social responsibility for which companies are held liable. Both aspects are essential ingredients for the necessary rethinking of the relationship between science and society. This is unfinished business – challenging and demanding.

3 Not on the Right Track…

I am surprised how conservative the public and scientific conversation on sustainability often is and how much time is being spent on the review (and defense) of the past and the present instead of time devoted to envisioning the future and rethinking ways and means of proceeding.

Precious time is being wasted by delaying possible change. Looking ahead towards the year 2050, all indicators on climate change; energy dependency, or biodiversity loss; and social equity and coherence show that societies and economies are on the wrong track, globally and that science is not playing its part in forging a sustainable future. This is a strong view, but it seems defendable.

The world population is expected to exceed nine billion people by 2050. Carbon constraints and restrictions in the use of natural resources challenge competitiveness. The geopolitical lineup is changing, and social equity is strikingly in disproportion beyond any reason. Solutions are not keeping pace with the scale of the challenge expressed by food loss and malnutrition, climate change, soil damage, and decline of biodiversity (to name only a few top issues). One example may show the complexity and scale of the challenge: To meet targets that would limit global warming to two degrees increase in worldwide average temperature, it is a fair assessment that pretty much all decades-long stable trends in carbon intensity must be broken. Carbon intensity needs to be reduced in a way that is unprecedented. The regular innovation rate would need to be x-times magnified and then continued on this scale for 35 consecutive years. All externalities assumed to be steady; this is far from even potentially realistic. To say the least, those trends require very serious efforts in rethinking concepts and ideas.

4 … But There Is Also Evidence for Change

Nevertheless, a lot is currently being done, and that is good. First, moving universities and research facilities tackles the sustainability agenda as mentioned above. Long gone are the days when research into sustainability immediately made the scientist an outsider, sometimes to the extent of even an academic pariah. Serious money is now being spent to foster research into social and ecological issues associated with production and consumption patterns, the agenda of, e.g., urbanism, energy, water, biodiversity, land management, and climate change. Most important are recent research clusters that look into the methods for excellence in sustainability research, and more and more researchers from all academic levels have gotten themselves involved with these issues.

In Germany, federal research funds for sustainability research have nearly doubled since 2005, together with an overall increase in funding of R&D and sector programs such as the Sixth Federal Energy Research Program. The Federal Government places emphasis on cooperation in research on sustainable development and on encouraging research and innovation in small- and medium-sized enterprises.

For Germany in particular, an ambitious commitment to science and advanced technology is the basis for strong industrial performance together with regulation assuring decent working conditions. This notion, inter alia, has been subject to the assessment and deliberations of the high-level peer review on German's sustainability performance. The peers have been mandated by the Federal Government. Chaired by Prof. Dr. Björn Stigson,[2] they conducted the review in 2013 and put emphasis on the science and research. Here is what they recommended: "There is a need for funding new ideas outside mainstream research that will ensure innovation in sustainability of land use and future cities, efficient use and re-use of resources, reliable climate forecasts and the impact of adaptation strategies. The topics of lifestyle, food and health also need a major boost in science and research programs. Advancing transformational knowledge for a sustainable future should be made a criterion for funding and for assessing scientific excellence. The issue of energy system integration and all energy-related topics (including energy production, sustainable mobility and sustainable buildings) should remain high on the agenda, as well as resource productivity and recycling, sustainable food strategies, organic

---

[2] Sustainability made in Germany, the Second Peer Review, Berlin Sept. 2013, downloadable from http://www.nachhaltigkeitsrat.de/uploads/media/20130925_Peer_Review_Sustainability_Germany_2013_02.pdf (accessed March 27, 2014).

farming, sustainable agriculture and lifestyles. We recommend increasing and expanding spending for projects as opposed to funding for institutions and structures, a need that we heard repeatedly in our stocktaking. We recommend fostering a systems approach in researching resource issues and in the earth sciences so as to take better account of the interconnectedness of nature as a system and the relation between humans, nature and technology. A better understanding of this interconnectedness is needed. This often involves encouraging transdisciplinary approaches and breaking out of the traditional silos of individual academic disciplines and their associated research programs. The nexus approach is one of the instruments that need to be further explored in this respect. For research policy, we recommend furthering the link between science and the national sustainability agenda, as far as this is appropriate. Just as the objectives of this agenda itself need to be based on engagement with business and all parts of civil society, so, too, we recommend that more use should be made of well-researched evidence and the co-designing of research programs. Co-designing efforts should involve civil society and practitioners in the formation of academic research goals and objectives. Innovation in more sustainable products and processes is crucial in all sectors. In both the public and the private sector, such innovations should in the first instance help to improve the originator's own sustainability performance but should then be assessed further in terms of how they could contribute to the sustainability performance of others, both within Germany and ultimately on a global scale."

Does that mean that science is already on track to meeting the sustainability challenge? Science, politics, and the business community are on the same page. A metaphor best describes the situation: The overall picture is that of a plateau filled with model lighthouse charges run by sustainability pioneers. On this plateau, you find high-end technical and cultural innovations, and all kinds of tracks and junctions, and the most up-to-date traffic light systems. But there is no clear way up to the summit, and the summit is where you need to go.

5 The Disconnect

One may conclude that funds and political regulation do not yet provide a big-enough impact to bridge the gap between what is necessary and what is possible. Certainly, this argument has substance, and an increase in funds and improved involvement of civil society may indeed be helpful. One may, however, also rethink the way the principle of self-organizing is implemented. It is a very important tool and, in the past, has proved a good approach to fostering innovation and critical thinking. When it comes to sustainability, it must again prove its merits, but this time, it has to develop out-of-area traction. Stakeholder interests outside of the formal academic system are a legitimate source for scientific agenda processes, and they might express opinions that are most relevant to the implementation of research budgets. This argument, too, has its justification.

Dedicated leadership and personal sustainability skills are needed on all levels. For private companies and public entities, the tools and means may be different, but responsibility for a common future is not. Thus, approaches should be centered on social responsibility, compliance, mitigation, and the license to operate. Value creation and innovation can and must be sourced from those approaches.

For mature economies, such as that of Germany, traditional growth patterns measured in gross output do not provide sound perspectives, but sustainability strategies do, and they must be based on scientific evidence. But conversely, one may ask what kind of understanding of its own role and performance science is following and how this translates into specific approaches to sustainability accounting and human resources management.

Observation is the methodological link between all empiric sciences. This will remain the working modality. The concept of sustainability requires a comprehensive approach to observation. *Observing the observers* is a relatively new and challenging idea. On first glimpse, this idea seems a bit generic or cloudy. But think of the issue of storing nuclear waste, with all its facets of scientific advice, malfunctioning, power structures, and organized protest, and the idea gains momentum immediately.

Transdisciplinarity is subject to self-organization of actors. If conducted properly it makes the scoping, design, performance, and communication of scientific excellence a part of transformational solutions. This is not specifically necessary for all kinds of research, but for a fair number of items, it is. The understanding of excellence for sustainability builds on disciplinary excellence, and, to be very clear, it does not replace disciplinary excellence. It rather requires a comprehensive reflection on the role of science, its structures, and the ways in which it may cooperate in a transdisciplinary mode. Explicitly, this is true for the upcoming implementation period of universe Sustainable Development Goals in all countries.

6 Iconic Game Changer

Nobel Prize awardees are often seen as near to changing the game as possible. More effectively, however, the political and societal perspective towards sustainability could capture the imagination and dedication of students, researchers, and the society as a whole. For example, the complete recycling of all waste, be it plastics, glass, and paper, or electronic waste and its rare earth components; turning carbon dioxide from undesired waste into usable raw materials; and combining sufficiency and efficiency strategies in the notion of green growth and sharing economies. Last, but not least, the German Energiewende can (and must, actually) be seen as a large-scale society lab for finding the sustainability trajectory.

Of course, this perspective requires increasing research in this area and developing an enhanced reporting system on sustainability performance that could make a difference.

Science is constitutive for the trustworthiness, credibility, and modality of sustainability trajectories. The future of science relies on the extent to which the society puts itself on the track towards an effective sustainable development. The dichotomy between science and society and between knowledge and practical change, although often repeated and evoked, might be a misleading concept in the long run. For "science for sustainability," the raison d'être lies in combination and cooperation. Combining externally recognized leadership with market-lead business performance and wise regulatory approaches is key. Most important are national sustainability strategies. If well sketched out they would provide room for leaders and advanced thinking. They would encourage leadership that does not wait for others to allow them to proceed and uncover what is already possible by today's standards. It is in the spirit of this meaning that sustainability science must be built into the academic mainstream.

Carl von Carlowitz, in his time, was called the Elector of Saxony's leading mining officer, and he was responsible for the ongoing creation of wealth and luxury goods that would add to the fortune of the Elector and the prosperity of the country. In today's framing, we would address him as a leading minister, and his concern would be how to continue essential ecosystem services in all three dimensions of sustainability.

MIX
Papier aus verantwortungsvollen Quellen
Paper from responsible sources
FSC® C105338

If you have any concerns about our products,
you can contact us on
ProductSafety@springernature.com

In case Publisher is established outside the EU,
the EU authorized representative is:
Springer Nature Customer Service Center GmbH
Europaplatz 3, 69115 Heidelberg, Germany

Printed by Libri Plureos GmbH
in Hamburg, Germany